THE

MEASUREMENT OF COLOUR

By the Same Author

The Perception of Light

Researches on Normal and Defective Colour Vision

Photometry and the Eye

The Rays are not Coloured

THE MEASUREMENT OF COLOUR

W. D. WRIGHT
A.R.C.S., D.Sc.
Professor of Applied Optics
Imperial College of Science and Technology

FOURTH EDITION

ADAM HILGER LTD
LONDON

ADAM HILGER LTD
60 ROCHESTER PLACE LONDON, NW1

SBN 85274 134 0

First edition 1944
Second edition 1958
Third edition 1964
Second impression 1968
Fourth edition 1969

PRINTED IN GREAT BRITAIN BY ROBERT MACLEHOSE AND CO. LTD, GLASGOW

PREFACE TO THE FOURTH EDITION

The main purpose of this book is to describe the principles, methods and applications of the trichromatic system of colour measurement. Not all colour measurement needs to be referred to this system, but since 1931, when the Commission Internationale de l'Eclairage (C.I.E.) first adopted a set of tables to define the colour-matching characteristics of a standard observer and establish a reference framework for the specification of colours, the trichromatic method has become fundamental to all others.

The most important formal developments in colorimetry since the third edition of this book was published have been the confirmation by the C.I.E. of the 1964 colour matching functions for large-field conditions of observation and the adoption of standard spectral energy distributions over a wavelength range from 300 to 830 nm to represent different phases of daylight. It is very valuable to have these daylight definitions, based as they are on new and extensive measurements of daylight, but full advantage of them will only be achieved when closely similar 'daylight' sources are available for use in the laboratory and factory as standard illuminants. The Colorimetry Committee of the C.I.E. has also recommended four alternative conditions for illuminating and viewing a surface whose colour is to be measured.

While these decisions are a sign of the continuing evolution of colorimetry, they are also an indication of the 'growing pains' from which the subject is suffering. The number of possible alternative conditions under which the colour of a surface can be measured has grown to the alarming matrix of $2 \times 6 \times 4\,(=48)$, depending on the choice of colour matching functions, illuminants, and illuminating and viewing geometry. (It could be even larger if we included more than three alternative phases of daylight.) Clearly some rationalization is called for, otherwise the subject will get out of hand.

Another rather disconcerting feature, especially for a newcomer to the field, is the massive tabulation of the 1931 and 1964 colour-matching functions prepared by the Colorimetry Committee of the

C.I.E. I realize that tables of values at close wavelength intervals, carefully smoothed and to a considerable number of decimal places, are necessary for certain sophisticated calculations and that once they are programmed in a computer, their use may not entail much extra work. Nevertheless, these tables do give a false impression of the level of accuracy to which colour matches can be made. We could also be in danger of forgetting that colour is something that we *see*, a fact which Professors Billmeyer and Saltzman have been at great pains to emphasize in their recent book *Principles of Color Technology*; moreover, we do not all see exactly alike. As a reminder of this I have thought it desirable to retain the diagrams showing the observer spread reported in the investigations which led up to the 1931 standard observer (Figs. 4.1, 4.2 and 4.3), and indeed to supplement them with a further diagram (Fig. 4.8) showing the spread among some of the observers who contributed to the 1964 standard observer.

The subject of colour has been greatly enriched by the publication in 1967 of a masterly reference book *Color Science* by Dr Wyszecki and Dr Stiles. Inevitably, this has called for a re-assessment of the material which should be included in a fourth edition of *The Measurement of Colour* and indeed of whether there should be a fourth edition at all. However, a case can be made out for text-books on the same subject which are written at different levels and from different points of view, and I hope that this new edition will make its own contribution to the better understanding and future development of colorimetry. The main changes from the third edition are new chapters on Metamerism and on Colour Difference Metrology and, in the final chapter, a more extended treatment of colorant formulation, a very lively industrial application of colorimetry at the present time. Other revisions and extensions have also been made at various points in the text, and additional tables have been included in the appendices.

A number of acknowledgments remain to be made and I should once again like to express my indebtedness to Mr Neville Goodman and Mr David Tomlinson of Adam Hilger Ltd, for their continued encouragement and support, and to my secretary, Mrs Jean King, for her assistance in the preparation of the text.

Acknowledgement is also due for the reproduction of many of the diagrams and tables as follows: The Physical Society for Figs.

3.3, 4.2, 4.3, 4.7, 6.3, 6.5, 9.2, and for Tables 5, 6, 7 and part of 8; the Optical Society of America for Figs. 6.4, 7.5, 7.6, 9.12 and 10.3; *Nature* for Figs. 10.6, 10.14, 10.21, 10.22 and 10.23; the University of Chicago Press for Figs. 2.3 and 2.5; the Royal Society for Figs. 4.1, 4.4 and 4.5; the *Journal of the Oil and Colour Chemists Association* for Fig. 10.9; the Paint Research Station for Fig. 10.8; the British Kinematograph Society for Fig. 8.15; the *Journal of the Society of Dyers and Colourists* for Figs. 1.11, 1.12, 10.2 and 10.4; John Wiley and Sons for Figs. 4.12, 4.13, and for Tables 8 (in part) and 12; the C.I.E. for Fig. 9.13; the *Journal of Applied Physics* for Fig. 10.15; Iliffe Books for Figs. 8.8, 8.9, 8.10, 8.11 and 8.13; the Massachusetts Institute of Technology for Table 9; the Interchemical Corporation for Plate 1; Messrs. H. K. Lewis for Fig. 2.1 and Plate 3; the Academic Press Inc. for Fig. 2.2; Vision Research for Fig. 2.9; Dr C. M. H. Pedler for Fig. 2.4; and the CIBA Foundation for Plates 2 and 6.

W. D. Wright

Applied Optics Section,
Imperial College of Science and Technology,
London.
March 1969

CONTENTS

page

PLATES

CHAPTER I

RADIATION IN THE VISIBLE SPECTRUM
ITS EMISSION, ABSORPTION, AND REFLECTION

THE NATURE OF LIGHT

Out of the whole gamut of electromagnetic radiations from gamma rays to radio waves, only a small band of radiations is capable of stimulating the eye. The limits of this band are ill-defined, since they depend on the amount of energy available, but for most practical purposes the visible spectrum may be regarded as terminating at a wavelength of $0{\cdot}38 \times 10^{-3}$ mm on the short-wave side, and at $0{\cdot}78 \times 10^{-3}$ mm on the long-wave side. However, with especially intense sources and with a completely dark-adapted eye, the boundaries may be extended at least down to $0{\cdot}35 \times 10^{-3}$ mm (even lower for some observers) and up to $0{\cdot}90 \times 10^{-3}$ mm, the limits apparently being set by absorption in the optic media at the short waves and by lack of sensitivity of the photochemical substances in the retina at the long waves.

Only radiations within the visible spectrum should properly be described as light radiations, although there is little physical difference between these and the radiations in the ultra-violet and infra-red regions. The extraordinary importance of light to the human race lies mainly in the fact that we possess in the eye a sense organ capable of reacting to light rays; yet this is a good and sufficient reason for dignifying them with a name of their own, in spite of their physical similarity to neighbouring radiations. In any case, when we refer to the measurement of light, we imply that the measurements relate to radiations in respect of their ability to stimulate the eye, and hence the amount of light, as such, outside the visible region of the spectrum is necessarily zero.

The unique physical characteristic of each radiation is its frequency of vibration; related to it, and, since the velocity *in vacuo* is the same for all radiations, inversely proportional to it, is the wavelength. The practical methods of radiating energy at different wavelengths vary very considerably from frequencies of one order to frequencies of another, as does the interaction of radiation with matter in its various forms; but these differences are rather differences in the properties of matter than of the radiations, which consist essentially of electromagnetic disturbances transmitted as displacements transverse to the direction of propagation. If the source of the radiation, or the original disturbance of the electric field, is within the nuclei of atoms, it gives rise to γ-rays having a frequency around 10^{19} per sec; lower frequencies of 10^{17} per sec have their origin in the inner electron shells of atoms and lead to the radiation of X-rays; light rays, of frequency about 10^{15} per sec, originate in the outer electron shells; infra-red rays, of frequency about 10^{13} per sec, in vibrations of atoms within the molecule and from the rotation of molecules; and so on.

For the problems considered in this book, we have to note three things about the nature of light: (i) light is a form of energy radiated as a transverse harmonic vibration over a frequency range between about 4×10^{14} and $7{\cdot}5 \times 10^{14}$ per sec, or between the corresponding wavelength limits $0{\cdot}75 \times 10^{-3}$ mm and $0{\cdot}4 \times 10^{-3}$ mm; (ii) the vibrations are electromagnetic disturbances in which the electric vector corresponds to the vibration considered in the simple harmonic theory; and (iii) the energy is radiated in the form of an extremely large number of elementary units of energy known as photons or light quanta, the energy of a photon being proportional to the frequency of the radiation.

The difficulties which have existed in understanding how wave and particle properties can be harmonized in a comprehensive theory of light need not concern us here. We make use of both aspects from time to time in colorimetry: the quantum concept provides the theoretical basis for determining the energy distribution in the radiation from a black body heated to some given temperature, it is involved in relating the spectral sensitivity of the retina to the absorption curve of the photochemical substances isolated from the retina, and chemists have to make use of it when analysing the absorption that occurs in a dye or pigment; on the other hand, calculations such as the loss of light by reflection in terms of the refractive index of the medium at which the reflection takes place are made with the aid of the electromagnetic wave theory. For the great majority of problems in colorimetry, however, light has to be regarded, physically, only as various amounts of energy being radiated per second (i.e. power) at various frequencies within the visible spectrum and propagated as transverse vibrations at a very high velocity. The physical identification of the different parts of the spectrum may be established either by the appropriate frequency or by the corresponding wavelength. Frequency is the characteristic which is independent of the medium within which the light is travelling, but wavelengths can be measured with greater absolute accuracy, since they do not depend for their determination on a knowledge of the velocity of light. The so-called 'wave-number', which is defined as the number of waves per centimetre, is proportional to the frequency yet does not suffer from its numerical uncertainty, and Stiles and Burch* reported their large-field colour matching functions on a wave-number scale. Wavelengths (symbol λ) are, however, more familiar to workers in colorimetry and have therefore been retained in this book.

Light can cause other reactions besides stimulating the eye. When it strikes the surface of an object, some fraction of the incident energy is generally absorbed and, if the surface is black and matt, the fraction will be large. The effect of the absorbed light is to heat the body. Again, light serves as the agent of numerous photochemical reactions, including the reaction which takes place in the retina, and can be recorded by the many types of photographic emulsion which are now available. Light can be converted into electrical energy by photoelectric action. This may take one of three forms: (i) in the photo-emissive cell electrons are emitted from a suitably prepared surface of one of the alkali metals, e.g. potassium or cæsium, under the impact of a beam of light and the action of an applied potential; (ii) light may cause a change in the electrical conductivity of certain materials, notably selenium, and thus lead to a change in the flow of a current through a closed circuit in which a suitably designed cell, such as a selenium cell, forms one of the components; this phenomenon of photoconductivity can also be used to generate an electrostatic image, as in xerography;† (iii) in the photovoltaic cell light falls on a

* W. S. Stiles and J. M. Burch: *Optica Acta*, 1959, **6**, p. 1.

† J. H. Dessauer and H. E. Clark: *Xerography and Related Processes* (Focal Press: London), 1965.

boundary surface between, say, cuprous oxide and copper, which possesses rectifier properties, and produces a potential difference sufficient to cause a small current to flow in a low-resistance circuit. Indirectly, light can be converted into electrical energy by focusing it onto the junction of two metals such as constantan and manganin, at which a potential is developed when the temperature is raised. Finally, light may be absorbed by certain crystals, such as activated zinc sulphide, and other chemical substances, and part of the energy re-radiated as light of a longer wavelength. This phenomenon is known as fluorescence.

LIGHT SOURCES IN COLORIMETRY

The most common method of making a body emit light is to heat it. As the temperature rises, the activity of the atoms and molecules in the substance greatly increases; their impacts become more frequent and more violent, and the resulting electromagnetic radiations increase in strength and spread to shorter and shorter wavelengths. Thus the traditional poker when heated in a fire at first radiates heat but no light. As it gets hotter, it begins to glow a dull red, and then a bright red, gradually becoming brighter and whiter as more energy, both relatively and absolutely, is radiated in the green and blue parts of the spectrum.

A few experiments suffice to show that the distribution of energy through the spectrum is a function principally of the temperature of the body rather than of the material of which it is composed. When a perfectly black body, i.e. a body that completely absorbs all the energy in a beam of light illuminating it, is heated, it can be shown from considerations of thermal equilibrium that the power radiated per unit area of its surface has a magnitude and a distribution through the spectrum depending only on its temperature. This distribution is given by Planck's law and typical curves for Planckian (black-body) radiators are shown in Fig. 1.1.

A close approach to a black body is not easily obtained, since most surfaces reflect some light and many show a certain amount of selective absorption, as indicated by their colour. This results, when such bodies are heated, in a departure from a true Planckian distribution of energy, although at no wavelength will the radiation of energy exceed that given by a black body heated to the same

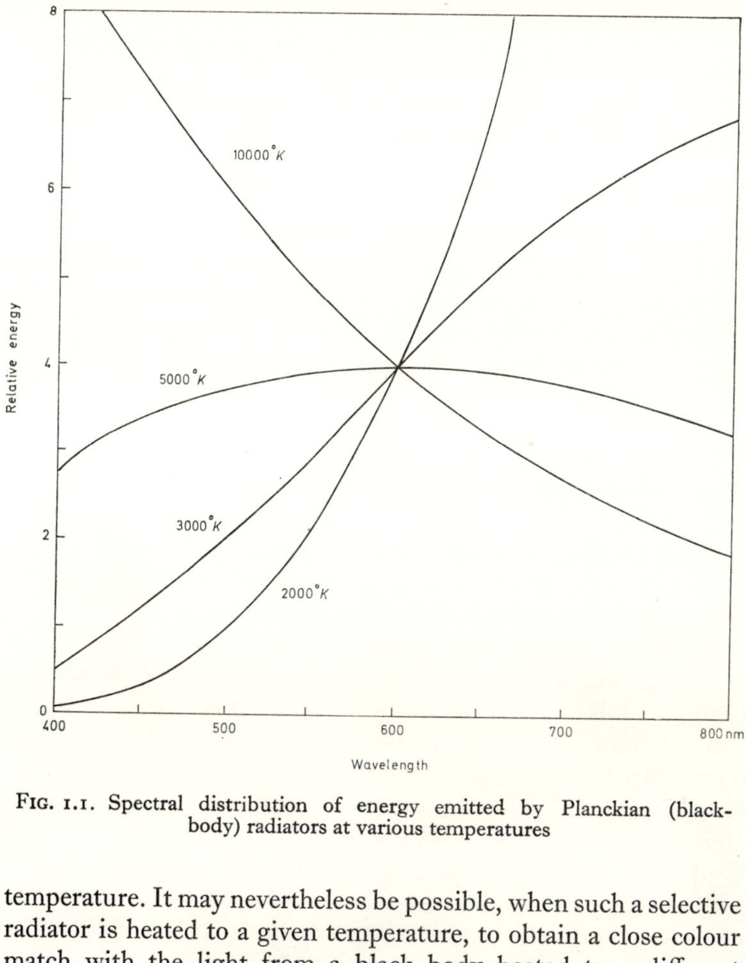

FIG. 1.1. Spectral distribution of energy emitted by Planckian (black-body) radiators at various temperatures

temperature. It may nevertheless be possible, when such a selective radiator is heated to a given temperature, to obtain a close colour match with the light from a black body heated to a different temperature. The selective radiator is then said to have a colour temperature equal to that of the matching temperature of the black body. This colour temperature may be either higher or lower than the true temperature of the heated material, according as the latter tends to radiate relatively more or less light at the blue end of the spectrum. The energy distribution need not be, and is indeed unlikely to be, identical to that of the black body at the stated colour temperature, since, as will be evident from later chapters, it is quite possible for two beams of light to give rise to the same colour sensation while possessing very different energy distributions. With materials such as tungsten, however, in which the selective absorption is not marked, the deviations from the Planckian distribution are small and may often be ignored. With some light sources such as the fluorescent lamp, where the energy distribution is irregular, it may not be possible to obtain a very close colour match to a black-body radiator. In that case, the lamp can be given a correlated colour temperature, which is the temperature of the black body nearest in colour to the lamp in question.

Sources of light such as the candle or the oil lamp are relatively yellow in colour because of the rather low temperature (about 1900° K) of the flame. The acetylene flame is hotter (2350° K) and whiter, while the tungsten filament lamp has a colour temperature ranging from 2400° K for a vacuum lamp to 3400° K for a gas-filled projection lamp or a tungsten-halogen lamp. The colour temperature of the arc crater of a carbon arc is about 3800° K* and the disk of the sun has a colour temperature of about 5000° K as seen through the earth's atmosphere. Typical energy distributions for a selection of these sources are shown in Fig. 1.2.

The most widely used source of light for industrial colour matching, and indeed for colour appraisal by the customer, is daylight. Only recently have reliable data become available on its spectral composition over a wide range of wavelengths from the ultra-violet to the far red end of the spectrum. Spectral energy distributions for different phases of daylight as reported by Henderson and Hodgkiss are illustrated in Fig. 1.3. These distributions clearly differ significantly from those of temperature radiators and on the basis of these and other measurements†, the Colorimetry Committee of the Commission Internationale de l'Eclairage

* M. R. Null and W. W. Lozier: *J. Opt. Soc. Am.*, 1962, **52**, p. 1156.

† S. T. Henderson and D. Hodgkiss: *Brit. J. App. Physics*, 1963, **14**, p. 125, and 1964, **15**, p. 947.

Y. Nayatani and G. Wyszecki: *J. Opt. Soc. Am.*, 1963, **53**, p. 626.

H. R. Condit and F. Grum: *J. Opt. Soc. Am.*, 1964, **54**, p. 937.

D. B. Judd, D. L. MacAdam and G. Wyszecki: *J. Opt. Soc. Am.*, 1964, **54**, p. 1031.

G. Kawakami, I. Ando and T. Ikemori: *J. Illum. Eng. Inst. Japan*, 1965, **49**, p. 521.

S. R. Das and V. D. P. Sastri: *J. Opt. Soc. Am.*, 1965, **55**, p. 319.

A. W. S. Tarrant: *Int. Farbtagung, Luzern* 1965 (Die Farbe), 1966, **2**, p. 689.

G. T. Winch, M. C. Boshoff, C. J. Kok and A. G. du Toit: *J. Opt. Soc. Am.*, 1966, **56**, p. 456.

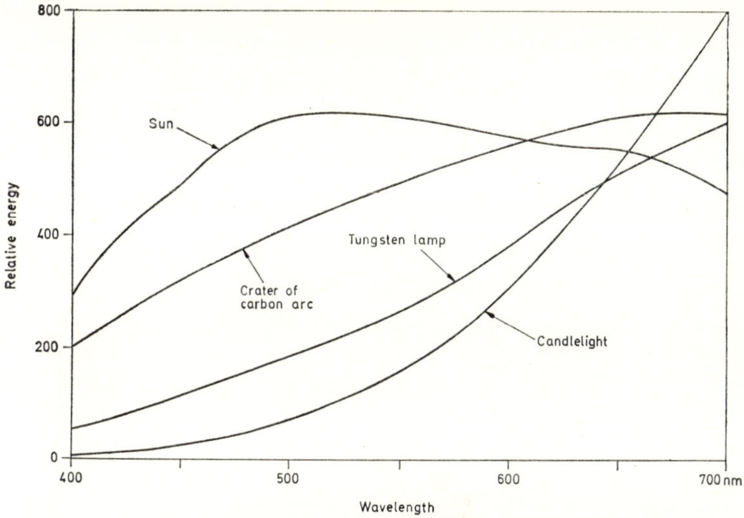

FIG. 1.2. Typical spectral-energy distribution curves for candlelight, a gas-filled tungsten lamp, the crater of a carbon arc, and the sun. The heights at which the individual curves have been drawn are quite arbitrary and bear no relation to each other

(C.I.E.) have defined standard spectral distributions to represent daylight at various correlated colour temperatures. For general use, a correlated colour temperature of 6500° K has been selected as typical of average daylight, and the spectral distribution for this source (D_{6500}) is tabulated in Table 2, Appendix 1. Distributions for D_{5500} and D_{7500} are also included in the Table. It now remains for the lamp manufacturer to produce lamps simulating these distributions as closely as possible, but even without these the tabulated data can be used to calculate colour specifications for standard daylight illumination provided the samples to be specified do not fluoresce.

Prior to the standardization of these new energy distributions, the C.I.E. had adopted in 1931 three standard illuminants based on work carried out by Davis and Gibson at the Bureau of Standards in the United States.* One of these, S_A, consists of a gas-filled tungsten lamp operating at a colour temperature of 2854° K on the international temperature scale as revised in 1948.

* R. DAVIS and K. S. GIBSON: *Bur. Stand. Misc. Pub.*, No. 114, 1931.

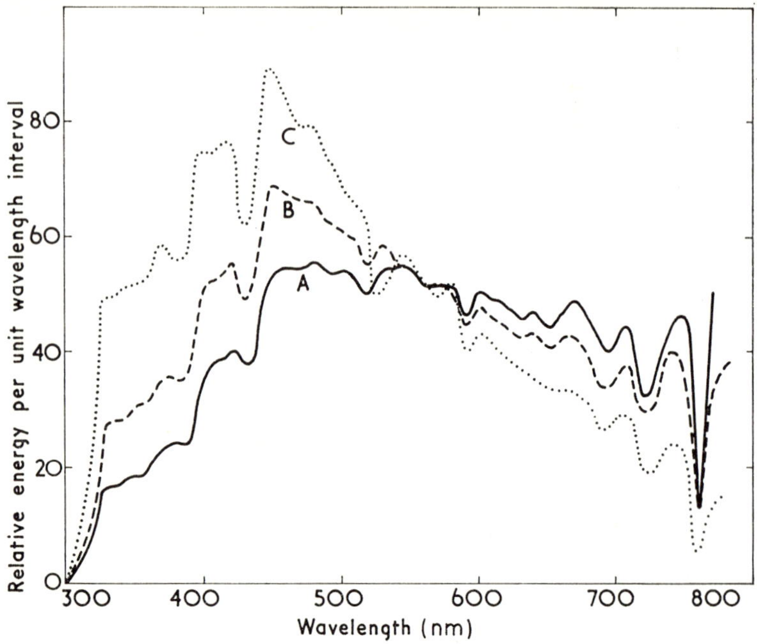

FIG. 1.3 Spectral energy distributions for three phases of daylight, as measured by Henderson and Hodgkiss (1963).
Correlated colour temperature: A—5630° K, B—7190° K, C—10500° K.

This provides a standard illuminant typical of average artificial tungsten lighting. The other two sources, S_B and S_C, are intended to be representative of sunlight and daylight, and are obtained by using S_A in conjunction with liquid colour filters of specified chemical composition and thickness, to give colour temperatures of approximately 4900° K and 6700° K respectively. The specification of the filters and the tabulated energy distributions of the three sources are given in Appendix I and Table 1.

It must be expected that in due course S_B and S_C will be discarded as standard sources in favour of D_{6500}. At the present time, the Tintometer Company must be almost the only instrument manufacturer who continues to supply the S_B and S_C filters as part of their instrumental accessories. How soon the tabulated data for the S_C distribution will be replaced by the D_{6500} distribution for colour specification by calculation will presumably depend on the readiness of industrial laboratories to accept some discontinuity in their records. Colorimetrically, there is no reason for any delay in switching to the improved daylight distribution, but changes in standards can obviously cause some commercial inconvenience.

In addition to light sources which produce their radiation by heat, caused either by the passage of electricity or by combustion, there are sources such as the neon lamp or the mercury lamp, in which the light is produced by electric discharge through gases. In this case the radiation of energy originates in the collision between the ions and atoms of the gas during the passage of the electric current. Under these conditions, a line spectrum is seen in the spectroscope corresponding to the natural frequencies characteristic of the gas being excited. Gas-discharge lamps are of great value in the laboratory as sources of monochromatic light and in the calibration of a spectrum scale in wavelengths. They have, however, little use as illuminants for colorimetry, although the colour-rendering properties of the mercury and sodium lamps may have to be borne in mind, since these lamps are employed in street lighting installations and elsewhere. When the gas pressure in the lamps is increased, there is a tendency for the spectrum lines to broaden as the radiation spreads to frequencies outside the natural frequency of the gas. A recent development has been the high-pressure sodium lamp, in which the radiation spreads well beyond the familiar yellow sodium lines to give a radiation which approximates to white in colour. In referring to gas-discharge lamps, it is necessary, therefore, to distinguish between lamps with the gas at high pressure and those at low pressure.

Another development of the gas-discharge lamp is the widely-used fluorescent lamp. In this, the inner glass wall of a mercury-discharge tube is lined with a suitable mixture of phosphors which are excited under the strong ultra-violet radiation that forms part of the mercury spectrum. The light from these lamps thus originates in part from the yellow, green and violet lines of the mercury discharge, but mainly from the fluorescence of the phosphors. Different spectral compositions can therefore be provided depending on the choice of phosphors, and a close approach to the colour of daylight can be secured if required. The presence of the mercury lines, however, makes it impossible for the spectral distribution to be a close approximation to that of daylight.

One of the most promising 'daylight' sources is the high-pressure xenon arc lamp.* This is a discharge lamp of very high intensity with a spectral distribution extending from the ultra-violet to the infra-red. It is unlikely to be used as a general illuminant because of its high intensity, the precautions under which it has to be run,

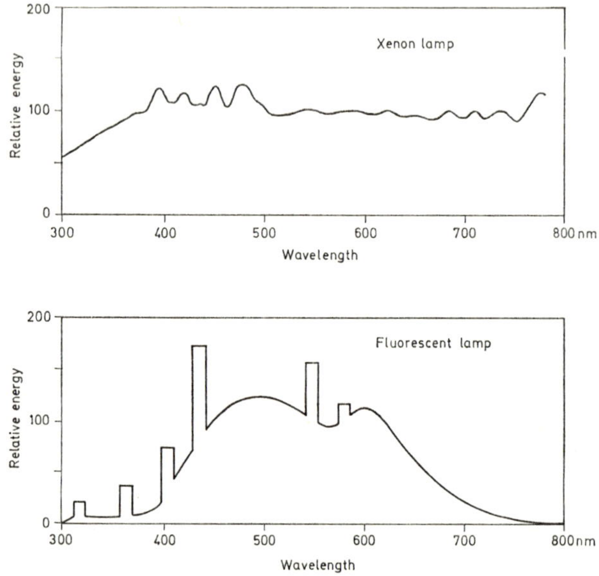

FIG. 1.4. Spectral-energy distribution curves for a xenon lamp and a typical fluorescent lamp

and the considerable amount of auxiliary equipment required for its operation. Nevertheless, as a laboratory source for colorimetry it is likely to be a strong contender for adoption, when used in conjunction with a filter, as the source to represent the new daylight illuminant D_{6500}. The energy distributions of typical xenon and fluorescent lamps are shown in Fig. 1.4.

The need for a daylight illuminant is two-fold. On the one hand, a standard illuminant with a defined ultra-violet content is required for the colorimetry of samples which fluoresce. These are now very

* H. G. FRÜHLING, W. MÜNCH and M. RICHTER: *Die Farbe*, 1956, **5**, p. 41.
A. BERGER and O. KOCH: *Die Farbe*, 1960, **9**, p. 259.
A. BROCKER and O. KOCH: *Die Farbe*, 1960, **9**, p. 267.

common owing to the widespread use of fluorescing pigments, dyes and optical bleaches. In the present state of our knowledge, it is virtually impossible to calculate the spectral composition of the light re-radiated in the visible spectrum through the process of fluorescence, even though we may be able to define the wavelengths and intensities of the ultra-violet and violet radiations that give rise to the fluorescence. Instead, it is necessary to measure either the spectral composition of the light from the fluorescing sample, or its integrated colour, when illuminated by a source giving as close a spectral distribution as possible to that of the standard source D_{6500}. It is primarily for this reason that there is so much urgency for the lamp manufacturer to produce a standard daylight illuminant as soon as possible. As already suggested, one likely source is the xenon lamp. Another possibility is a tungsten lamp with a daylight filter, used in combination with another source contributing the required amount of ultra-violet radiation. Until a standard daylight illuminant is available, there must be a considerable degree of imprecision in the colorimetry of fluorescing samples.

The second need for a daylight illuminant is for general purpose colour matching and colour inspection in industry. Here the precise imitation of the standard daylight distribution may be less critical, but even so the representation of the ultra-violet and visible radiations must be reasonably close. What is required is the provision of high level, uniform illumination over a considerable area, and ideally it should be possible to illuminate the whole of the colour matching room with standard daylight. The old S_B and S_C sources were never suitable for the purpose because of the inconvenience of liquid filters. One alternative has been to use blue glass filters specially designed to approximate to colour-temperature conversion filters, i.e. to filters having a linear relation between log τ_λ and $1/_\lambda$, where τ_λ is the transmission of the filter at wavelength λ.* By use of a suitable thickness of such a filter in conjunction with a tungsten lamp, a spectral distribution of almost any colour temperature can be obtained, but the higher the temperature, the thicker and bluer the filter and the greater the light absorption. The wattage of the lamps has therefore to be increased if an adequate illumination level is to be maintained at

* H. P. GAGE: *J. Opt. Soc. Am.*, 1933, **23**, p. 46.

the higher colour temperatures. Another disadvantage, of course, is the limited ultra-violet radiation from the tungsten lamp.

The more common practice is to illuminate the colour matching room, or perhaps a special colour-matching cabinet, with a combination of fluorescent lamps, including probably an ultra-violet lamp, and tungsten lamps. The level and uniformity of illumination from such an installation can be very good even if the spectral distribution differs significantly from daylight. Nevertheless, if the lamps are specially selected for their colour-rendering properties, this arrangement is likely to be entirely satisfactory for the vast majority of industrial colour-matching and colour-inspection problems. At the moment only the xenon lamp seems to issue a serious challenge to this type of installation.

THE ABSORPTION OF LIGHT

When a beam of llght, as represented by the arrows in Fig. 1.5, is travelling through a medium, a certain amount of energy may be lost owing to absorption or scattering. If the light travels a distance l in the medium in going from A to B, and if the light flux is represented by F_0 when the beam is at A and by F when it has arrived at B, then F will be less than F_0 by an amount depending on the distance l and on the degree of scattering and absorption occurring in the medium. The reduction due to scattering depends on the size of the particles producing the scattering relative to the wavelength of light, on their shape, on their refractive index relative to that of the surrounding medium, etc. Exact analysis is often impossible, so that particular problems may have to be solved empirically by making reasonable approximations.*

The loss of light by absorption alone can be evaluated without difficulty once certain constants of the absorbing medium are available, although its interpretation in terms of the structure and nature of the medium is still a matter of great intricacy. So far as the visible spectrum is concerned, the absorption of light produces a change in the electron energy of the atom or molecule of the absorbing substance, and causes an electron transition from one energy level to a higher one. In vapours and gases at low pressures, the absorption spectrum may therefore appear as

* W. E. K. MIDDLETON: *Vision through the Atmosphere* (University Press: Toronto), 1952.

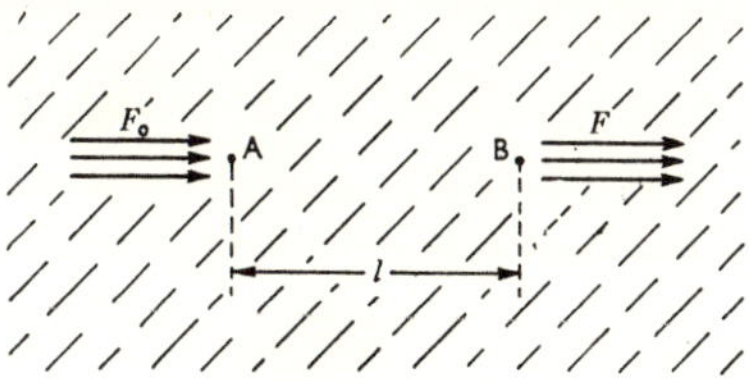

FIG. 1.5

discrete lines, or as very fine lines packed very close together, but in liquids and solids the spectral absorption is blurred out into broad bands owing to the disturbing influence of molecules in close proximity to each other.

The difficulty of predicting the absorption characteristics of a substance is very great,* and the physicist who deals with colour needs to be reminded that the chemist has a truly gigantic problem to solve before he can deduce the spectral absorption curve of a material from a knowledge of its molecular make-up. Still less can he be expected to produce, to order, a substance with a particular absorption curve, however urgently the physicist may need it, or however high a price a manufacturer may be prepared to pay for it.

With a given substance, on the other hand, the position is different, since once the absorption is known for a given thickness and, with liquids, for a given concentration, the absorption can in general be found for other thicknesses and concentrations. There are two laws, Lambert's law and Beer's law, by means of which such calculations can be made.

Lambert's law states that the fraction of light which is absorbed by a substance is independent of the intensity of the incident light. It follows that, for light of any given wavelength, each successive layer of the medium absorbs an equal fraction of the light passing through it. Hence, this exponential decay for a monochromatic beam can be expressed by

$$F = F_0 . e^{-al}$$

where a is a constant depending on the medium and is known as the absorption coefficient of the substance. The law cannot be

* See, e.g., E. J. BOWEN: *The Chemical Aspects of Light* (University Press: Oxford), 1949,

extended in that form to a beam of white light or any other broad band of wavelengths, since the quality of such light will change in its passage through the medium. An alternative expression of the law can be obtained by making use of the extinction coefficient, K, as defined by Bunsen and Roscoe. K is defined as the reciprocal of the thickness, measured in centimetres, of the medium through which the light must travel to reduce the intensity of the light to one-tenth of its initial value, so that, for any other thickness l, the flux transmitted is given by

$$F = F_0 \, . \, 10^{-Kl}$$

Beer's law states that the absorption is directly proportional to the number of molecules of absorbing substance through which the light passes. In practice, accurate measurement shows that this law is often not exactly obeyed, owing possibly to some chemical change in the absorbing molecules rather than to any invalidity of the law. When an absorbing substance is dissolved in a non-absorbing medium, the absorption, according to Beer's law, will be proportional to the concentration. If c is the concentration, Beer's law may be written

$$F = F_0 \, . \, \alpha^{-cl}$$

where α is the absorption factor for the substance and is equal to F_0/F for unit thickness and unit concentration. Since the extinction coefficient K is proportional to the concentration, Lambert's law can be converted into Beer's law by writing ϵ for K/c, giving

$$F = F_0 \, . \, 10^{-l\epsilon c}$$

When c is expressed in gramme-molecules of absorbing substance per litre of solution, ϵ is known as the molecular extinction coefficient, since it is a measure of the absorption due to a single molecule.

Another term widely used in connection with the absorption of a medium is the density D, which is defined by the equation

$$D = \log_{10} \frac{F_0}{F}$$

Since this is equivalent to the equation

$$F = F_0 \, . \, 10^{-D}$$

it follows that D is related to the molecular extinction coefficient by the relation

$$D = \epsilon c l$$

Hence the density for a given wavelength is proportional to the molecular extinction coefficient for that wavelength. Density is a particularly convenient measure of the absorption of a medium, because, when the light travels through several absorbing substances, the overall density at any wavelength is equal to the sum of the individual densities for the same wavelength.

For colorimetric purposes the variation of absorption with wavelength is usually demonstrated in one of three ways. In the first, the fraction of the light transmitted at each wavelength in the visible spectrum is plotted against wavelength. An example of this is illustrated in Fig. 1.6(*a*), in which the transmission curves for three concentrations of a hypothetical solution are drawn. The concentrations are in the ratio of 1 : 2 : 4. In Fig. 1.6(*b*) the same data are recorded in terms of density; thus, if the transmission at some wavelength and concentration is τ, i.e.

$$\tau = \frac{F}{F_0}$$

then the value of D plotted in the second diagram is

$$D = \log_{10}\left(\frac{F_0}{F}\right) = \log_{10}\left(\frac{1}{\tau}\right)$$

In Fig. 1.6(*c*), the curves record the variation with wavelength of $\log_{10} D$; i.e. the values of $\log_{10}\{\log_{10}(1/\tau)\}$ are plotted against λ for each concentration. This last form has the interesting and valuable property that the shape of the curve is a property of the medium alone, being independent of both thickness and concentration. Thus, since

$$D = \epsilon c l$$

then $$\log_{10} D = \log_{10} \epsilon + \log_{10} c + \log_{10} l$$

Hence, as c and l are both independent of wavelength, a change in either by a given factor will result simply in a constant arithmetical increase or decrease of $\log D$ for all wavelengths. In the example illustrated in Fig. 1.6, the concentration is increased by a

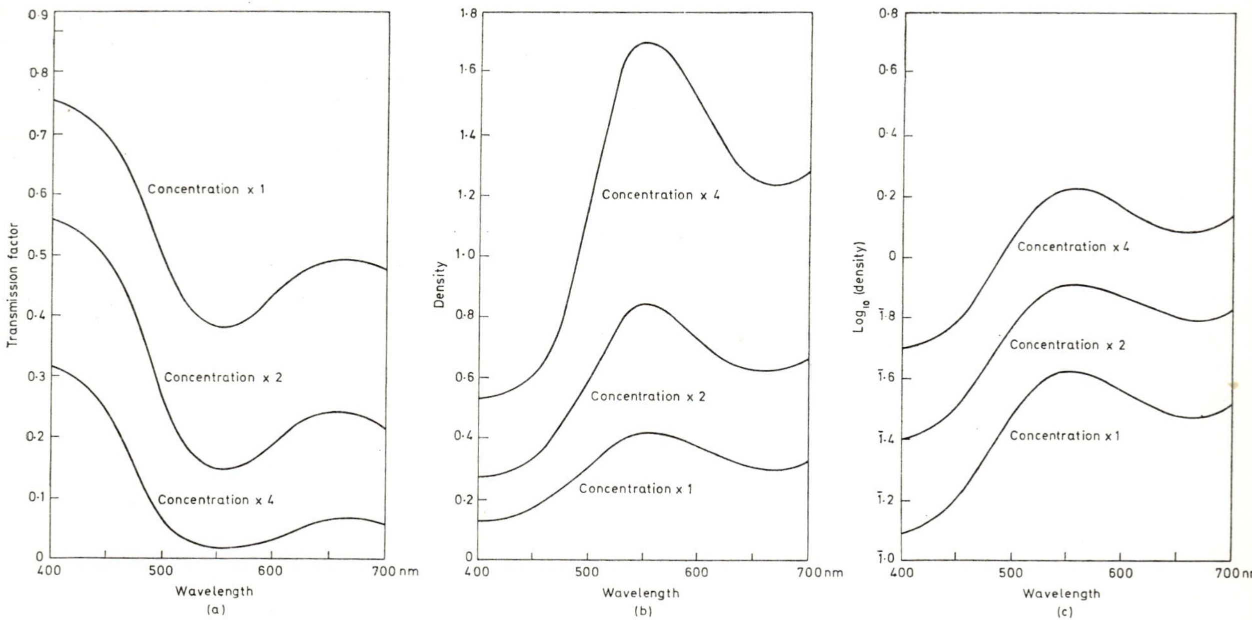

(*a*) (*b*) (*c*)

FIG. 1.6. The spectral absorption of a hypothetical solution at three concentrations, plotted in terms of (*a*) the transmission factor, (*b*) the density, and (*c*) $\log_{10}$ density.

factor of 2 between successive strengths, and hence each curve in Fig. 1.6(*c*) is displaced vertically from its neighbour by $\log_{10} 2$, i.e. by 0·3.

When two or more colouring media exist in the same solution, absorption occurs as if each substance were acting independently, always assuming that there is no chemical interaction between them. If c_1, c_2, c_3 . . . are the respective concentrations of the substances and ϵ_1, ϵ_2, ϵ_3 . . . their molecular extinction coefficients at a given wavelength, their density at that wavelength is given by

$$D = l(\epsilon_1 . c_1 + \epsilon_2 . c_2 + \epsilon_3 . c_3 + \ldots)$$

THE REFLECTION OF LIGHT

When light is travelling in a certain medium A and strikes a surface dividing A from a new medium B, a fraction of the light is reflected back again through the original medium A, while the remainder enters into B. The amount of light reflected depends on the nature of the two media, while the fate of the light entering B depends on the nature of B. Thus metals, which exhibit very high reflection when polished, do so very largely owing to the presence of free electrons, to which their good electrical conductivity is also due, and some 70, 80 or 90 per cent of the incident light may be reflected, while the remainder, which enters into the material, is very rapidly absorbed. Metallic reflection is also characterized by the fact that if the metal is coloured, the specular reflection will also be coloured, which is, in general, not the case for the light specularly reflected from the polished surface of a transparent medium. Most transparent substances possess few free electrons and are consequently poor conductors; they reflect only a small proportion of the incident light, while the light refracted into the medium will be transmitted freely, except where some degree of absorption of particular regions of the spectrum occurs, as described in the preceding section. In that event the medium appears coloured.

When light is incident on the polished surface of a transparent medium, the reflected beam will in general exhibit some degree of polarization. If i is the angle of incidence and r the angle of refraction (Fig. 1.7), and if a is the amplitude of each of the components (i.e. the electric vectors of the electromagnetic theory) of the incident beam resolved in and at right angles to the plane of

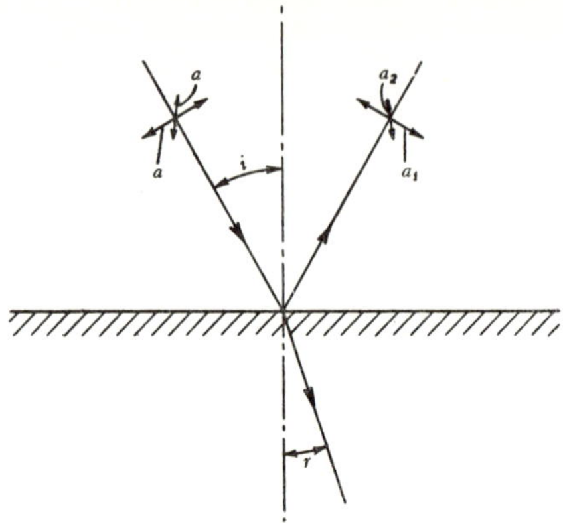

FIG. 1.7

incidence, the amplitudes, a_1 and a_2, in these two planes after reflection are given by Fresnel's equations; thus, for the component whose vibration lies in the plane of incidence

$$a_1 = a \cdot \frac{\tan(i - r)}{\tan(i + r)}$$

and for the vibration perpendicular to the plane of incidence,

$$a_2 = a \cdot \frac{\sin(i - r)}{\sin(i + r)}$$

The intensities of the two components are, of course, proportional to the squares of the amplitudes. For nearly normal incidence, i and r become very small, the law of refraction reduces to $n = i/r$, where n is the refractive index of the medium, and the ratio of the amplitude of the reflected beam to that of the incident beam becomes $(n - 1)/(n + 1)$.

The corresponding ratio of the intensities is then $[(n - 1)/(n + 1)]^2$, which, for a refractive index of 1·5, gives the result

$$\left(\frac{0{\cdot}5}{2{\cdot}5}\right)^2 = 0{\cdot}04$$

i.e. the intensity of the reflected beam is 4 per cent of that of the incident beam.

When $i + r = 90°$, $a_1 = 0$, so that the vibrations of the reflected beam all take place in a direction normal to the incidence plane corresponding to the a_2 component. This occurs when $\tan i_p = n$, i_p being known as the polarizing angle; for glass, this has a value of about 57°.

When the light, before reflection, is in the medium of higher refractive index, then for angles of incidence greater than a critical angle, all the light is reflected, none being transmitted. The critical angle of incidence i_c is given by

$$\sin i_c = \frac{n}{n'}$$

where n is the index of the less dense medium, n' that of the medium with higher refractive index. For a glass-air surface, the critical angle within the glass is of the order of 41°.*

Polished surfaces are of less interest in colorimetry than diffusely reflecting surfaces, such as those of paper, fabric or paint. When light is scattered back from these diffusing materials, it appears to have come from the topmost layer of the surface, and the fact that the majority of the rays have travelled some distance into the surface before reflection is by no means apparent. This illusion is aided by the solidity of the material itself, which to all appearances consists of a solid block of matter; the fact that it is composed of a fine structure of particles or fibres may be revealed only with the aid of a magnifying glass or microscope.

If we wish to discover how and why a so-called opaque surface reflects light, we have to apply the facts of reflection and absorption to the structure of the surface concerned. Consider, first, the example of a white diffusing surface, such as snow. Snow is formed by the crystallization of water vapour in the atmosphere and consists of a variety of complex hexagonal crystals some 0·01–0·5 in. across. When light falls on a layer of snow, 3 or 4 per cent is reflected back by the uppermost surfaces of the top

* The matters discussed in the last few paragraphs can be studied in greater detail in any standard textbook on physical optics, e.g.:

R. S. LONGHURST: *Geometrical and Physical Optics* (Longmans, Green: London), 2nd Ed., 1967.

F. A. JENKINS and H. E. WHITE: *Fundamentals of Optics* (McGraw-Hill: London), 1957.

R. W. DITCHBURN: *Light* (Blackie: London), 2nd Ed., 1963.

S. G. LIPSON and H. LIPSON: *Optical Physics* (Cambridge University Press) 1969.

layer of crystals, but the random arrangement of the crystal faces causes these first reflected pencils to be returned in a great variety of directions. Most of the light, however, is refracted into the top crystals and travels through them until it strikes their lower faces. Here again the greater part will be refracted more deeply into the snow, but in some crystals (usually only a small proportion of the total) it may happen that the angle between the lower face and the light falling on it is such that the light is totally reflected. For those individual pencils, all the energy is redirected, possibly along a path that will ease its ultimate escape through the top of the snow again. The majority of the rays travel into lower layers of the crystals, at each face giving up more of their energy by random reflections, and themselves changing direction at each refraction. Ultimately, the light in each original pencil will become so attenuated by repeated reflections that its remnant, if not totally reflected, becomes insignificant. All the reflected elements are themselves subject to additional reflections and refractions until practically the whole of the original light re-emerges into the air, with the exception of that which is absorbed in its passage through the ice and converted into heat. The rays which eventually emerge from the surface will be so numerous and so random as regards their direction and point of emergence, that each point in the surface will in effect act as if it were radiating light in all directions. It is for this reason that we are not conscious of looking into the interior of the surface, although the light has actually come from within.

The effectiveness with which snow diffuses the light arises primarily from the large number and variety of crystals which go to form the snow and from their random arrangement in the snow. There is no one direction in which more of the light is reflected than in any other, and a layer of snow therefore forms an almost perfect diffusing surface and will therefore appear equally luminous from all directions of view. Again, because of the random reflections which occur, there is no question of any polarization by reflection, there being as many rays with vibrations in one plane as in the plane at right angles.

The whiteness and lightness of snow arise because the crystals are very transparent and because there are a vast number of them. The amount of light reflected will obviously depend on the number of reflections that take place as well as on the fraction of light reflected at each surface; hence larger and fewer crystals would lead to reduced reflections and, since the light would have to travel, on the average, through a greater thickness of the medium between one reflection and the next, a greater opportunity would exist for absorption to occur. A thin layer of snow does not appear perfectly white, because an appreciable fraction of the light can escape through the lower side of the layer and be partially absorbed by whatever surface lies below. Again, with snow which has half melted, water will tend to fill the interstices between the crystals and thus reduce its diffusing power and whiteness. This follows because the difference between the refractive index of the crystals and that of water is very small compared with the difference between the indices for ice and air, and so, in accordance with Fresnel's equations, a much smaller fraction of light will be reflected at the ice-water than at the ice-air boundaries.

The whiteness of snow represents a rather simple case of diffuse reflection, in that snow crystals, although quite complicated (and indeed beautiful) structures, are relatively large compared with the wavelength of light. With crystals or particles less than about 5μm in diameter, the light is diffused not only by reflections and refractions according to the Fresnel equations, but also by diffraction and scattering. This means that the particles in effect become new sources of light and may redirect the radiation over considerable angles. The smaller the particle, the greater the light scatter for diameters down to about 200 nm, where the scatter is a maximum; it is still, however, a function of the refractive index of the particle relative to that of the surrounding medium. With even smaller particles, the scattering begins to diminish again and eventually the particles behave as if they were in solution in the medium.

We are now in a position to state the properties which a substance must possess if it is to have a perfectly white, perfectly diffusing surface, or if, on the other hand, it is to be a black matt surface, reflecting a minimum amount of light. For the former, the material must consist of finely divided particles arranged in an entirely random manner, absorbing the minimum amount of light both at their surfaces and within them, and embedded in a medium having as large a difference in refractive index from that of the material itself as possible. This means that the surrounding

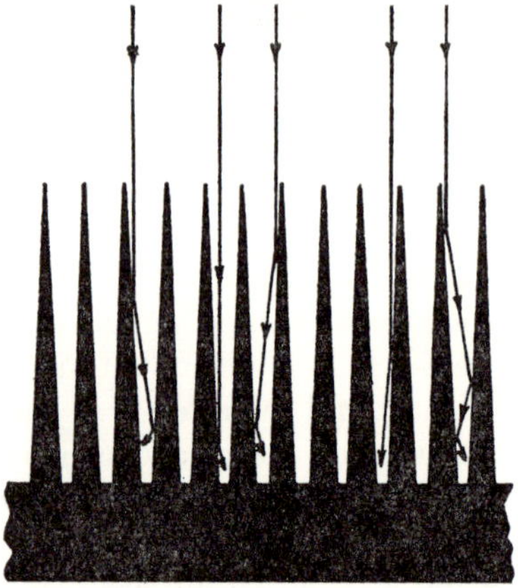

FIG. 1.8. Ideal structure for a matt black surface

medium should normally be air. For a black, matt surface, the material should be highly opaque or absorbent, and if it were not for the front surface reflections, large particles would be better, to avoid numerous reflections. But to reduce the front reflections and make the surface as nearly matt as possible, the particles should preferably be tiny and needle-shaped, arranged with their points jutting out normal to the surface. Then, as shown in Fig. 1.8, the light reflected from the outer surfaces of the particles is nearly all redirected farther into the material, where it will be largely absorbed. The surface needs an open structure and needs, as it were, to be made porous to the light—to admit the light and absorb it and let only a minimum escape again. In this way, black velvet reflects very little light although the individual fibres do not themselves possess matt surfaces. Again, lampblack, or finely divided platinum or silver is very matt and black because of its fine state and the open spaces between the highly absorbent particles into which most of the reflected light is directed and subsequently absorbed. Then the petal structure of certain flowers is so delicate, fibrous and absorbent that there is virtually no top-surface reflection. A practical black surface on the principle of Fig. 1.8, described to the author by Professor S. Tolansky, consists of some hundreds of tightly packed razor blades. Another effective optical light trap can be made from two pieces of polished black glass and a black diffusing surface, arranged as shown in Fig. 1.9.

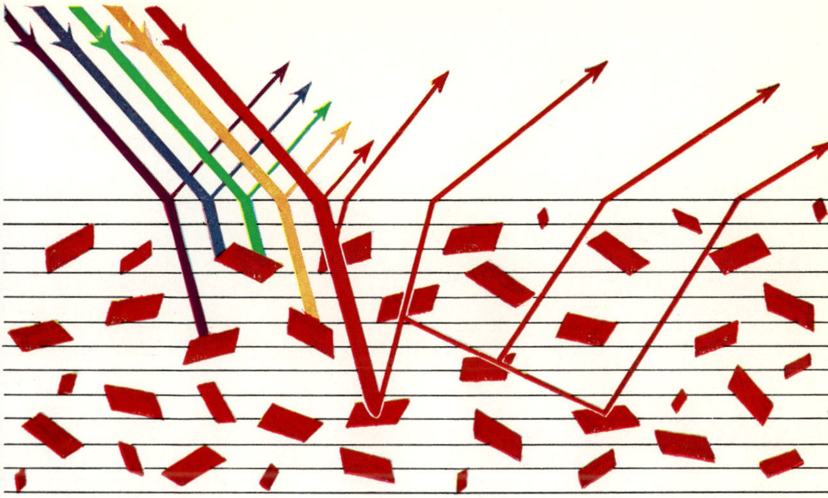

The absorption, scattering, and reflection of light in a layer of pigment (see page 24)

PLATE 2

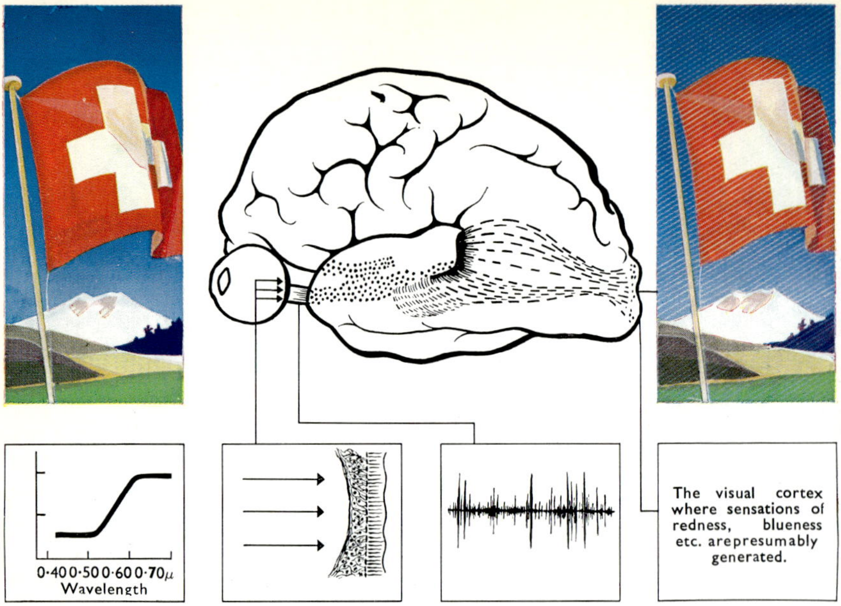

Spectral composition of light from the flag.

Light entering the colour receptors (cones) in the retina.

Nerve impulses travelling along optic nerve fibres to brain.

Stages in the process of perception
(see page 31)

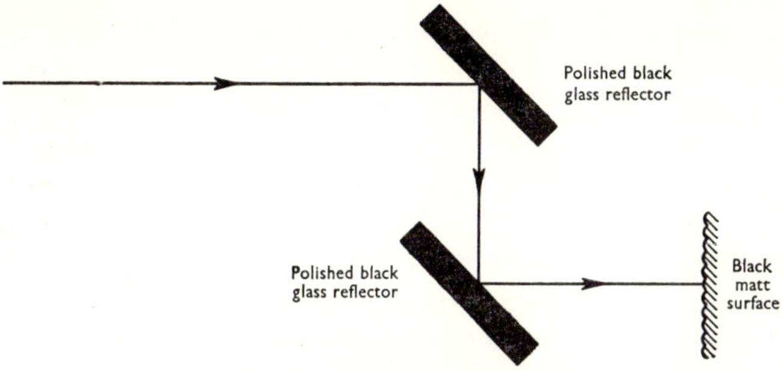

FIG. 1.9. An optical light trap

The preparation of a perfectly white, perfectly diffusing surface is of great importance in photometry and colorimetry. Such a surface is required as a standard against which the luminance factor of other surfaces can be measured; it may form an essential element in a photometer and may contribute to the successful operation of an integrating sphere when used as a lining to the inner wall of the sphere. There are a number of surfaces which approximate to the ideal required, such as white blotting paper, white opal diffusing glass, and matt white celluloid, and these may often be usefully employed. But a closer approach to perfect diffusion and reflection is attained by titanium oxide, magnesium carbonate, barium sulphate, and magnesium oxide, and it is the last of these which, although providing a somewhat fragile surface, has been generally employed as a standard diffusing surface. It can be prepared by allowing the magnesium oxide smoke from a clean strip of burning magnesium ribbon to deposit on a cold surface inverted a few inches above the flame. The most complete reflection is secured when the surface on which the oxide is deposited consists of a polished silver plate and when the oxide layer is built up to a thickness of about 2 mm. Under these conditions the surface is a very perfect diffuser and, when fresh, reflects about 98 per cent of the incident light.* Alternatively, disks of magnesium oxide powder may be prepared in a suitably designed press, and although the luminance factor may be lower by 1 or 2 per cent compared to a freshly smoked surface, the pressed disks are relatively easy to prepare and are claimed to be more stable and

* W. E. K. MIDDLETON and C. L. SANDERS: *J. Opt. Soc. Am.*, 1951, **41**, p. 419.

reproducible. The effectiveness of magnesium oxide as a diffuse reflector is due to the extremely small particles which form the surface, the absence of absorption, and the high refractive index. Photographs taken with the electron microscope† show that the particles of magnesium oxide smoke consist of very minute, regular cubes, 200 nm in width, a size which incidentally produces an optimum amount of scattering.

The use of magnesium oxide as a standard white reference surface has come under increasing criticism in recent years through the recognition that its reflection may vary from preparation to preparation by 1 or 2 per cent and is likely to fall off by a similar amount after a few days. Following an enquiry to determine the views and experience in different countries‡ the Colorimetry Committee of the C.I.E. has recommended that the hypothetical perfectly white, perfectly diffusing surface should be adopted as the reference standard in the colorimetric specification of opaque samples, instead of magnesium oxide. At the C.I.E. meeting in Washington in 1967, January 1st, 1969, was proposed as the date when this change should come into operation.

In the colour industries, and particularly in the paint industry, the light-scattering properties of a pigment layer determine its opacity or hiding power, and thus the particle size and refractive index of a pigment are of the utmost importance. The colour of the pigment will be governed by the absorption of some wavelengths more than others, the absorption taking place either on the surface of the particle or within it, or both, depending on the opacity of the particle.§ Similarly, with a dyed fabric, the multiple reflections and scattering between and within the fibres cause successive absorption of some part or other of the spectrum, and the emergent light will thus appear coloured. The depth of colour of a surface will depend in part on the extent to which the light penetrates into the surface, and this in turn will be governed by the dimensions, structure and refractive index of the material of which the surface is composed. The penetration of a beam of light into a pigment layer is illustrated schematically in Plate 1.

In addition to the effect on its colour, the physical structure of

† R. D. HEIDENREICH: *Phys. Rev.*, 1942, **62**, p. 291.
‡ *C.I.E. Compte Rendu*, 1963; Colorimetry E-1.3.1.
§ D. R. DUNCAN: *J. Oil and Colour Chem. Assoc.*, 1949, **32**, p. 296.

a surface will have a profound effect on its texture also. The variety of surface textures is immense, but we will for the moment consider the two divisions of glossy and matt surfaces. A glossy paint, for example, is one in which the medium forms a smooth and shiny surface from which 3 or 4 per cent of the incident light is specularly reflected. This top-surface reflection will have the same colour, generally white, as the incident illumination, while the remaining 96 per cent of the incident light, which penetrates into the pigment layer, will become coloured by the successive absorptions at the pigment particles. This coloured component will eventually emerge from the surface as a diffusely reflected beam, and by choosing the appropriate angle of illumination and viewing, the specular component can be reflected away at an angle and the colour of the surface can be seen in its full depth. For this reason, glossy paints can be deeper in colour than matt paints. It must, however, be remembered that in normal viewing a glossy top surface is likely to reflect light from the surroundings into the observer's eye and to that extent to modify the colour of the surface.

With a matt surface, the top-surface reflection will be diffusely distributed in all directions by the random orientation of the minute facets forming the top layer. Nevertheless, a major part of the light will still penetrate into the pigment layer, where it is partially absorbed and partially reflected to emerge in all directions as the coloured component of the reflected light. It is now, however, impossible to separate the two components, so that the body colour of the paint tends to be slightly desaturated by the admixture of the top-surface reflection; with a truly matt surface, the colour is also independent of the directions of illumination and viewing.

Nearly all surfaces encountered in practice are neither perfectly glossy nor perfectly matt, but the immense variety of surface textures virtually defies any precise classification. Names such as *egg-shell finish*, *distinctness-of-image gloss*, and *absence-of-bloom gloss* show the type of verbal descriptions that are attempted, while polar curves such as those in Fig. 1.10 provide a first optical description of the reflection characteristics of a surface. A complete specification requires very much more than this, since account has to be taken of the spread of a given pencil of light within the surface, the variation both in the colour and in the reflection characteristics from point to point of a surface, the depth of the

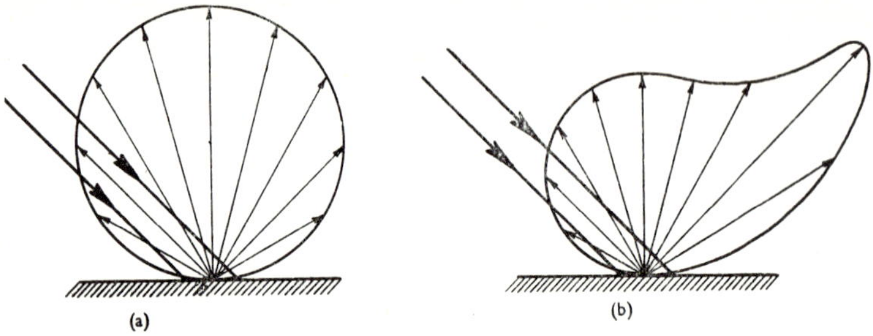

FIG. 1.10. Polar curves showing the intensity distribution of the light reflected (*a*) from a perfectly diffusing surface, and (*b*) from a surface possessing some degree of specular reflection, when the incident illumination makes an angle of 45° with the normal.

surface structure, etc.* Where there is regularity in the pattern of a surface, as in a woven material, this again will introduce its own characteristic appearance.

For ordinary colour measurement, many of these variables have to be ignored and the method of illumination illustrated in Fig. 1.11(*a*), with the light incident at 45° to the surface and collected normally, was recommended by the C.I.E. in 1931 as standard practice unless otherwise stated. This ensures that any specular component from a glossy surface will be reflected well out of the path of the measured light, but the inverse arrangement of normal incidence and 45° reflection, Fig. 1.11(*b*) would be equally effective in this respect. For various reasons but particularly shortage of light, other ways of collecting the light are frequently used, and four further examples are illustrated in Fig. 1.11(*c*), (*d*), (*e*) and (*f*).

It has generally been assumed that the same value for the luminance factor will be obtained either for 45° incidence and normal collection or the other way round, but this has now been shown to be incorrect if there is some partial polarization in the incident light.† The reason for this is illustrated in Fig. 1.12 which shows the different percentages of the light reflected from the top surface of a glossy sample for the two conditions of illumination and for light plane polarized both with its vibration direction in the plane of the diagram and at right angles to it. For 45° incidence, the top

* W. D. WRIGHT: *J. Oil and Colour Chem. Assoc.*, 1953, **36**, p. 537.
† S. JOHNSON, D. H. PHILLIPS, A. R. ROBERTSON and W. D. WRIGHT: *J. Soc. Dyers and Colourists*, 1963, **79**, p. 731.

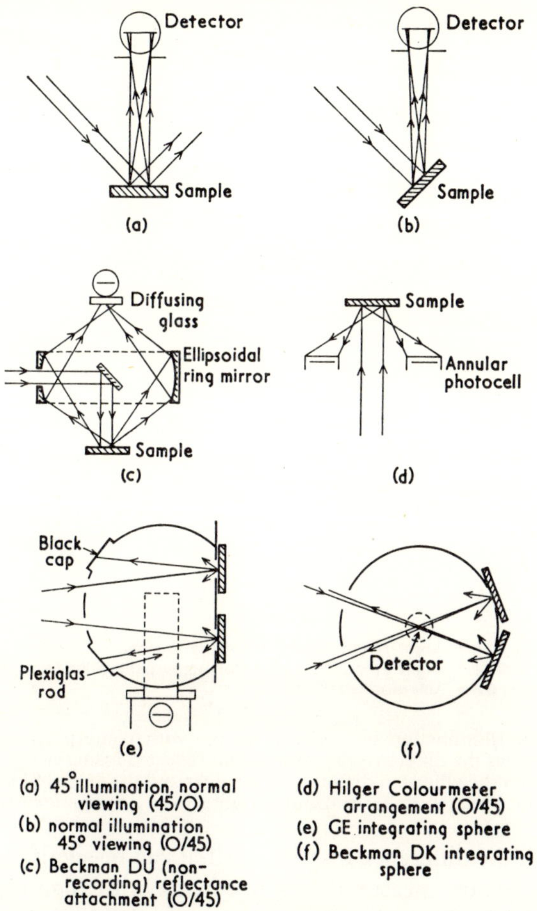

FIG. 1.11. Various methods of illuminating a surface and collecting the diffusely reflected light employed in colorimeters and spectrophotometers. (Johnson, Phillips, Robertson and Wright, *S.D.C. Symposium*, 1963)

surface reflection differs markedly for the two states of polarization and this means that the illumination of the pigment layer beneath will also be affected. As a consequence the luminance factor when measured relative to a matt diffusing surface such as magnesium oxide will also be a function of the polarization. On the other hand, for normal incidence, the luminance factor will be independent of the polarization.

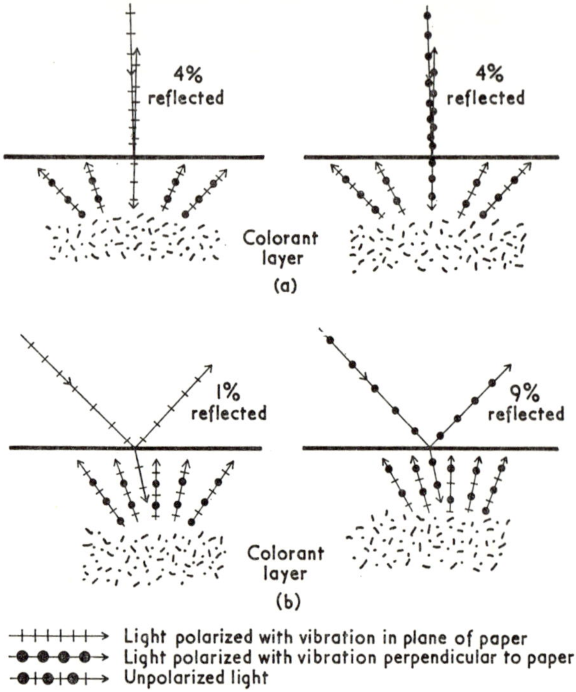

FIG. 1.12. Illumination of a glossy sample with plane polarized light. Diagram shows the different fractions of light reflected from the top surface depending on the vibration direction of the light and the angle of incidence. (Johnson, Phillips, Robertson and Wright, *S.D.C. Symposium*, 1963)

Although the difference may be small, it can be significant when spectral reflection measurements are made with monochromatic light derived from a prism monochromator. With an instrument such as the author's double monochromator described on p. 247, partial polarization will occur at each of the eight refractions through the dispersing prisms. Since the normal/45° arrangement will give the correct result irrespective of the state of polarization, it should certainly be used where polarization of the incident light is present.

At the 1967 C.I.E. meeting, four alternative illuminating and viewing conditions were recommended in place of the one condition recommended in 1931. These are (*a*) 45° illumination and normal viewing, (*b*) normal illumination and 45° viewing, (*c*) diffuse illumination and normal viewing, (*d*) normal illumination and diffuse viewing. The Colorimetry Committee also set limits to which these angles of incidence and viewing should be met and to the angular cone of light used to illuminate the sample. These are given in Appendix I, part 2. Instruments intended for measurements made in accordance with C.I.E. recommendations should therefore be designed within these tolerances.

CHAPTER 2

THE PERCEPTION OF LIGHT AND COLOUR

THE FUNCTIONS OF THE EYE

The essential function of the eye is to impart to an observer knowledge about the external world. Except under conditions of extreme darkness, all objects radiate or reflect some measure of light from the visible spectrum, and it is through its sensitivity to these radiations that the eye has acquired its ability to see. First, then, the eye is a detector of light. Next, its sensitivity to the direction from which the light has come leads to spatial location, to vision of form and ultimately to the perception of fine detail. Sensitivity to different amounts of light leads to the perception of a greater or less degree of contrast between an object and its background; sensitivity to the quality or wavelength of the light has developed into colour perception; while the combined operation of form vision, contrast perception and colour vision leads to the recognition of the texture of a surface. Other vitally important functions of the eyes include the detection of movement and the perception of depth.

The information which the eyes transmit to the brain is interpreted, on the basis of the observer's previous experience, into detailed knowledge about the objects within his field of view. This knowledge is co-ordinated with the information derived through the other sense organs to form a perceptual pattern which serves as the basis for his subsequent thought and action. Colour is only one of a number of qualities that go to make up this pattern, and the colour sensation of which the observer is aware is a function of the physical quality of the light, the physiological processes in retina and brain, and the psychological interpretation of the physiological response, modified to a greater or less extent by association with other characteristics of the visual sensation and with other non-visual aspects of the perceptual pattern. In all ordinary observations, the ultimate colour perception is mentally projected into the external world and regarded as a property of the object in view (see Plate 2).

THE OPTICAL SYSTEM OF THE EYE

A horizontal section of the eyeball is shown in Fig. 2.1. Light from the external world is admitted into the interior of the eye

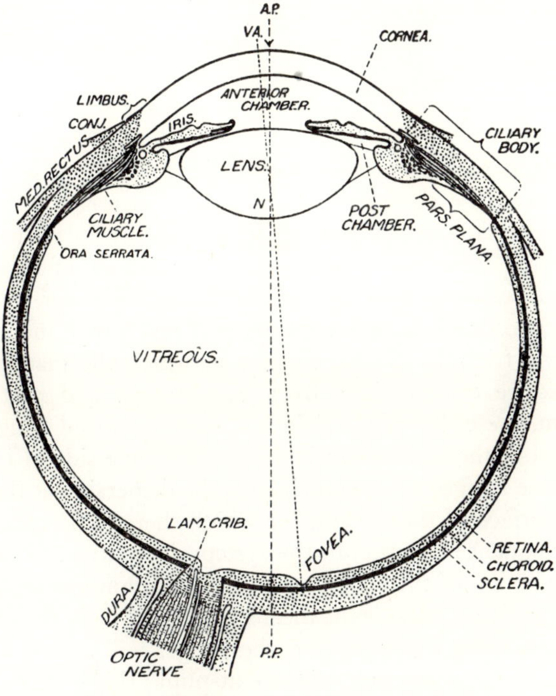

FIG. 2.1. A horizontal section of the human eye (From Wolff's *Anatomy of the Eye and Orbit*)

through the cornea, a transparent segment of rather shorter radius than the rest of the eyeball. The refraction which takes place at the cornea is sufficient, when combined with the additional refractive power of the crystalline lens within the eye, to focus the light onto the inner wall of the eyeball. Immediately in front of the anterior surface of the crystalline lens is the iris, a fibrous membrane with a central circular aperture which contracts and expands as more or less light is incident on the eye and which to some extent compensates for sudden changes in the illumination level. The refracting power of the crystalline lens is itself capable of adjustment through the action of the ciliary muscles attached by ligaments to the circumference of the lens, a greater convexity of the lens permitting near objects to be sharply imaged, and *vice versa.* The regions behind and in front of the crystalline lens are filled with transparent media known as the *vitreous* and *aqueous humours* respectively.

In comparison with well-designed systems of glass lenses, the optical correction of the lens system of the eye cannot be rated very high, particularly on account of the complete absence of any correction of the chromatic aberration.* Fortunately, the effects of this aberration are in some way suppressed by the retina or brain to an extent sufficient to make us quite unaware of its existence in normal seeing. Under special conditions, however, the lack of colour correction becomes apparent and may prove inconvenient. Thus in a trichromatic colorimeter in which the matching field is a mixture of radiations from widely separated parts of the spectrum, the edges of the field will show coloured fringes if the exit-pupil of the instrument is displaced to one side of the visual axis. These fringes are due mainly to the dispersion at the steeply curved surface of the cornea, for which there is no correction at any subsequent refraction in the system, in contrast to the achromatizing action of the combined crown and flint glasses in a telescope objective. Unless some means is adopted to fix the observer's head when using a colorimeter, the fringes will appear whenever his eye moves to one side of the exit-pupil; this is more troublesome to a person who has had little experience of colour matching than to the experienced observer, but with neither is it conducive to the highest precision in matching. Again, the difference in focus

* A. Ivanoff: *Les Aberrations de l'Oeil* (Revue d'Optique: Paris), 1953.
G. Wald and D. R. Griffin: *J. Opt. Soc. Am.*, 1947, **37**, p. 321.
R. E. Bedford and G. Wyszecki: *J. Opt. Soc. Am.*, 1957, **47**, p. 564.

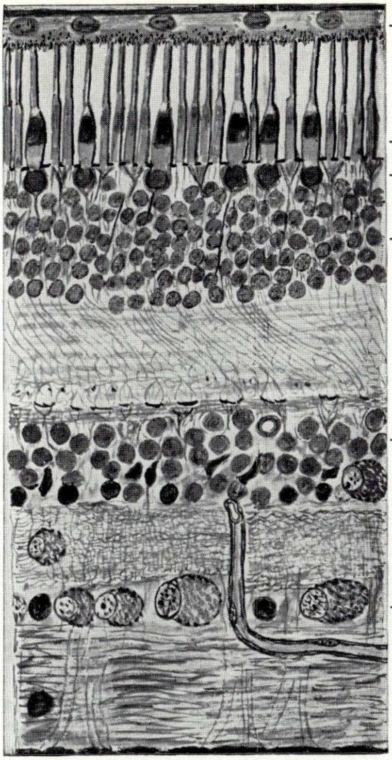

A vertical section of the retina
(see page 33)

(From Wolff's *Anatomy of the Eye and Orbit*)

between the far red and far violet parts of the spectrum has made it necessary, in small-field experiments with a colorimeter, to add a special lens to introduce sufficient over-correction to bring the differently coloured images sharply in focus simultaneously.*

Some longitudinal spherical aberration is also present, but the shape of the refracting surfaces and the higher refractive index of the core of the crystalline lens provide a measure of correction, which varies however from individual to individual and with the state of accommodation.† So far as the off-axis aberrations are concerned, the location of the iris with respect to the cornea is such that the amount of coma and astigmatism introduced at the cornea is not likely to be excessive, while the curved wall of the eyeball on which the image is formed tends to fit the large curvature of field which the lens system of the eye inevitably introduces. As it happens, however, the structure of the retina is such that fine detail can be resolved over only a very small area called the *fovea*, which is located near the centre of the retina and defines the visual axis of the eye; the detail of an extended object can therefore be studied only by scanning the object with the eye so as to direct the axis onto successive points of interest and thus image them on the fovea.

THE RETINA

The conversion of light energy into nervous energy takes place in the light-sensitive retina which lines the inner wall of the eyeball and which comprises several layers of neurones and their interconnections, as shown in Plate 3. The actual light receptors, the rods and cones, are located in the layer most remote from the cornea, so that the light has first to pass through more or less transparent layers of ganglion and bipolar cells to reach the receptors. Between the retina and the wall of the eyeball, there exists a black pigment, the pigment epithelium, which serves to absorb unwanted light and which may play some part in the visual

* W. D. Wright: *Researches in Normal and Defective Colour Vision* (Kimpton: London), 1946.

† A. Ivanoff: *Les Aberrations de l'Oeil* (Revue d'Optique: Paris), 1953.

reaction. The tips of the rods and cones just make contact with the pigment.*

There is general agreement that the initial reaction in the retina is photochemical, and extensive information is accumulating about the chemical nature of visual purple and its derivatives. This is the light-sensitive substance which can be extracted from the rod receptors and which decomposes in the light and is regenerated in the dark. A similar photochemical cycle is assumed to occur in the cones although far less is known about the chemistry of the cone substances. As we shall see, the rods function mainly at low illumination levels, while the cones are the receptors which operate at higher levels and which are primarily responsible for the perception of detail and colour. This division of function is the basis of the long-standing Duplicity Theory of vision, but the validity of such a clear cut distinction between the rod and cone processes has recently been challenged by Pedler.†

Studies with the electron microscope have revealed a fine lamillar structure within the rods and cones as seen in Fig. 2.2. It may be that the light-sensitive material is distributed over these layers or perhaps the structure has some more physical purpose analogous to a waveguide. However that may be, it must be assumed that the photochemical decomposition within the receptors generates a

* There are a number of books that may be consulted for more detailed information on the structure, photochemistry and electrophysiology of the retina, e.g.:

W. S. Duke-Elder: *System of Ophthalmology*, vol. I (Kimpton: London) 1958.
S. L. Polyak: *The Retina* (Chicago University Press: Chicago), 1941.
S. L. Polyak: *The Vertebrate Visual System* (Chicago University Press: Chicago), 1957.
R. Granit: *Receptors and Sensory Perception* (Yale University Press: New Haven), 1955.
E. Wolff: *The Anatomy of the Eye and Orbit* (Lewis: London), 5th Ed., 1961.
Documenta Opthalmologica, vol. III, 1949. (Contains papers read at the Colour Vision Conference, Cambridge, 1947.)
Coloquio Sobre Problemas Opticos de la Vision (Instituto de Optica: Madrid), 1953.
H. J. A. Dartnall: *The Visual Pigments* (Methuen: London), 1957.
R. A. Weale: *The Eye and its Function* (Hatton Press: London), 1960.
G. S. Brindley: *Physiology of the Retina and Visual Pathway* (Edward Arnold: London), 1960.
CIBA Foundation Symposium on Colour Vision (Churchill: London), 1965.
J. J. Sheppard: *Human Color Perception* (Elsevier: New York and London), 1968.

† C. H. M. Pedler: *CIBA Foundation Symposium on Colour Vision* (Churchill: London), 1965, p. 52.

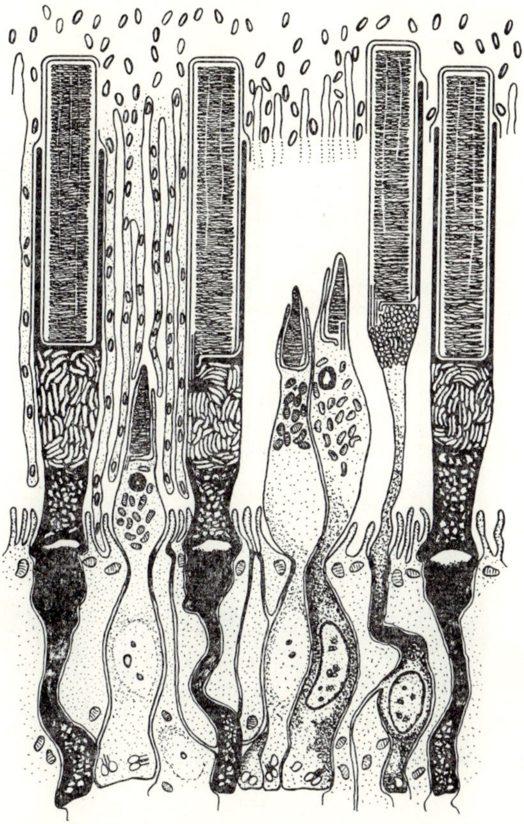

FIG. 2.2. The structure of the different types of photoreceptor in the frog retina (after S. E. Nilsson, *J. Ultrastruct. Res.*, 1964, **11**, 147)

change of electric potential which is possibly amplified within the receptors before being transmitted through to the bipolar and ganglion cells. At some point, possibly in the ganglion layer, the electrical signal is converted into nerve impulses which are then transmitted along the fibres of the optic nerve to the brain.

The structural intricacies of the retinal tissue as analysed by Polyak using the light microscope are illustrated in Fig. 2.3. At the first synaptic layer contact is made between the base of the receptors and the protoplasmic processes, known as *dendrites*, at the receiving end of the bipolar cells. The place of contact has a

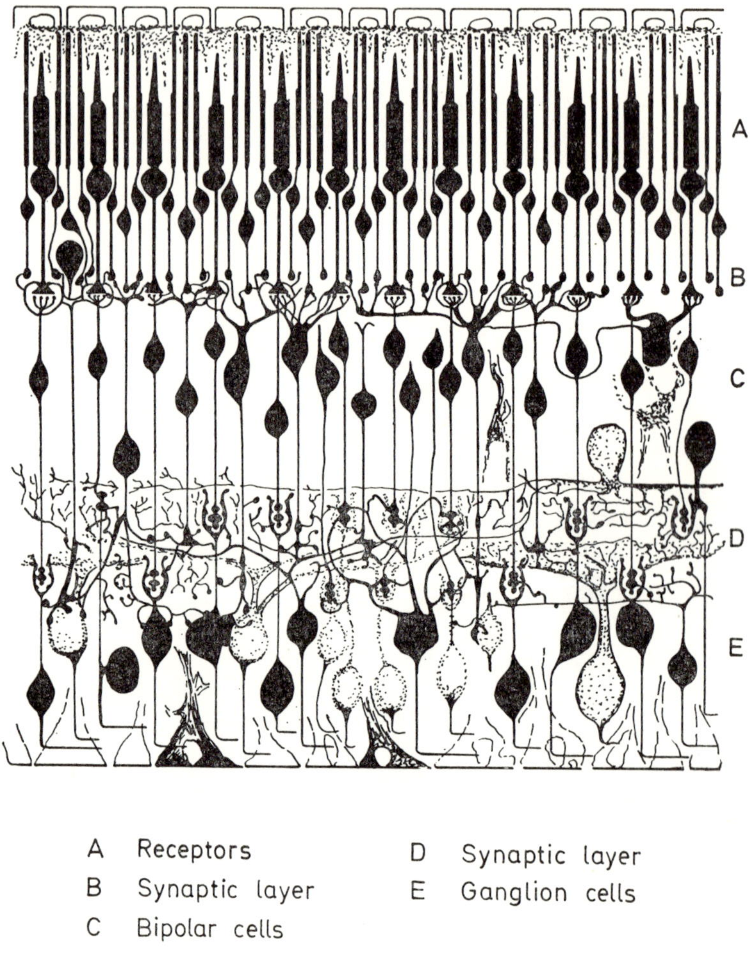

FIG. 2.3. The primate retina showing the receptors and the synaptic layers linking the receptors with the bipolar and ganglion cells (after Polyak)

polarity which facilitates the passage of the electrical signals in one direction, namely into the body of the bipolar cell and thence into the slender axis cylinder, which divides into numerous branches to form the discharging end of the cell, called the *teledendron*. The teledendrons of the bipolar cells make contact with the dendrites of the ganglion cells, into which the signals pass; from there they travel through the nucleus of the ganglions into the fine axis cylinders, which become the fibres leading to the optic nerve.

In the foveal area of the retina the central region is populated only by cones, and Polyak reports the presence here of so-called midget bipolar cells which provide a very direct route from cone to nerve fibre with a minimum of opportunity for lateral spreading to neighbouring receptor units. The high resolving power of this area can then be accounted for by the one-to-one connection between cone and nerve fibre. Outside the central pit of the fovea, more diffuse forms of connection occur and in the peripheral retina quite extensive arborization is found at the synaptic junctions. There is no doubt that a considerable degree of convergence and summation takes place outside the fovea, not only between groups of cones, but also between rods and cones. While this must inevitably lead to a loss of resolving power it can also lead to enhanced light sensitivity. Horizontal and amacrine cells provide a further possible source of lateral interaction between retinal areas.

The description of the retinal tissue as given in the two previous paragraphs and illustrated in Fig. 2.3 represents what we may describe as the classical picture of the retina derived from the light microscope. Complex as this picture appears, the electron microscope reveals a still higher order of complexity, especially in the synaptic layers. Fig. 2.4 is a schematic diagram due to Pedler* based on his own studies, in which he reports an intricate network of very fine processes running across the synaptic layers and providing a vastly greater opportunity for interaction between the signals from individual receptors than had previously been imagined. Pedlar considers, in fact, that this network is responsible for processing the visual information recorded by the receptors and transforming it into a code suitable for transmission along the optic nerve. This concept of the retina as a computer or data-processing device fits in neatly with modern ideas of the brain as a computer, since in its structure the retina is a morphological expansion of the brain.

Returning to the receptors themselves, the functional differences between the rods and cones is on the whole more clearly defined than their known physiological differences. Rods are

* C. M. H. Pedler: *CIBA Foundation Symposium on Colour Vision* (Churchill: London), 1965, p. 52. *Spectrum*, 1965, No. 16, p. 9.

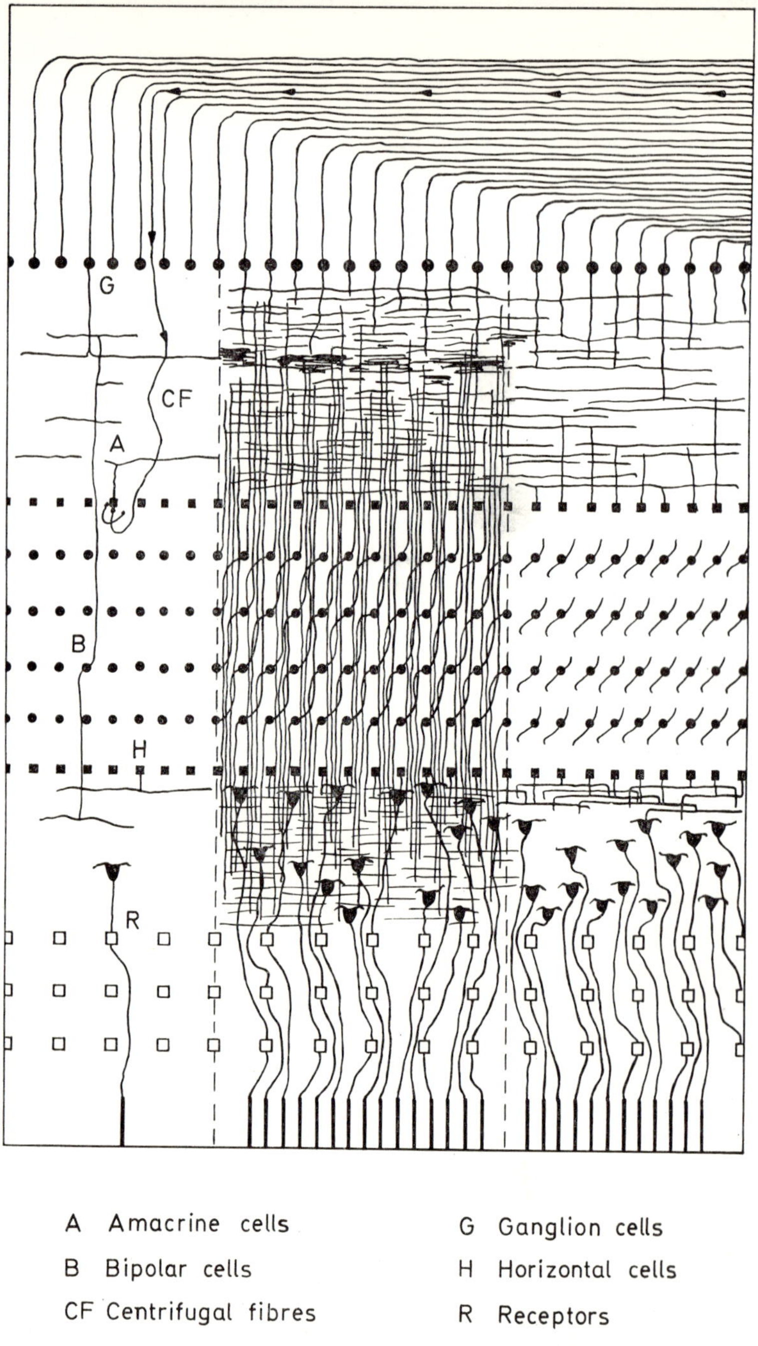

FIG. 2.4. The neural network of the vertebrate retina as analysed by Pedler from studies with both the light and electron microscopes

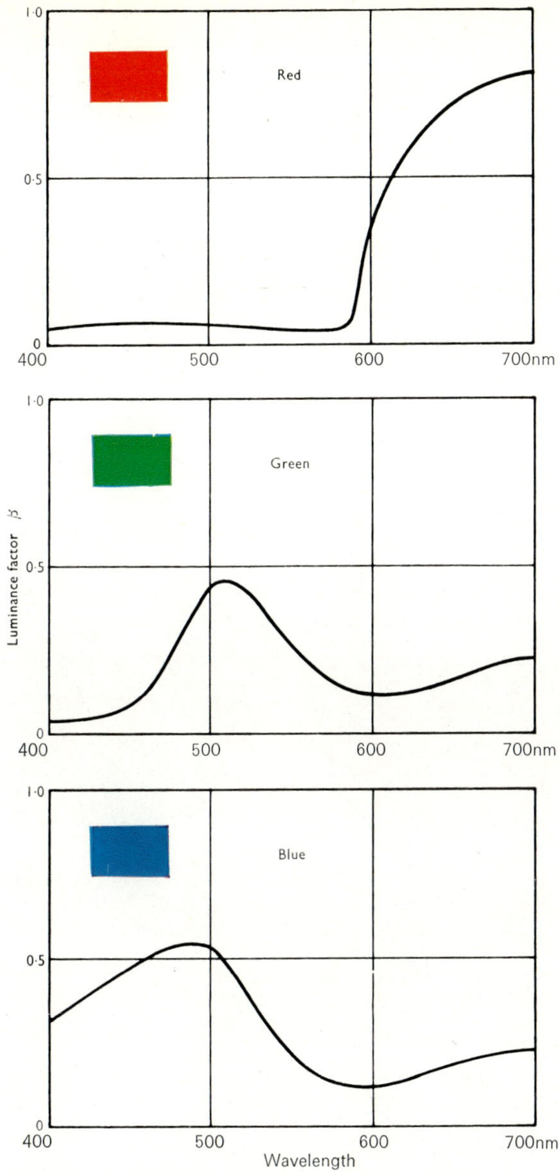

Spectral reflection curves for typical
red, green, and blue pigments (see page 47)

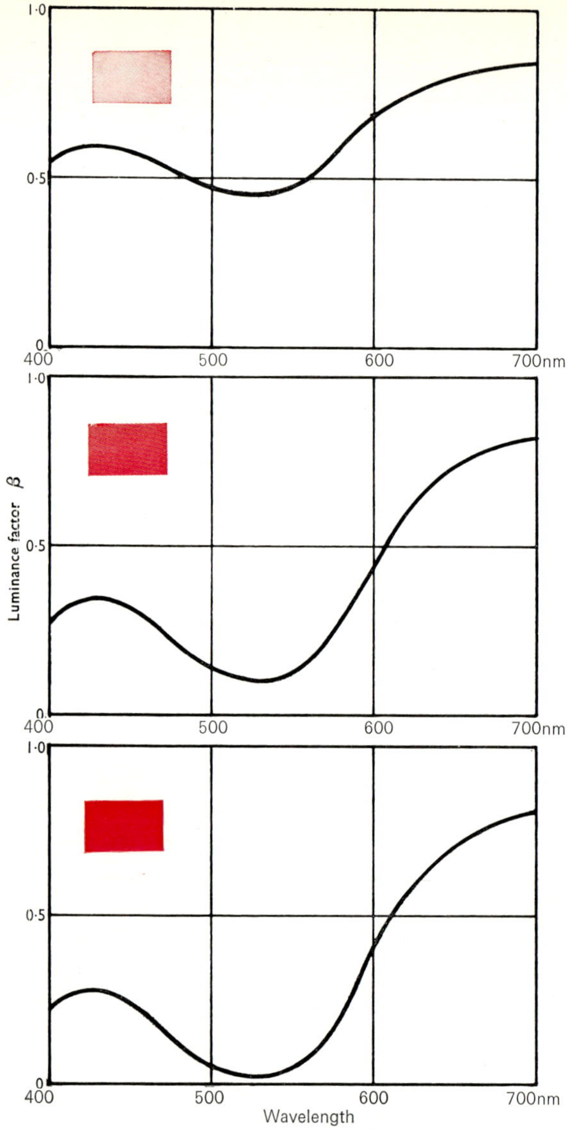

Effect of decreasing saturation on the spectral reflection of a magenta pigment (see page 47)

essentially the receptors for low-level (scotopic) vision, while the cones operate at high (photopic) illumination levels. There is, however, an overlapping range where both rods and cones are functioning simultaneously (mesopic vision). Scotopic vision is in monochrome, photopic vision is coloured. The cones, in particular the foveal cones, are the receptors which give high resolution, while the rod system sacrifices visual acuity for light sensitivity.

As their names imply, the rods and cones differ in shape, although in the fovea itself the cones, while most highly developed, become elongated into a very fine rod-like shape, no doubt to acquire minimum diameter and maximum resolving power (Polyak). The distribution of rods and cones across the retina is also very significant. The central area, the fovea, is populated only by tightly packed cones, the completely rod-free area extending over a diameter of some 1° to 2°, as measured by its angular subtense in the visual field. With increasing distance from the fovea, the cone/rod ratio steadily decreases until the peripheral zones of the retina are populated almost entirely by rods. In consequence, we find the visual acuity of the central area very high, that of the periphery low; colour discrimination is maximal in the centre of the field of view, and is almost non-existent at the margin of the field; at high intensities, when the rods appear to lose the greater part of their activity, the area most sensitive to light is the central area; at low intensities, the fovea becomes a 'blind spot', and maximum light sensitivity occurs some 10° from the fovea.

Another probable source of variation of sensitivity across the retina is the yellow *macular pigment*, which most, but not all, authorities believe to be present in the retinal tissue around the foveal area. Its distribution within the fovea itself is uncertain, and its presence in the central area where the neurone layers have largely been pulled to one side (Fig. 2.5) is particularly in doubt. The position is complicated by the fact that small-field observations suggest that the blue receptors may also be absent from this area, and it is difficult to disentangle the evidence for the absorption of blue light by the yellow pigment and for insensitivity to blue light owing to the absence of blue receptors.*

One of the most significant differences between the rods and

* G. WALD: *J. Opt. Soc. Am.*, 1967, 57, p. 1289.
W. D. WRIGHT: In the press.

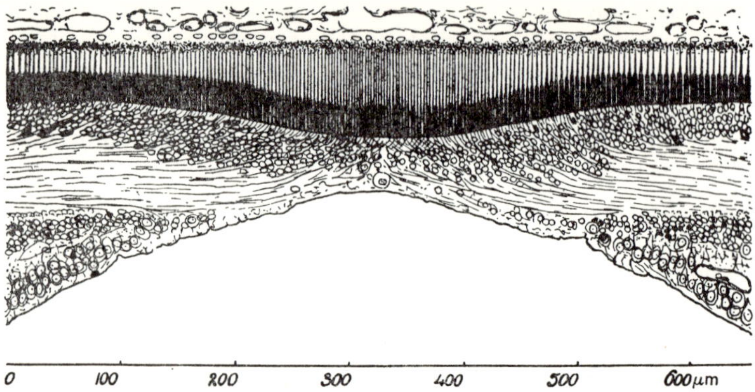

FIG. 2.5. Diagram of a section through the central fovea of the human eye (From Polyak's *The Retina*)

cones is in their spectral sensitivity. This may be recorded subjectively by finding the energy required at each wavelength in the visible spectrum to produce a response of constant magnitude; for the cones, the observations have to be made under photopic conditions or by using the foveal area of the retina, while the rod sensitivity is obtained at near-threshold illuminations using the extra-fovea. The spectral sensitivity curve is then given by plotting the inverse of the energy against wavelength. Since scotopic or rod vision is colourless, the determination of the scotopic curve presents no difficulty in principle, but with the photopic or cone curve, the observer has a more ambiguous task because of the intrusion of the colour. This question is elaborated further in Chapter 3, and we need list here only the different observing techniques which can be used, namely a threshold technique, a direct luminosity match between the monochromatic light and a fixed comparison field, a step-by-step matching procedure through the spectrum, a colour-matching technique and a flicker technique.

Fig. 2.6 shows the scotopic and photopic curves as standardized in 1951 and 1924 respectively by the C.I.E., in which V_λ and V'_λ (see p. 64) are plotted linearly against wavelength. The scotopic curve was based primarily on an investigation by Crawford,* in which the observer made a photometric match between the monochromatic light and a white comparison field of low luminence, but

* B. H. CRAWFORD: *Proc. Phys. Soc., B*, 1949, **62**, p. 321.

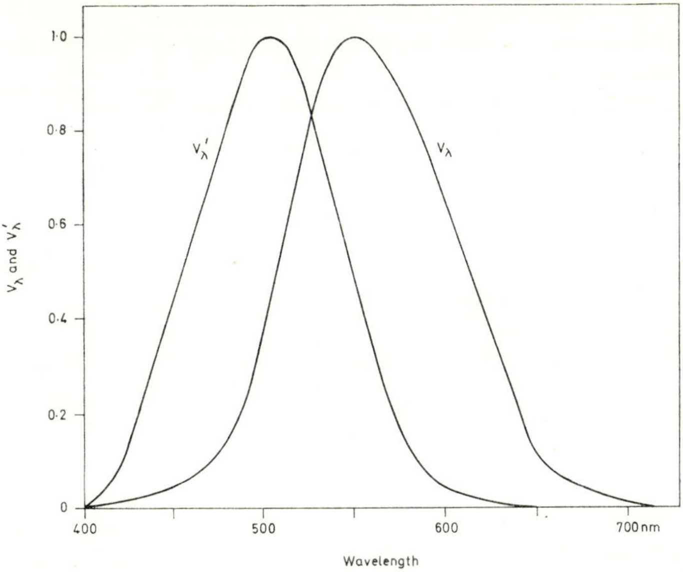

FIG. 2.6. The spectral sensitivity curves of the eye for photoptic vision (V_λ curve) and scotopic vision (V'_λ curve), as defined by the C.I.E. in 1924 and 1951 respectively. In photometric terms, these curves are given by plotting against wavelength the relative spectral luminous efficiencies of monochromatic radiation for photopic and scotopic vision

the 1924 photopic curve was a composite curve built up from several American investigations in which the flicker technique predominated. The curves show that the rods have their maximum sensitivity at a wavelength of 507 nm and the cones at one of about 555 nm, a result which is in general agreement with objective, electro-physiological data.* The difference between the two curves is responsible for the change in relative luminosity of coloured surfaces as the illumination is reduced, a change known as the *Purkinje effect*, in which the most notable feature is the darkening of red surfaces.

It is reasonable to suppose that the shape of the sensitivity curve is determined primarily by the absorption curve of the photochemical substance secreted in the receptors. This has been confirmed for

* R. GRANIT: *Receptors and Sensory Perception* (Yale University Press: New Haven), 1955.

the rods, since the absorption curve of visual purple extracted from the rods has a very close similarity to the scotopic curve, especially when a correction is made for the quantum efficiency of the light.* There is still no comparable data on visual substances in the cones, although this is no doubt due to their occurrence in smaller quantities or weaker concentration and so to the greater difficulties of their extraction.

Most problems in colorimetry are related to surfaces seen under high illuminations, for which the photopic curve applies, although certain problems do involve lower retinal illuminations. The colour aspect may virtually disappear in such problems, which then reduce to one of scotopic photometry. This will apply to luminance levels below about 10^{-4} candela/ft^2, while the photopic condition holds above about 10^{-1} candela/ft^2. In mesopic vision between these two levels, the sensitivity curve changes from the scotopic to the photopic curve so that in this region photometric and colorimetric calculations are fraught with great difficulty and uncertainty of interpretation,† and problems within these limits must be studied individually. Our concern in this book is almost entirely with the higher illumination levels, although even at these levels we cannot completely ignore the existence of the rods and their influence both on colour matching and on the subjective appearance of a colour when seen some distance from the centre of the visual field. As we shall see, allowance had to be made for rod intrusion in deriving the new colour matching functions for 10° fields.

We must now consider how the eye can distinguish one wavelength from another under photopic conditions. The perception of colour necessarily implies that within the structures illustrated in Fig. 2.3, processes must exist which are capable of reacting differently according to the wavelength of the incident light. It is conceivable, although scarcely possible, that each monochromatic radiation might be associated with a particular type of nerve impulse which has to be analysed and interpreted in the brain as a

* F. Crescitelli and H. J. A. Dartnall: *Nature*, 1953, **172**, p. 195.

† See, e.g.:

W. D. Wright: *Photometry and the Eye* (Hatton Press: London), 1949.

H. A. E. Keitz: *Light Calculations and Measurements* (Philips Technical Library: Eindhoven), 1955.

D. A. Palmer: *Nature*, 1966, **209**, p. 276; *Vision Research*, 1967, **7**, p. 619.

particular colour; in that event, the colour-mixture relations which are known to exist between spectral stimuli would have their physiological counterpart in the brain. There is, however, good reason to believe that there are only three types of receptor among the cone processes, and that the whole gamut of colour sensations is derived from variations in the magnitude of the responses in these three systems. The most important support for this theory, which was originally postulated by Young and elaborated by Helmholtz, is the fact that, in order to reproduce a colour, it is necessary to take only three radiations, such as a red, a green and a blue, and mix them in the correct proportions. There is no need to amplify the experimental phenomena here, since the greater part of the remainder of this book is devoted to that purpose; the essential conclusions to be drawn from this key experiment are that colour vision is a function of three variables and that the physiological processes must at some point sub-divide into three channels.

A great deal of ingenuity has been expended in the past to discover the nature of this sub-division. There are a number of possibilities, of which perhaps the simplest is to assume that there are three kinds of photochemical substance secreted in the cones, with the absorption maxima of the substances located at different regions of the visible spectrum. No such substances have yet been isolated from any one retina, neither has histological examination revealed any difference in structure between one cone and another, but there is considerable activity in this field. Thus, Polyak* considers that there is a basis for a triple mechanism in the different types of bipolar cell which he has observed in the retina, while Le Gros Clark† has reported evidence for three types of fibre in the lateral geniculate body, which may be regarded as a kind of relay-station between the retina and the visual cortex; Granit,‡ in his electro-physiological studies, has found cone receptor units in mammalian eyes which possess widely different spectral sensitivities; Wald, Dartnall§ and others have been re-

* S. L. Polyak: *The Retina* (Chicago University Press: Chicago), 1941.

† W. E. Le Gros Clark: *J. Anat.*, 1941, **75**, p. 225.

‡ R. Granit: *Receptors and Sensory Perception* (Yale University Press: New Haven), 1955.

§ G. Wald: *Ann. Rev. Biochem.*, 1953, **22**, p. 497.
H. J. A. Dartnall: *Brit. Med. Bull.*, 1953, **9**, p. 24.

sponsible for extensive studies of the photochemistry of the retina; Inglestam* has investigated the extent to which waveguide principles might be applied to the retinal receptors leading perhaps to a mechanism for wavelength discrimination; while Rushton and his colleagues and Weale† have shown that with a suitably designed optical system, the light reflected back from the retina may be used in a very remarkable manner to record objectively the changes in photochemical concentration in a living eye.

Most recently, the micro-spectrophotometer has been used by Marks, Dobelle and MacNichol‡ and by Brown and Wald§ to measure the spectral absorption of individual cones in a human retina kept active for some hours after death. The evidence which this technique is yielding tends to confirm completely Young's original concept of three types of colour receptor, but for the moment we must receive the results with some caution. Apart from the biological hurdles to be overcome in perfecting such a technique, there are optical difficulties as well, as Enoch¶ has reminded us. Nevertheless, this is potentially the most direct and convincing approach yet devised for confirming the existence of three separate colour receptors in the retina.

Further subjective evidence for the existence of a limited number of receptor types will be considered later in this chapter, but for the purpose of technical colorimetry we can postulate three more-or-less distinct processes within the cone systems of Fig. 2.3, all of them sensitive to wide wavelength ranges but with maximal sensitivities at three separated regions of the spectrum. Various sensitivity curves have been proposed from time to time as a basis for colour perception and a typical set derived by Thomson and Wright|| is shown in Fig. 2.7, although there is as yet no general agreement as to the correctness of one set of curves or another.

* E. Ingelstam: *Problems in Contemporary Optics* (Istituto Nazionale di Ottica: Florence), 1956, p. 640.

† W. A. Hagins and W. A. H. Rushton, *J. Physiol.*, 1953, **120**, p. 61 *P*.
R. A. Weale: *J. Physiol.*, 1953, **122**, p. 322.
W. A. H. Rushton: *J. Physiol.*, 1956, **134**, p. 11; *J. Colour Group*, 1965, No. 4, p. 22.
R. A. Weale: *The Eye and its Function* (Hatton Press: London), 1960.

‡ W. B. Marks, W. H. Dobelle and E. F. MacNichol: *Science*, 1964, **143**, p. 1181.

§ P. K. Brown and G. Wald: *Science*, 1964, **144**, p. 45.

¶ J. M. Enoch: *J. Opt. Soc. Am.*, 1966, **56**, p. 833.

|| L. C. Thomson and W. D. Wright: *J. Opt. Soc. Am.*, 1953, **43**, p. 890.

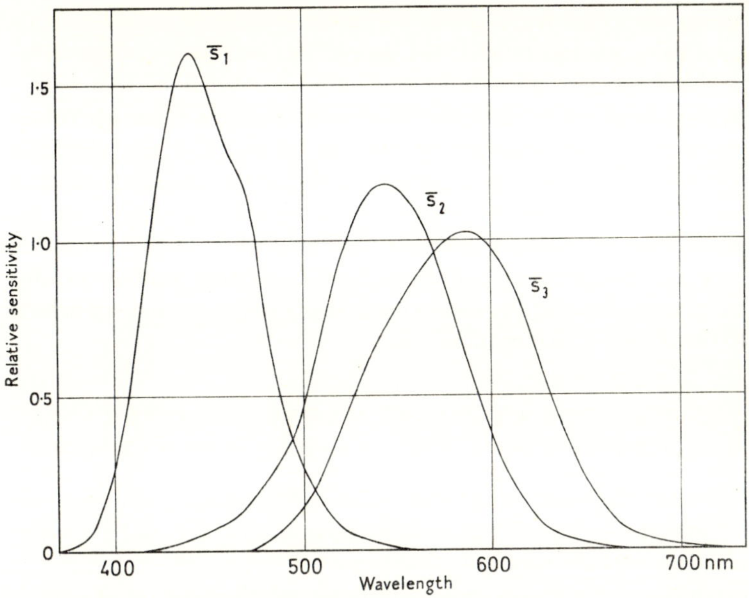

FIG. 2.7. Typical set of colour sensitivity curves on the three-components theory, as derived by Thomson and Wright

We then assume that, when light strikes the retina, the reaction in each receptor process will be a function of the distribution of the energy in the stimulus and the spectral sensitivity of the receptors. How each response preserves its identity as it travels to the brain has yet to be demonstrated and this is a question which demands an answer whatever theory of colour vision is accepted. There may, indeed, be some re-coding of the neural response even before the colour information leaves the retina. An indication that this may be the case is provided by recordings of nerve impulses in the lateral geniculate nucleus of the macaque monkey. Thus De Valois, Abramov and Jacobs* have recorded responses from cells in the nucleus when the eye has been exposed to flashes of monochromatic light of various wavelengths. The cells were found to divide into two general classes, one class apparently responsible for the transmission of intensity information, the other class for colour. The 'colour' cells responded by excitation or inhibition

* R. L. DE VALOIS, I. ABRAMOV and G. H. JACOBS: *J. Opt. Soc. Am.*, 1966, **56**, p. 966.

according to the wavelength, and were coupled in opponent pairs by their response to red and green, and yellow and blue. This certainly suggests that the colour information may be transmitted to the visual cortex in accordance with Hering's opponent colour theory.* The process would still, however, be a three-variable process. When the three responses arrive at the visual cortex, they may be regarded as the physiological stimuli responsible for the production of a colour sensation, although the ultimate sensation may be modified, as indicated in the opening section of this chapter, by interaction with other sensory elements that contribute to the complete perceptual pattern.

THE RELATION BETWEEN SPECTRAL COMPOSITION AND VISUAL SENSATION

Although the colour appearance of a light source or a surface will be affected by the state of adaptation of the eye and by other physiological and psychological factors, it is possible and useful to establish some broad generalizations relating the colour of a sensation with the spectral composition of the light causing the sensation. Thus white light, for example, is generally associated with a light source radiating over the whole of the visible spectrum with no one region greatly in excess over any other region. Experiments on colour mixture reveal that this is by no means the whole truth, and examination of the energy distribution for typical sources of white light, such as those given in Fig. 1.2, show how widely the statement has to be interpreted, but it remains nevertheless a useful generalization.

When the energy distribution loses these characteristics, the light in general appears coloured. The colouring power of a dye or pigment is nothing more and nothing less than the power to reduce the energy in certain parts of the spectrum relative to the remainder. If we measure the spectral quality of the light reflected from different coloured surfaces, we find that red surfaces reflect highly at the long-wave end of the spectrum, but the reflecting power falls off in the region of 580 nm and remains comparatively small through the rest of the spectrum; a yellow surface usually

* E. HERING: *Outline of a Theory of the Light Sense*, trans. by L. M. HURVICH and D. JAMESON (Harvard University Press), 1964.

absorbs heavily at the short-wave end up to about 480 nm, but reflects well from 550 nm onwards; greens reflect maximally between 480 nm and 560 nm and absorb at either end of the spectrum; blues reflect well in the short-wave region up to, say 530 nm and absorb elsewhere; purples reflect light from each end of the spectrum, but absorb in the middle regions. These descriptions are obviously over-simplified and give no indication of the varied types of absorption shown by individual dyes or pigments, but their general aptness is illustrated by Plate 4, in which reflection curves for typical red, green and blue pigments are shown. White surfaces, it should be noted, can be defined more precisely than white light sources; thus a surface is white when it possesses a high luminance factor at all wavelengths through the spectrum. When one part of the spectrum is absorbed more than another, the surface becomes tinted, and when the luminance factor at all wavelengths is reduced, the surface appears grey.

If we pursue the matter further to discover why one part of the spectrum is generally predominant for a surface of one colour, and another region for a different colour, we are naturally led to a consideration of the spectrum itself. When the spectrum is examined, it is found, of course, that the long waves from, say, 630 nm upwards are red, from 590 nm to 630 nm orange, from 570 nm to 590 nm yellow, from 550 nm to 570 nm yellow-green, from 510 nm to 550 nm green, from 480 nm to 510 nm blue-green, from 450 nm to 480 nm blue and from 400 nm to 450 nm violet. The gradual transition from one hue to the next makes the boundaries between them ill-defined, and their exact limits are to some extent a matter of opinion. Once the relation between any given wavelength and its associated colour sensation has been recognized, it is a natural corollary to expect that, when a number of radiations are united in a single beam, the colour of the beam will be characterized mainly by the colour of those radiations which are predominant; hence the redness of the pigment illustrated in Plate 4(*a*), the greenness of (*b*) and the blueness of (*c*).

While the hue of a pigment may be similar to that of a particular spectral radiation, its depth of colour will depend on its concentration. This is illustrated in Plate 5 which gives the reflection curves for three densities of a magenta ink. As the density of the ink increases, so the colour becomes deeper or more saturated. This can be understood in terms of the spectral reflection curve, since the top curve for the palest of the three inks does not drop below a level of some 40 or 50 per cent reflection. This means that there is a massive white component on to which the excess red and blue reflections are superimposed. With the higher concentration or density, this white component diminishes almost to zero, leaving a spectral composition consisting of broad reflection bands in the blue and red and hence a strong magenta coloration.

So far, then, we have correlated the physical composition of a beam of light with the visual sensation experienced by the observer to the extent that, broadly speaking, the hue is determined by the position in the spectrum of those radiations that are markedly stronger than the remainder, while the saturation is governed by the degree to which these radiations predominate over the others. Later discussion of the laws of colour mixture will reveal that the situation is in fact far more complex than this, but from the facts we have considered and as a first description of the colours we commonly see, the inferences we have drawn may be regarded as legitimate.

We have now to examine the bearing which the total amount of light has on the quality and intensity of the visual sensation. Changes in the amount of light may occur under three distinct conditions, in which the corresponding changes in visual appearance are not necessarily identical. The first is that of the self-luminous object, including the illuminated field of an optical instrument such as a colorimeter or a photometer; the second that of a diffusely reflecting surface seen under various levels of illumination; the third that of a diffusely reflecting surface in which the reflected light can be reduced by making the surface darker.

For self-luminous objects, the more intense the source, the greater its luminosity. When two sources which have the same spectral composition are compared side by side, the one emitting the greater amount of radiant flux per unit area per unit solid angle will appear the brighter. The apparent brightness will admittedly be affected by the state of adaptation of the eye, about which more will be said later in this chapter, but for a given state of adaptation we can assume that the greater the physical energy, the greater the intensity of the visual sensation.

On the other hand, the change in the appearance of a diffusely reflecting surface with change in illumination level is generally small. Non-luminous surfaces are judged almost entirely by comparison with surrounding surfaces; hence an object observed, say, in a hall illuminated by a hundred lamps will alter little in appearance if ninety of the lamps are switched off, since the change in illumination affects all the objects in the room equally. It is true that we are conscious of the lower illumination in the room as a whole, but by some process of mental adjustment we allow for this in judging the appearance of a surface, so that a sheet of white paper still looks white and not grey, an orange surface continues to look orange and not brown, and so on. Moreover, after a few minutes in the hall, the light sensitivity of the retina adjusts itself to the reduced illumination and the hall may look almost as bright as before. For this condition, therefore, there is in general little correlation between amount of light and visual sensation; the important physical characteristic is the fractional, and not the absolute amount of light reflected by the surface, and this is what we perceive.

The third condition we have to consider is that of the diffusely reflecting surface in which the amount of reflected light is reduced by what we may conveniently describe as 'blackening'. We are not here concerned with the process by which the surface is dulled or blackened, but we will make the assumption that the blackening is a strictly neutral process in which the relative energy distribution of the light reflected from the sample is not disturbed. We are now faced with a most interesting phenomenon. If we take, say, a white surface and blacken it so that it is now only one-tenth as effective a reflector, the surface will appear to be a dark grey; if, on the other hand, we had kept the surface white but had reduced the illumination to one-tenth of its initial value, the surface would still have looked white, although the physical stimuli entering the eye would have been identical in the two cases. In the same way, an orange surface that is blackened looks brown, a green surface blackened may become an olive-green, and so on. We may conclude that where surfaces are seen in relation to each other and under a recognizable system of illumination, the lightness or darkness of a given surface is closely correlated with the fraction of the incident light reflected by the surface but has little correlation with the absolute amount of light in the reflected beam.

THE GEOMETRICAL REPRESENTATION OF SURFACE COLOURS

Surface colours occupy such an important place in the world of colour and are so little subject to change in appearance with change in level or, surprisingly enough, with moderate change in quality of illumination, that it is convenient to develop a three-dimensional structure within which any surface colour can be located by its appearance, i.e. by its lightness, hue and saturation. A number of solids have been described from time to time, but they have had, in the main, the same general features, which are illustrated in Fig. 2.8.

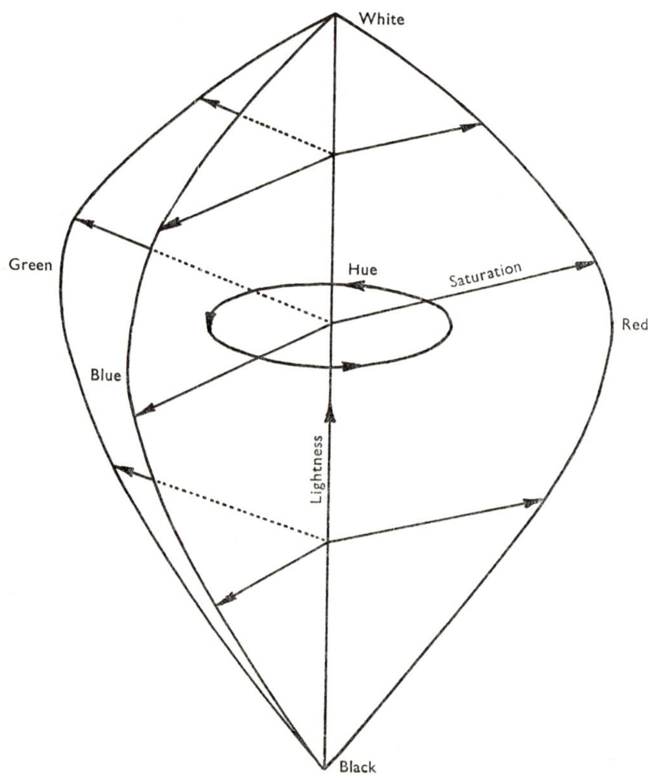

FIG. 2.8. The representation of surface colours in a three-dimensional figure

The central vertical axis represents the locus of greys, with black at the lower and white at the upper extremity. The distance from black to white fixes the scale of the solid. Colours of any one hue can be located in a vertical plane with the axis as one of its limiting ordinates. The height of a sample above the 'black' level will be determined by its lightness, while its distance from the black-and-white axis indicates its saturation. If the plane is assumed to rotate around the axis, it will pass through successive hues of red, orange, yellow, etc.

The exact distribution of the colours in a solid might be made on the basis of any one of a number of criteria. It might be decided that the most saturated colours which can be produced should lie on a circle around the central axis; or that complementary colours should lie on opposite sides of the axis; or that the colours produced by additively mixing any two colours together should lie on the straight line joining them; or perhaps that any short length in the solid, no matter in what region or in what direction it be situated, should correspond to the same subjective difference in colour. These conditions are not mutually compatible, so that the criterion adopted must be chosen in relation to the purpose for which the solid is required. One aim may be to produce a framework within which the samples in a colour atlas might be located, another may be to establish the principles of colour harmony, which are naturally related to the appearance of colours and not to their physical specification. If, however, an attempt is made to impose too many conditions on the colour solid, it will become overconstrained. The solid cannot in itself serve as a fundamental system of colour specification, since its shape has to be determined by the position in which given colour samples have to be placed; in other words, the framework is defined by the samples, not the samples by the framework. If the samples were to fade or become discoloured, there would be no reliable indication of this, other than by comparison with freshly prepared samples, and hence the standardization would depend on the successful repetition of chemical prescriptions and not on the reproduction of a particular quality of light stimulus. On the other hand, given a physically defined system of specification such as that described in Chapter 4, each sample in the colour solid could be measured and specified in terms of the standard system and then located in the solid according to its subjective appearance; the solid might then serve as a most useful secondary reference system in which to describe a colour. However, if a colour solid is required for purposes of numerical specification, some agreed and defined quality of illumination must be adopted; the relative positions of the samples may not alter very perceptibly with illumination, but some shift must be expected.

A more detailed discussion of surface colours, colour charts and the colour solid is given in Chapter 7.

ADAPTATION AND CONTRAST

Reference has already been made to the existence in the retina of two kinds of receptor, the rod and the cone, which operate respectively at low and high illumination levels. The existence of these two systems in itself ensures that the eye can operate over a wide range of illuminations, but the sensitivity of each system also adapts as the illumination rises or falls. This is believed to occur in part through the varying photochemical concentration as the light-sensitive material is decomposed in the light and regenerated in the dark, but probably also in part through adaptation in the neural processes, and by summation and convergence in the visual pathways.

Subjectively, adaptation is a continually recurring experience whenever we go from a higher to a lower illumination, and *vice versa*. Faint lights become more visible when we are dark-adapted, and corresponding changes in apparent brightness occur with sources well above the threshold: thus, a motor-car headlight seen during the day does not appear unduly bright, but when seen by the dark-adapted eye at night will appear quite glaring.

Changes in colour sensitivity are less frequently noticed, although they, too, are continually occurring and may be quite large. A yellow patch, for example, can be made to look green by first adapting the eye to a red stimulus for a short time. The physiological explanation that can be given must in this case be more tentative, but in essence we assume that adaptation to red light reduces the sensitivity of the red receptor system more than that of the green or blue. If the yellow stimulus initially gave rise to fairly equal activities in the red and green processes, then after adaptation of the eye to red light, the green response would predominate. Effects of a similar character occur when the eye is adapted to other coloured radiations.

Less striking, but nevertheless important, changes are also induced in the appearance of an object through contrast with neighbouring areas in the field of view. Thus, the yellow of a primrose is deepened by contrast with the surrounding green leaves, and the sky immediately above the horizon appears especially light in contrast to the darker sea below. There are probably a number of factors which contribute to the production of contrast phenomena: histological and electro-physiological studies provide evidence for some interaction between neighbouring retinal areas; the continual movement of the eyes means that an area of the retina which at one moment receives light from one part of the field is at the next stimulated by a different region; the light scattered within the eye from a bright area leads to a haze of light over other areas; perceptual processes within the brain itself may enhance differences in colour, just as the height of a tall man and the shortness of a small man are enhanced when the two are seen side by side. We must note, however, that in a colour pattern, very striking and unexpected colour changes can occur. These changes are a function of the fineness of the pattern and the angular subtense of the areas of the pattern at the observer's eye, as illustrated in Plate 6, facing p. 86. As Evans has rightly pointed out,* these effects have not received the attention they deserve, yet they are of very real interest to our understanding of colour perception and of quite vital importance to the use of colour in design.† Recent studies by Gindy and by Melkman,‡ however, have included quantitative measurements of the colour changes by a method of subjective colour measurement, and from their analyses it seems likely that scattered light in the eye, chromatic aberration of the optical system of the eye, simultaneous

* R. M. EVANS: *An Introduction to Color* (Wiley: New York. Chapman and Hall: London), 1948.

† P. WARBURTON and G. V. LUND: *J. Textile Inst.*, 1956, **47**, p. 305.

‡ S. S. GINDY: *Techniques for subjective colour measurement and their application to colour contrast phenomena*, 1963, Ph.D. Thesis, London University.

M. J. MELKMAN: *A Study of Colour Contrast Effects and their Measurement Using the Binocular Matching Technique*, 1967, Ph.D. Thesis, London University.

contrast, eye movements and local adaptation, all contribute to the effect.

Colour adaptation plays an important role in maintaining the colour constancy of objects seen under different qualities of illumination. The colour of daylight and the colour of tungsten light are very different when seen side by side, but objects in a room show little change in appearance whether they are seen during the day by the light from the window or at night by the light from tungsten lamps. The yellowness of the tungsten illumination and the blueness of daylight are largely compensated by corresponding degrees of specific yellow and blue adaptation in the two cases. Nevertheless, exact compensation rarely occurs, and with light sources such as the mercury lamp with its line spectrum, the colour distortion may be considerable. The fluorescent lamp with its combination of line spectrum and phosphor spectrum Fig. 1.4, poses special colour rendering problems, as discussed in more detail in Chapter 10.

In subsequent chapters we shall elaborate the system of colour measurement based on matching a colour by the appropriate mixture of three specified radiations, and it is essential that we should understand the limited relation which this specification has to the appearance of the colour as seen under various conditions of adaptation. Most important of all is the fact that experiment shows that colour matches remain valid over a wide range of adaptation levels and conditions of contrast, at least for 2° fields; only when the eye is subjected to glaring conditions of pre-adaptation does a colour-match break down.* If this were not so, a colour specification would be virtually useless, and it is an essential assumption of colorimetry that the equivalence between the trichromatic mixture and the test colour holds over all normal conditions of viewing. Physiologically, we can deduce from this colour match stability that while the individual sensitivity levels of the three receptor mechanisms may be independently affected by the adaptation, the shape of their spectral sensitivity curves remains effectively unchanged. Yet this stability of matching also means that a colour specification cannot take account of the changes in appearance of a colour induced by adaptation, and it is for this

* W. D. Wright: *J. Physiol.*, 1936, **87**, p. 23.
G. S. Brindley: *J. Physiol.*, 1953, **122**, p. 332.

reason that we must regard a tristimulus specification as a measure of the quality of the stimulus rather than of the quality of the sensation.

This may be regarded as a limitation of colour measurement, and it must be admitted that we are at present a long way from establishing a system on which colour appearance can be specified exactly, yet experience shows that colorimetry as at present practised has many applications and is, indeed, an essential stage towards the specification of colour appearance.

VARIATION OF COLOUR PERCEPTION ACROSS THE RETINA AND AMONG INDIVIDUALS

In view of the varying structure of the retina from fovea to periphery, it is not surprising to find that the perception of colour varies from one area to another. Even within the fovea there are significant differences, as revealed, for example, by the threshold sensitivity to light of different wavelengths. Thus Stiles* has shown that, while the very centre of the fovea is an area of maximum sensitivity to red wavelengths, there is a sharp drop there in sensitivity to blue wavelengths. Colour matching with small fields subtending 20 minutes of arc or less shows that the foveal centre is in fact dichromatic,† since all colours can be matched by a mixture of only two stimuli. It is probably true that this dichromatism, which is similar in general characteristics to the colour defect known as tritanopia, is associated with the smallness of the field rather than the retinal area used to view it,‡ yet the deficiency of the foveal centre is especially striking in view of its otherwise high acuity and discriminating power. The evidence certainly suggests that the foveal centre is devoid of blue receptors, and therefore differs in this respect from other retinal areas.

* W. S. Stiles: *Doc. Opthal.*, 1949, **3**, p. 138.
G. Wyszecki and W. S. Stiles: *Color Science* (Wiley: New York and London), 1967, p. 572.

† A. König: *S. B. Akad. Wiss. Berl.*, 1894, 21st June, p. 577.
E. N. Willmer and W. D. Wright: *Nature*, 1945, **156**, p. 119.

‡ H. Hartridge: *Nature*, 1945, **155**, pp. 391 and 657.
L. C. Thomson and W. D. Wright: *J. Physiol.*, 1947, **105**, p. 316.
G. Wald: *J. Opt. Soc. Am.*, 1967, **57**, p. 1289.

Peripheral colour vision has long been known to differ from foveal colour vision, but it is only comparatively recently that extensive quantitative studies have been undertaken.* These show, among many other results, the desaturation of the colours perceived peripherally as compared with their foveal appearance, a reduced colour discrimination, a tendency towards dichromatic vision, and an enhanced blue sensitivity. Such effects can be explained in part by the admixture of rod vision with cone vision, and by the reduction in the density of the cone population in the peripheral retina. The colour receptors in the periphery may also have characteristics which differ from the cone characteristics in the fovea, and it may well be that peripheral vision corresponds to an earlier stage of evolution of the colour sense.

One of the most striking features of extrafoveal vision is the speed and extent of the local adaptation which occurs when a test light is viewed steadily on a given area of the peripheral retina. This can lead to the complete fading of a visual image and has been the subject of intensive study by Clarke,† who has named the fading Troxler's effect. It is the peripheral equivalent of the effects studied by Ditchburn and others‡ at the fovea, using special devices attached to the observer's eye to produce a stabilized retinal image at the fovea. In the case of the periphery its poorer visual acuity implies that effective stabilization can be secured by less complicated optical arrangements, and normal foveal viewing of a fixation target is generally sufficient.

With large areas of colour, both the foveal and peripheral receptors are in action simultaneously, but the relative contributions of the two groups of receptors to the determination of the overall colour appearance are not easily determined. Eye movements, of course, ensure that any given object is scanned by the fovea as well as being seen by indirect extra-foveal vision, but it is surprising that we are not more frequently aware of the desaturated quality of peripherally perceived colours. Only with certain dichroic dyes does a uniformly-coloured area appear subjectively non-uniform and this is presumably due to the

* M. Gilbert: *Proc. Phys. Soc.*, *B*, 1950, **63**, p. 83.
R. A. Weale: *J. Physiol.*, 1953, **119**, p. 170.
J. D. Moreland and A. Cruz: *Optica Acta*, 1956, **6**, p. 117.

† F. J. J. Clarke, *Optica Acta*, 1960, **7**, pp. 219 and 355; 1961, **8**, p. 121.

‡ R. W. Ditchburn, *Nature*, 1963, **198**, p. 630.

absorption of the blue end of the spectrum by the yellow macular pigment, Fig. 2.9.* A reddish patch, sometimes referred to as *Maxwell's spot*, can then be seen, but even this rapidly disappears as local adaptation compensates for the absorption.

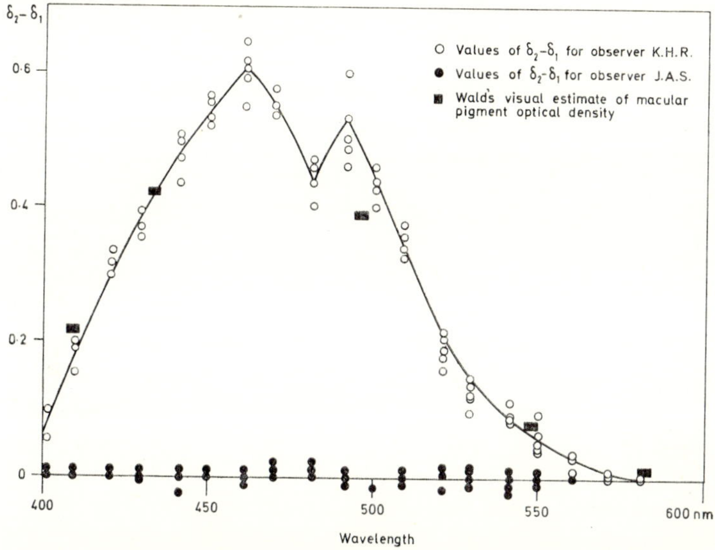

FIG. 2.9. The optical density of the yellow macular pigment for a heavily pigmented subject (K.H.R.) compared with a subject (J.A.S.) with little or no pigment. (Ruddock)

Maxwell's spot, which with many observers shows interesting internal structure, has been the subject of considerable study† and is one of the major and most intractable causes of variation in colour vision between one observer and another. The extent to which this can affect colour matching is discussed in Chapter 4 and is illustrated in Fig. 4.3, but evidence is also given there to show that observers may differ in respect of the spectral sensitivity of the receptor systems themselves, as well as in the density of the yellow macular pigment. The crystalline lens of the eye also be-

* K. H. RUDDOCK: *Vision Research*, 1963, **3**, p. 417.

† See, e.g., G. L. WALLS and R. W. MATHEWS: *New Means of Studying Color Blindness and Normal Foveal Color-Vision* (University of California Press: Berkeley and Los Angeles), 1953; K. H. RUDDOCK: *Vision Research, 1963*, **3**, p. 417; and J. A. SPENCER: *Brit. J. Physiol. Optics*, 1967, **24**, p. 103.

comes yellow with age and this can be a significant cause of differences of colour vision between observers.*

These variations have probably given rise to more trouble in visual colorimetry than any other difficulty, and one of the major aims has been to devise methods of measurement in which the effects of individual variations of colour vision are reduced to a minimum. The adoption of standard observer data has made the problem easier to the extent that the goal to be aimed at is precise and definite, and by using physical methods of spectrophotometry or photoelectric tristimulus colorimetry, the observer problem is avoided. The Donaldson six-stimulus colorimeter (p. 226) provides the closest approach to a visual solution of the problem.

Yet if standard colour measurements have largely to be made with physical apparatus, a vast number of colour matches are still made by eye every hour of the day and every day of the week. These matches range from those made by a customer in a shop to those of a trained colourist in a dye-house, and observer differences can be very significant in these situations. Also, some observers are much more critical in their judgement than others, apart altogether from any pigmentary and other differences of vision. For this reason, tests have been devised from time to time to determine the ability of a person to detect small differences, and the most recent and elaborate of these is the *colour aptitude test.*† This is a colour-matching test in which a number of samples have to be matched against a battery of reference colours. The differences are very small and, if the time is limited, it is virtually impossible to make a completely correct series of judgements. The final score is thus a measure not only of the visual discrimination of the observer, but also of his confidence in making the judgements. It therefore provides a critical test of the candidate's ability to make colour matches under exacting conditions.

DEFECTIVE COLOUR VISION

In addition to differences of colour vision and discrimination

* R. A. Weale: *The Aging Eye* (Lewis: London), 1963.
F. S. Said and R. A. Weale: *Gerontologia*, 1959, **3**, p. 213.
K. H. Ruddock: *Vision Research*, 1965, **5**, p. 47.

† F. L. Dimmick: *J. App. Psych.*, 1946, **30**, p. 10. *J. Opt. Soc. Am.*, 1956, **46**, p. 389.

among normal observers, there are some eight per cent of males and a much smaller percentage of females who possess more or less grossly defective colour vision. The borderline between excessive variation from the normal and the initial stages of defective vision is ill-defined, but the outstanding characteristic of 'colour blindness', to use a familiar but misleading description of the defect, is an inability to detect colour differences which can be readily distinguished by the normal observer. In addition, there is generally a pronounced tendency to record matches in a colorimeter which are patently wrong to a more normal eye, while in some cases the photopic sensitivity curve differs considerably from that shown in Fig. 2.6. A defective observer cannot be employed for making visual measurements with a colorimeter, partly because he is unable to repeat his matches with sufficient accuracy and partly because his average reading is likely to depart widely from the true value. He is also likely to be handicapped in colour work generally, since his description of a sample may not tally with that given by a normal person, nor will he be able to recognize errors in colour quality so readily.

In the more serious cases, amounting to at least 30 per cent of all colour-defective observers, colour matching is reduced to a function of two variables instead of three. Such dichromats as they are called can match colours merely by mixing two radiations taken from either end of the spectrum, from which it can be reasonably deduced that only two forms of receptor system are functioning. There are three main types of dichromat—the protanope, deuteranope and tritanope—although the last type is very much rarer than the others.* Both the protanope and the deuteranope have poor colour discrimination in the red-to-green part of the spectrum, whereas the tritanope is unable to distinguish blues from greens; the spectral sensitivity of the protanope is also much reduced at the long-wave end of the spectrum, so that reds appear darker than in normal vision. There are also partially defective observers known as anomalous trichromats who are intermediate between the dichromats and normal trichromats, and, at the other extreme, the very rare monochromats who have no colour discrimination at all.† Classification of the various forms of defect

* W. D. Wright: *J. Opt. Soc. Am.*, 1952, **42**, p. 59.
† R. A. Weale: *J. Physiol.*, 1953, **121**, p. 548.

has been described in detail elsewhere,* but we may note here that in the more common forms the most striking feature is a reduced ability or complete inability to distinguish colours in the red-yellow-green range.

Tests for colour defects may, broadly speaking, be carried out with one of two objects in view. They may be undertaken, on the one hand, as a purely scientific investigation to discover the incidence and nature of the defect and, on the other, to decide whether an individual is capable of undertaking certain types of work in which colour discrimination is involved. The latter may include not only occupations requiring the correct recognition of signal lights, as on the railways or at sea, but also many trades and professions, such as dyeing, printing, paint, geology, cartography and photography, as well as industries such as radio engineering, where colour coding is widely used. A most valuable report on this subject has been published by the Physical Society;† in it examples from many industries are described, and the need for adequate testing, both for the benefit of the employee and for that of the employer, is emphasized.

To test whether a man is fitted for a certain type of work, it is perfectly legitimate to give a trade test in which he is asked to perform operations similar to those he would have to carry out in practice. Two provisos only need be made in regard to this form of test. First, the conditions of the test should be similar to those encountered in the works; thus, if the wiring of colour-coded cables sometimes has to be checked when the cables are dirty or the lighting is poor, then the tests should also be made under these conditions. Secondly, it must be recognized that the results of the test do not necessarily indicate whether a person is or is not colour defective, but only whether he can perform certain operations. Hence, tests which are adequate as trade tests are not necessarily very suitable for clinical diagnosis.

Complete diagnosis of colour-defective vision calls for the measurement of the photopic sensitivity curve, spectral mixture

* D. B. Judd: *J. Opt. Soc. Am.*, 1943, **33**, p. 294.
W. D. Wright: *Brit. Med. Bull.*, 1953, **9**, p. 36. *Ann. Roy. Coll. Surgeons Eng.*, 1957, **20**, p. 177. *The Rays are not Coloured* (Adam Hilger: London), 1967.

† *A Report on Defective Colour Vision in Industry* (Physical Society: London), 1964.

data and wavelength discrimination with the aid of fairly elaborate spectroscopic equipment.* For most purposes, however, a simpler test is adequate. One well-known type of test consists of a series of cards on which a variegated pattern of coloured dots is printed to form a figure or letter of one colour outlined against a background of a different colour. The observer with normal colour vision can generally distinguish the figure with ease, while the defective observer may either fail to see any number or see a different number. The success of the test depends on the proper choice of colours and their arrangement in the pattern, one of the most effective series being the Ishihara Charts published in Japan. Other examples of this kind of test are the Swedish Bostrom test, the American Optical HRR test, and the Dvorine test also produced in America. The relative effectiveness of these different tests has been examined in a survey of some 500 subjects by Belcher, Greenshields and Wright.†

Another form of test, especially useful for classifying the type of defect, is the Nagel anomaloscope. This consists of an optical system incorporating a direct-vision spectroscope prism and arranged to provide a small photometric field, one half of which is illuminated with yellow light from the spectrum, while the other consists of a mixture of spectral red and spectral green. The observer is required to vary the proportion of red to green on the one side and to vary the intensity of the yellow on the other, until a match between the two is secured. The nature of the defect is indicated by the need for excess red or excess green in the match and by the accuracy to which observations can be repeated.

A third type of test, of general application where the recognition of coloured signals is involved, consists of a lantern in which a pair of apertures can be illuminated by light of various colours and luminance. In the Martin lantern,‡ a modern development of the test, two sizes of aperture are available, of diameter 0·2 in. and 0·02 in., and, as normally used, each aperture can be illuminated in red, green or white light. Provision is also made for

* W. D. Wright: *Researches on Normal and Defective Colour Vision* (Kimpton: London), 1946.

† S. J. Belcher, K. W. Greenshields and W. D. Wright: *Brit. J. Ophthal.* 1958, **42**, p. 355.

‡ L. C. Martin: *Brit. J. Ophthal.*, 1939, **23**, p. 1, and 1943, **27**, p. 255.

dimming one or other of the two apertures. In making the test, the observer has to name the colours he sees when viewing the apertures from a distance of 20 ft and when they are illuminated in various combinations of colours. The test is, in effect, a trade test, but it also serves as an effective means of detecting small degrees of deficiency.

Another very interesting test is the Farnsworth-Munsell 100 Hue test,* in which four series of colours taken from the Munsell atlas (see p. 183) and embracing the whole colour circle, are mounted in Bakelite caps in such a way that a half-inch circle of colour is exposed. The terminal colours of each series are left in position, while the intermediate colours are shuffled together; the candidate then has the task of replacing them in the correct hue sequence. The steps between one colour and the next are sufficiently small so that, although a normal observer should be able to arrange the colours correctly, one or two mistakes would not be an indication of a colour defect. A defective observer is likely to make a good many mistakes in one part of the hue circle or another, and by an appropriate method of scoring, a good indication of the type and magnitude of the defect can be obtained.

No test of this kind, however, is infallible and it is good practice to use at least two, or even three, tests in any serious investigation or selection. However, when a person is first shown to be colour defective, he need not feel embarrassed about it; it is not a disease and, from his point of view, it is an advantage to know about his vision, since he can then choose a career in which his poor colour discrimination is unlikely to be a handicap.

* D. Farnsworth: *J. Opt. Soc. Am.*, 1943, **33**, p. 568.

CHAPTER 3

THE PRINCIPLES OF PHOTOMETRY AND COLORIMETRY

THE CONCEPTS AND CONVENTIONS OF PHOTOMETRY

The feature which distinguishes photometry from radiometry and is responsible for its peculiar conceptual difficulties is that radiometry is concerned merely with the measurement of radiation in terms of physical units of energy or power, while photometry has to take account of the effectiveness of the radiation in stimulating the eye and producing a visual sensation. The apparent simplicity of photometry, as suggested by the elementary experiments of our schooldays, is deceptive, and indeed the orthodox interpretation of much of photometry is now being challenged except where the sources being compared are of very nearly the same colour and are viewed under closely similar conditions of observation.

To a very considerable degree, photometry is based on a system of somewhat arbitrary conventions. These conventions are necessary if photometry is to be of use to those who buy and sell light, but unless their conventional character is recognized, photometric data may be misused and the integration of photometry with colorimetry may be misunderstood. Consider, for example, the meaning of such a simple statement as: 'The luminous flux in a beam of light is *n* lumens.'

Its interpretation is found in part, of course, in the definition of the standard source. Since 1948, the standard has been defined as a Planckian radiator (black body) at the melting point of platinum (2042° K): the luminance of such a source is specified as 60 candelas/cm^2, and the luminous flux radiated in unit solid angle by a source having an intensity of 1 candela is, by definition, 1 lumen. A standard source having an area of $n/60$ sq. cm would therefore radiate a flux of *n* lumens in unit solid angle, if we neglect the purely geometric complications associated with the finite area of the source.

This interpretation of a flux of *n* lumens follows directly from the definition of the standard source without any conceptual difficulties whatever. But suppose the light is of a different colour from the rather yellowish white of the standard source; it might, for instance, come from a gas-filled tungsten lamp or a daylight fluorescent lamp or be quite strongly coloured, as in a signal light. If we go back to first principles and examine the procedure by which measurements are made on a photometer bench, we shall recognize that the basic operation consists, in principle, of adjusting the amount of light from a standard source until equality of illumination, as judged by eye, is achieved on the two sides of the photometer field. A subjective element has thus been introduced, and if the test and standard sources are significantly different in colour, the observer will have the task of mentally abstracting and equating the luminous attributes of the two halves of the field. Assuming this can be done, the statement that the beam under test has a flux of *n* lumens would then mean that *n* lumens from the standard source produced as luminous a sensation as the beam under test. This, in fact, is the fundamental concept underlying photometry. However, the observation itself may not be easy to make and is in any case a function of the spectral sensitivity of the observer's eye; hence a standard procedure has to be agreed in order that lamps which differ in colour from the standard source can be measured and specified without ambiguity.

Two conventions have been adopted to provide the necessary photometric bridge. The first was the adoption in 1924 of the V_λ curve, defining the relative contributions which each wavelength makes in an equal-energy spectrum to the total light flux. The second was the assumption that the total light flux in a beam is given by the arithmetic sum of the individual flux contributions of the spectral components in the beam. These two conventions are implicit in the definition of the luminous flux F in a beam of light as given by the equation

$$F = K_m \int P_\lambda \cdot V_\lambda \cdot d\lambda$$

where P_λ is the amount of the radiation in watts between the wavelength limits λ and $\lambda + d\lambda$, V_λ is the relative luminous efficiency at wavelength λ, and K_m is the luminous efficiency in lumens per watt of the monochromatic radiation ($\lambda = 555$ nm) for which V_λ has its maximum value of 1·0.

Armed with this equation, we are able to calculate the transmission factor τ of a colour filter when interposed in a beam of light having a known spectral composition $P(\lambda)$; thus

$$\tau = \frac{K_m \int \tau_\lambda \, . \, P_\lambda \, . \, V_\lambda \, . \, d\lambda}{K_m \int P_\lambda \, . \, V_\lambda \, . \, d\lambda}$$

where τ_λ is the transmission of the filter at wavelength λ. Only physical measurements are involved in this derivation of τ, so that it is possible to produce beams of coloured light of known 'standard' lumen content based on the original standard source and the conventions just described, without any subjective judgement having to be made. Such beams can then be used as comparison lights against which test beams of closely similar colour can be matched and measured with greatly reduced observational difficulties and uncertainties.

Alternatively, the conventions of heterochromatic photometry can be applied by means of a physical photometer incorporating a photocell, or photocell-filter combination, with a spectral sensitivity curve similar to the V_λ curve. Provided that the photocurrent is a linear combination of the elementary currents initiated by the individual monochromatic components of the beam under test, such a physical photometer can be used to measure the light flux in a test beam relative to the standard source, and will give an answer in accordance with the standard conventions of photometry, whatever the colour of the test light.

There are, in fact, a number of established procedures by which conventional photometric data can be derived. Unfortunately, it now seems clear that the results obtained in this way are not always in accord with what we should expect; in particular, evidence is accumulating to show that when a beam of strongly coloured light is measured by one of these techniques and is found, to have a flux of say, n lumens, then in general the coloured light produces a significantly greater luminous sensation than a flux of n lumens from the standard source.*

* C. L. SANDERS and G. WYSZECKI: *C.I.E. Compte Rendu*, 1963.

This possibility should not in itself be a cause for surprise in view of the complex nature of the visual processes and of visual perception. In part the explanation could lie in the realm of psychology, since we may from experience have come to associate a certain coloration with a high or a low luminosity, an association which we may subconsciously carry over incorrectly to other conditions of observation. Then again we may find it difficult to dissociate the vividness of a colour from its luminosity. There may, too, be a failure of additivity in that part of the physiological process which is responsible for the luminosity element in coloured stimuli. Whether or not this is the case, the effect in itself is equivalent to a breakdown in photometric additivity for mixtures of stimuli of different colours, *when the mixtures lead to the comparison of stimuli that are themselves different in colour*. This most important reservation must be remembered, since in colorimetry we are also concerned with the matching of stimuli of different spectral compositions, but in such measurements complete identity of the sensations is achieved, i.e. there is a match in respect of both colour and luminosity. If additivity failed significantly in this type of comparison, the whole subject of colorimetry would virtually collapse.

It is open to question whether the failure of conventional photometry to give results in accordance with subjective impressions is of serious practical consequence. It could be argued that we are rarely interested in knowing the accurate photometric equivalence of lights which differ strongly in colour, e.g. red and green signal lights. On the other hand, the embarrassing situation can arise that a fluorescent lamp A of one colour may look brighter than a lamp B of slightly different colour, although the lumen output of B as measured photometrically is greater than that of A.

Obviously, photometric quantities must be interpreted with considerable caution when they are concerned with coloured stimuli, although this caution has also to be extended to other aspects of the measurement of light. The author has discussed this matter at length elsewhere* and we need only emphasize here the frequent need to take account of adaptation, contrast, retinal position, etc., in assessing the visual impression that a given stimulus may produce. Special considerations also apply to photometry in the

* W. D. WRIGHT: *Photometry and the Eye* (Hatton Press: London), 1949.

Purkinje range of illumination levels between photopic and scotopic levels, and when dealing with point sources and flashing lights. Colorimetry, too, has its special difficulties under these conditions, and although this book is primarily concerned with objective colour measurement as applied to fairly normal viewing conditions, reference will be made from time to time to these fascinating subtleties.

LIGHT FLUX, LUMINANCE AND LUMINANCE FACTOR

The three basic photometric quantities which enter most frequently into colorimetric calculations are light flux, luminance and luminance factor. Light flux (or luminous flux) has already been discussed in the previous section and defined by the equation

$$F = K_m \int P_\lambda \, . \, V_\lambda \, . \, d\lambda$$

Its unit is the lumen. Under normal conditions of viewing we are not conscious of the light flux itself as part of the visual image, since we usually perceive illuminated objects rather than the light that does the illuminating. But the flux is important because it determines the level of illumination, and in image-forming optical systems, including the eye, we are very much concerned with the light flux as focused on screen, photocell, photographic emulsion or retina and recorded by these various light-sensitive devices.

Luminance, which is now the internationally accepted term in place of photometric brightness, determines the light being radiated per unit area of a surface. It applies both to self-luminous sources and to surfaces that reflect or transmit light. It can be defined as the intensity in a given direction per unit area of the surface and is expressed in candelas per unit area. The numerically equivalent quantity—flux per unit area per unit solid angle —is also commonly regarded as the luminance. It may be defined algebraically by the equation

$$L = \frac{F}{a \, . \, \omega}$$

where F is the flux radiated in a given direction within the small solid angle ω by an element of the surface of effective area a.

It is then measured in lumens per unit area per unit solid angle.

Any typical scene at which we may be looking is made up of a highly complex pattern of light and colour, but each area could be analysed into elements having a given luminance and chromaticity. Luminance is therefore something which enters directly as a component of the visual image. Luminance is also fundamental to the discussion of the photometry of optical systems since it is an invariant throughout a system, if reflection and absorption losses are neglected.

The luminance factor of a surface under given conditions of illumination and viewing is defined as the ratio of its luminance to that of a perfect white diffuser under the same conditions of illumination and viewing. The symbol for luminance factor is β and its unit is a simple ratio unless expressed as the percentage luminance. The luminance factor is a quantity which is much easier to measure than the reflection factor of a surface, which is defined as the ratio of the total light reflected to the total incident light. It is also of much greater visual significance, since the lightness of a surface when viewed from a particular direction is governed by the light reflected in that direction and not by the total reflection. Moreover, the luminance factor can be measured for a variety of directions of illumination and of viewing, to give a family of polar curves defining the reflection characteristics of the surface and thus distinguishing between glossy and matt surfaces and all the intermediate stages of semi-specular reflection.

The appearance of coloured surfaces has already been discussed in Chapter 2, where it was noted that very considerable changes in illumination level could occur without any very marked change taking place in the appearance of the objects illuminated. This constancy is primarily the result of the balance set up between the adaptation of the eye and the average luminance level of the scene under regard. For this reason luminance is of less perceptual significance than luminance factor, and although from time to time the absolute luminance has to be specified in colorimetric problems, it is the luminance factor that is usually given when specifying surface colours.

We shall see in Chapter 4 that a system of colorimetry has been adopted in terms of three stimuli (X), (Y) and (Z), and these have been so chosen that the photometric value of the light is given entirely by the magnitude of (Y). As a consequence of the form in which tabulated data have been prepared for colorimetric calculations, the luminance factor of a surface is given directly from these computations by the tristimulus value, Y, except that $Y = 100\beta$. It has thus become common practice for the colour of a surface to be specified in terms of its chromaticity co-ordinates, x and y, and the tristimulus value Y. Here, Y and β specify the same property of a surface, although the use of β is much to be preferred when purely photometric questions are under consideration.

THE EXPERIMENTAL BASIS OF TRICHROMATIC COLORIMETRY

There are two fundamental observations on which trichromatic colorimetry depends. The first is that a colour can be matched by a suitable mixture of three selected radiations, and the second that if two colours are matched in turn by mixtures of three radiations, the two colours together, when mixed additively by suitable optical means, will be matched by the sum of the two mixtures similarly combined.

There is nothing corresponding to this for sound, since the ear does not synthesize the aural response in the way the eye integrates the visual response. When a chord of music is struck, the listener, especially if he is trained, can identify the individual notes comprising the chord. The quality of the sound cannot be reproduced by striking a limited number of standard notes in some suitable combination. The same sound can usually be matched only by playing the same notes on a similar type of instrument, and an adequate specification of the chord would require an analysis of the fundamental frequencies and their harmonics in the chord.

With colour, the phenomenon of three-colour mixture and matching has been attributed to the existence of a triple-receptor system in the eye, as already discussed briefly in Chapter 2, and on this hypothesis the spectral composition of the stimulus can have no detailed counterpart in the pattern of the visual response, once the initial photochemical absorption has taken place in the retina. It is for this reason that no amount of training will enable the colour expert to identify in detail the spectral components in a heterogeneous beam of light.

It might be thought that the measurement of colour would require the measurement of the physiological activity in the retina, but nothing as ambitious as this is at present within the reach of existing physiological techniques. However, an adequate system of colour measurement can be developed in terms of the amounts of three defined radiations which, when mixed, will produce a complete colour match of the sample under test. Two comments should, however, be made. First, the system is as satisfactory as it is only because a trichromatic match is reasonably stable over a wide range of adaptation conditions; if this were not so, a colour specification would have such a limited significance as to be virtually useless. Secondly, it has to be recognized that the specification defines the three-stimulus mixture that will produce the same colour sensation as the test stimulus, without actually defining the sensation itself. It is perhaps surprising that a system of colour measurement which does not claim to measure the appearance of the colour can have the wide applications which have been found for it and which are discussed, in particular, in Chapter 10. Of course, there is, under normal viewing conditions, a fairly close correlation between the trichromatic specification of a stimulus and its appearance, but we should be making false claims for colorimetry if we did not acknowledge that a colour specification merely expresses an equivalence between physical stimuli in their capacity to arouse a colour sensation, and that the specification itself is a physical specification of stimuli, not of a sensation.

A few observations with an additive colorimeter quickly demonstrate the basic facts of trichromatic colour mixture. The matching stimuli usually employed are saturated red, green and blue radiations, and with these, a yellow can be produced by mixing the red and green, a blue-green by mixing the blue and green, and a purple by mixing blue and red. White will result from the correct combination of all three stimuli, and intermediate hues of varying degrees of desaturation can be obtained by combining the stimuli in the appropriate proportions. Those accustomed to mixing pigments or dyes may be unfamiliar with these effects, since pigments and dyes control the quality of the light entering the eye by the absorption or subtraction of certain parts of the spectrum and not by their addition. There is no inconsistency between the two methods of mixing, as will become apparent from Chapter 8, but whatever agency is responsible for determining the composition of a beam of light entering the eye, its spectral components are ultimately combined additively when focused on the retina, and it is for this reason that the phenomena of *additive* colour mixture must provide the foundation for a system of colour measurement.

The triple character of colour mixture can be demonstrated with a trichromatic colorimeter quite unequivocally. The mathematical additivity of colour matches, however, requires quantitative confirmation of the following kind. We will suppose that two coloured lights, C_1 and C_2, have been matched in turn in a trichromatic colorimeter incorporating red, green and blue matching stimuli (R), (G) and (B).* The readings on the instrument will give a measure of these stimuli on some arbitrary linear scale, and we can suppose the values for the two matches to be respectively u_1, v_1, w_1 and u_2, v_2, w_2. These matches can be represented algebraically by the expressions

$$c_1 \,.\, (C_1) \equiv u_1 \,.\, (R) + v_1 \,.\, (G) + w_1 \,.\, (B)$$

and

$$c_2 \,.\, (C_2) \equiv u_2 \,.\, (R) + v_2 \,.\, (G) + w_2 \,.\, (B)$$

where c_1 and c_2, the amounts of C_1 and C_2 that have been matched, are expressed in arbitrary units which, for our present purpose, we do not need to define.

The test of additivity then consists in combining the two lights, C_1 and C_2, to provide a new test colour, $c_1 \,.\, (C_1) + c_2 \,.\, (C_2)$, and matching the combination on the colorimeter. Let the readings for this match be u_3, v_3, and w_3. Then

$$c_1 \,.\, (C_1) + c_2 \,.\, (C_2) \equiv u_3 \,.\, (R) + v_3 \,.\, (G) + w_3 \,.\, (B)$$

If additivity holds, we shall expect to find that

$$u_3 = u_1 + u_2$$

$$v_3 = v_1 + v_2$$

and

$$w_3 = w_1 + w_2$$

Experimental tests of the law were made in the pioneer work of König and others, but its inclusion in the principles of colour mix-

* See p. 81 for an explanation of the significance of the brackets and the use of the equivalent sign $\equiv$ in the colour matching equation.

ture postulated by Grassmann as long ago as 1853* has tended to make its acceptance axiomatic. Within the last few years, however, more exacting checks have been carried out by Blottiau and by Trezona,† while the recent investigation by Stiles on the colour-matching functions ‡ has incidentally included a further examination of additivity.

In these recent studies, there has been some indication that additivity may not hold exactly even for a 2° field, at least for the blue stimulus, but when proper account is taken of the discrimination limits at each stage of the matching sequence, then departures from additivity, if they occur at all, appear to lie within a discrimination step. At the same time, the direction in which additivity appears to fail shows a fairly consistent trend, and it has been suggested that this may be due to scatter of the colour-match settings about a point which is not the mid-point of a discrimination step. Something of this kind could happen if colour-matching laws were arithmetic and additive, while colour discrimination was more a function of the logarithm of the stimulus.

Nevertheless, for all practical and technical purposes it is legitimate to assume that additivity does hold for foveal colour matching using 2° fields. For larger fields the situation is more uncertain because of the unusual adaptive properties of the extrafoveal retina, and because both the rods and the cones are being stimulated, as discussed by Clarke.§ Rod intrusion is itself a source of additivity failure and, if present under conditions of industrial colour matching, could restrict the relevance of colorimetry in such problems.

The essential principles of three-colour measurement were familiar to Maxwell, Helmholtz, König, Abney and others, but in the early days there were a number of obscure features and imagined limitations that required clarification. There was, for example, confusion between the specification of a colour in terms of the three fundamental physiological responses of the Young-Helmholtz theory and its measurement in terms of a mixture of three physically defined radiations; the possible arithmetical

* H. Grassmann: *Poggendorf's Ann.*, 1853, **89**, p. 69 and *Phil. Mag.* (4), 1854, **7**, p. 254.

† F. Blottiau: *Rev. d'Opt.*, 1947, **26**, p. 193.

P. W. Trezona: *Proc. Phys. Soc.*, *B*, 1953, **66**, p. 548, and 1954, **67**, p. 513.

‡ W. S. Stiles: *Optica Acta*, 1955, **2**, p. 168 and 1963, **10**, p. 229.

§ F. J. J. Clarke: *Optica Acta*, 1963, **10**, p. 257.

manipulations of the data were not appreciated, in particular the freedom with which values recorded in terms of one set of stimuli could be transformed to a different set, the consequent liberty to choose reference stimuli for a colour specification different from the radiations used experimentally in colour matching, and so on. Two papers by Ives* provided the first detailed solution of colour-mixture problems on the trichromatic system, followed by two papers by Guild,† who reduced the problems of colour-mixture to two dimensions by separating the amount of light from its quality, and applied the ordinary rules of plane geometry to their solution.

Since then, many papers have been published elaborating different aspects of the system and discussing variations in the method of calculation. Two investigations by Guild and the author‡ to determine the relative amounts of three specified radiations required to match the colours in the spectrum for 2° foveal vision were used in 1931 by the Commission Internationale de l'Eclairage (C.I.E.) to standardize the colour-mixture characteristics of a normal observer and to adopt a standard framework for colour specification. More recent studies by Stiles§ and by Speranskaya|| have been used by the C.I.E. to tabulate colour matching functions for large-field, 10°, viewing conditions. It will be useful to give here a brief account of the instrumental and observational background to these investigations.

The purpose of the two earlier researches was to supplement the 1924 V_λ curve by providing additional colour-mixture data to enable the V_λ curve to be broken down into its three constituent red, green and blue mixture curves. From these curves the colour specification of any stimulus with a spectral composition $P(\lambda)$ can be calculated by a process of multiplication and integration exactly analogous to the calculation of light flux using the simple V_λ

* H. E. Ives: *J. Frank. Inst.*, 1915, **180**, p. 673, and 1923, **195**, p. 23.

† J. Guild: *Trans. Opt. Soc.*, 1924, **26**, p. 95, and 1924, **26**, p. 139.

‡ J. Guild: *Phil. Trans. Roy. Soc., A*, 1931, **230**, p. 149.
W. D. Wright: *Trans. Opt. Soc.*, 1928, **30**, p. 141.

§ W. S. Stiles: *Optica Acta*, 1955, **2**, p. 168; *Physical Society Yearbook* (London), 1955, p. 44; *N.P.L. Symposium on Visual Problems of Colour* (H.M. Stationery Office London), 1958, p. 211; and J. M. Burch: *Optica Acta*, 1959, **6**, p. 1.

|| N. I. Speranskaya: *N.P.L. Symposium on Visual Problems of Colour* (H.M. Stationery Office, London), 1958, p. 317; *Optics and Spectroscopy*, (Trans. by Opt. Soc. Am.), 1959, **7**, p. 424.

curve. These calculations will be elaborated in the next section and in the next Chapter.

The essential observations that had to be made were a series of colour matches of the spectral colours by means of some form of trichromatic colorimeter. Guild* used an instrument operating on the principle illustrated in Fig. 3.1. The light from the diffusing

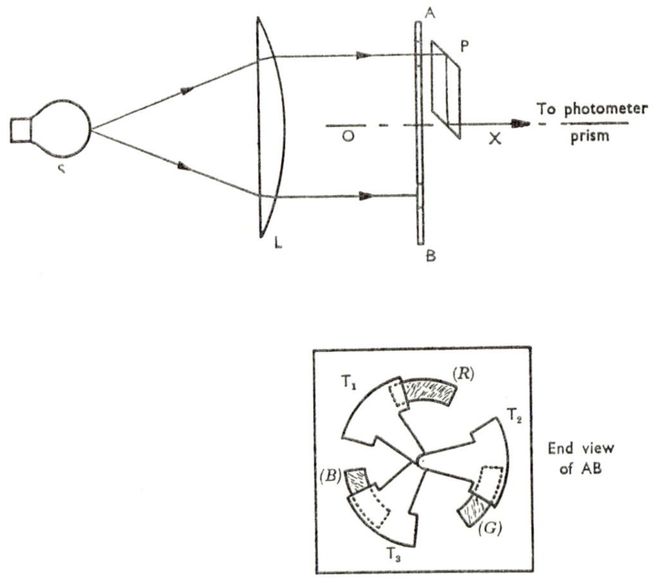

FIG. 3.1 The colour-mixing system of the Guild colorimeter

bulb of a lamp S is collimated by a large condensing lens L and projected as a broad beam onto the end of the instrument AB. An end view of AB is shown in the lower diagram and is seen to contain three sector-shaped openings in which red (*R*), green (*G*) and blue (*B*) colour filters are mounted. Inside the instrument a prism P is mounted immediately behind the colour apertures and is rotated rapidly about an axis OX. In this way the light passing through each filter is swept into the system and reflected into one half of a photometer prism and thence into the observer's eye. The light received by the observer thus consists of a series of rapid pulses of red, green and blue radiation, which, owing to the per-

* J. GUILD: *Trans. Opt. Soc.*, 1925, **27**, p. 106.

sistence of vision, are perceived as a steady beam of a uniform colour whose chromaticity and luminance are governed by the openings in front of the three filters in AB. These openings are controlled by radial shutters T_1, T_2 and T_3, which are adjusted by the observer until the mixed beam matches the test colour; the angular setting of the shutters then gives a measure of (R), (G) and (B) in the mixture.

The spectral colour to be matched was provided by an independent double-monochromator; light of wavelength λ emerging from the exit-slit was projected into the other half of the photometer prism, but for reasons which will be apparent later, it was necessary to desaturate the spectral light by admixture of a small amount of one or other of the three instrument stimuli. This desaturation was achieved by passing a pencil of light through a separate red, green or blue filter and reflecting it into the spectral beam by means of a plane-glass reflector.

In the author's experiments, an instrument operating on the principle illustrated in Fig. 3.2 was used.* A parallel beam of light is incident on the upper half of the dispersing prism D and focused by the lens L to form a spectrum in its focal plane. In the spectrum, three roof-reflecting prisms (R), (G) and (B) are located and, after two reflections at their adjacent faces, the light entering each prism is reflected back and retraces its path through the lower half of L and D. On emerging from D the light is reflected by the right-angled prism P, lying below the level of the incident light, into the photometer prism. Three monochromatic radiations are thus isolated and recombined by normal prismatic dispersion, the wavelengths selected being governed by the positions of (R), (G) and (B) in the spectrum. The band of wavelengths reflected by each prism need not be especially narrow, since the purity of the light ultimately reaching the eye is controlled by the diameter of the exit-pupil (not shown in the diagram) which is in conjugate focus to the prisms, and not by the width of the prisms themselves. The intensity of each of the beams in the mixture is controlled by photometer wedges T_1, T_2 and T_3 mounted in front of the prisms (R), (G) and (B) and through which the light passes twice.

* W. D. WRIGHT: *Trans. Opt. Soc.*, 1927, **29**, p. 225, and *J. Sci. Inst.*, 1939, **16**, p. 10.

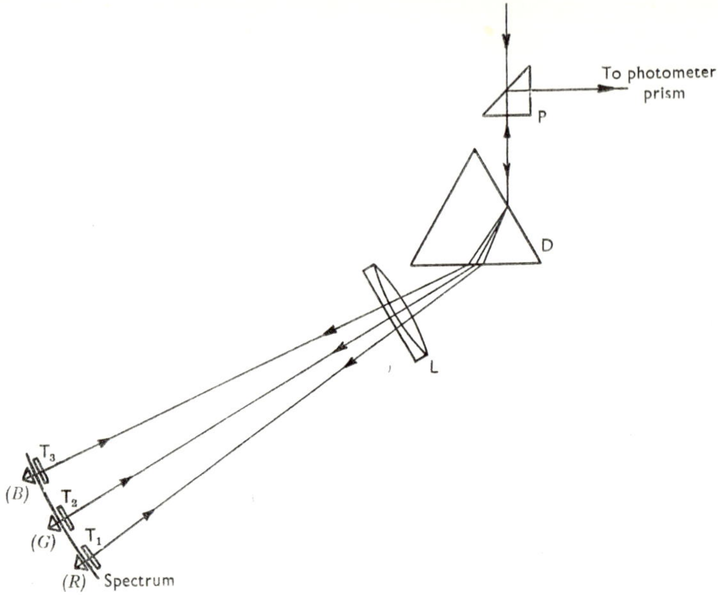

FIG. 3.2. The colour-mixing system of the Wright colorimeter

Fig. 3.2 actually shows only half the system, since the monochromatic test light to be matched was derived from a second spectrum taken from the same source and prism system. Two roof prisms were located in this second spectrum, one to provide the test light λ and the other a small amount of (R), (G) or (B) to act as desaturator.

Both Guild's and the author's instruments provided a field of view subtending 2° at the eye. With larger fields, the macular area of the retina may be seen as a subjective non-uniformity centred around the visual axis and moving about as the point of regard is transferred from one half of the field to the other. This projection of the macular area is not noticeable as a rule in looking at an extended area of colour under everyday conditions, because local adaptation of the retina tends to compensate for the absorption in the yellow macular pigment. But in a colorimeter the spectral compositions of the two halves of the field may be quite different even though they have the same colour appearance, and in consequence the effect of the pigment absorption will be different for the two sides of the field. Adaptation is too slow a process to provide the rapid local compensation that would be necessary to eliminate this difference as the eye scans from one half of the field to the other; hence with a large field of view, a match can be made either for the central 2° or 3°, with the surround a mismatch, or the surround can be matched, leaving the central area a mismatch.

A further reason for the 2° field in these earlier investigations was consistency with the 1924 V_λ curve with which they were to be combined. This curve was based on 2°-field conditions, because Ives* had shown that reasonable agreement could be secured in this way between the direct comparison and flicker methods of heterochromatic photometry. Nowadays, however, many people would consider that inconsistency between the direct comparison and the flicker methods under large-field conditions was an argument for abandoning the flicker data rather than for artificially restricting the observing conditions in order to obtain consistency.

Another argument for colorimeters with 2° fields was that, in the measurement of dark samples, the illumination might be so low that the visual response would have included a significant rod component if the field had been extended beyond the rod-free area. Visual instruments were then in common use for colour measurement and since the samples being measured were usually likely to be seen under photopic conditions, it was obviously desirable that the measurements should relate to those conditions, and thus that cone vision alone should be involved in the match.

More recently the recognition that much industrial colour matching is undertaken between large samples has led to a demand for additional colour matching data which apply to large-field conditions. At the National Physical Laboratory this demand has been met by Stiles† and his colleagues using a trichromator of new design and exceptional flexibility.

Their aim has been to extend the observations made by Guild and by the author, using both 2° and 10° field sizes and with the colour-matching functions recorded in energy units. The V_λ curve

* H. E. Ives: *Phil. Mag.*, 1912, **24**, pp. 149, 352, 744, 845 and 853.
See also J. Guild: *Proc. Opt. Convention*, 1926, Part I, p. 61, for a most informative discussion of the outlook prevalent at that time.

† W. S. Stiles: *Physical Society Year Book*, (London), 1955, p. 44; *Optica Acta*, 1955, **2**, p. 168.

has been computed by summing the three colour-mixture curves after they have been weighted by suitable factors, rather than relying on the direct determination of the V_λ function by one or other of the methods of heterochromatic photometry, although as a subsidiary investigation the V_λ curve has also been determined directly.

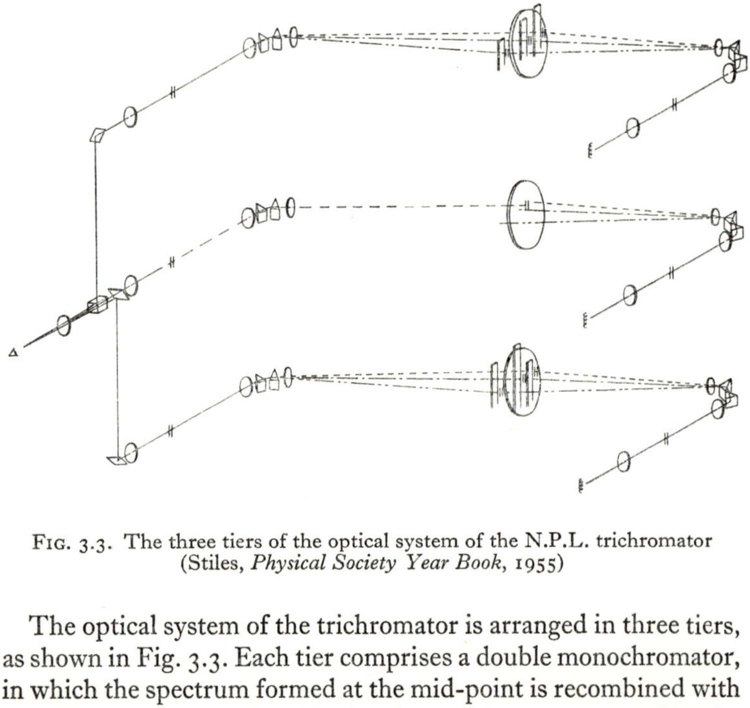

FIG. 3.3. The three tiers of the optical system of the N.P.L. trichromator (Stiles, *Physical Society Year Book*, 1955)

The optical system of the trichromator is arranged in three tiers, as shown in Fig. 3.3. Each tier comprises a double monochromator, in which the spectrum formed at the mid-point is recombined with the aid of a field lens placed in the plane of the spectrum and by the passage of the light through the second dispersing system. The central tier provides the test wavelength, the upper tier the three matching stimuli and the lower tier the desaturating stimulus.

The beams from the top and bottom tiers are reflected into the photometer prism and the three exit slits are then imaged at a common point in space. The subject's eye is brought into position so that his pupil centre coincides with this point and the two halves of the photometer prism are thus seen filled with light. This arrangement avoids any vignetting that might occur with fields subtending 10° or more if the observer had been required to place his eye immediately behind a small aperture to define the exit pupil. A further additional feature that had to be introduced was a rotating wheel of lenses to draw the slit images out into a small band about 2 mm square. This improved the uniformity of the field of view, which can otherwise show striations if the light is focused into a narrow slit as it passes through the cornea and crystalline lens.

The dimensions of the optical components are such that either a 2° or a 10° matching field can be provided, each with a 14° surround field of the same colour and luminance. The intensities of the stimuli are controlled with photometer wedges in the plane of the spectrum, and by an ingenious combination of magnetic clamping devices, the three wedges may be moved individually or two or three at a time. A variety of matching procedures thus becomes possible. For normal work the light source of each tier is a tungsten-ribbon lamp, but to increase the intensity at the ends of the spectrum, a zirconium or a xenon arc lamp can be substituted as the source in the middle tier.

The difficulties of large field observations that were anticipated thirty years earlier and which led to the adoption of the 2° field size, namely non-uniformity due to the macular absorption and desaturation due to the intrusion of rod vision, have necessarily been encountered by Stiles. The non-uniformity has been met by asking the observer to ignore the central 2° or 3° and to make a match for the outer zone of the field. This can certainly be done, but it does mean that the fovea, where colour discrimination is most acute, has no influence on the observations. The desaturation caused by the rod contribution has slightly affected the colour-mixture curves in certain parts of the spectrum, and a method of allowing for this has been worked out.

The large-field observations made by Speranskaya* at the Mendeleef Institute of Metrology, Leningrad, were obtained using a Demkina filter colorimeter. To provide the test field, light from the exit slit of a double monochromator was projected on to a white magnesia plate at 45° incidence to form a coloured spot some 6 mm in diameter. This spot was in turn placed at the focus of an

* N. I. Speranskaya: *Optics and Spectroscopy* (Trans. by Opt. Soc. Am.) 1959, 7, p. 424.

auxiliary lens system so as to illuminate one half of the colorimeter field and then be imaged in the exit pupil of the colorimeter. The colorimeter field had an angular size of 10° with a vertical dividing line between the two halves of the field, and was seen against a black background. To avoid the difficulty of the Maxwell spot, a central area of the field of about 2° subtense was blacked out but had in its centre a bright fixation point. The observations thus referred to an annular zone of the retina surrounding the fovea.

Desaturation of the test field was accomplished by the addition of light from a purple filter of the same type as the filters used in the colorimeter. The amount of desaturation was controlled with a photometer wedge so that the mixture of the spectral test colour with the purple desaturation could just be matched by only two of the colorimeter matching stimuli. The level of illumination at which the matches were made was relatively low, in spite of the use of a 500-watt tungsten lamp with the monochromator, because of the introduction of the magnesium oxide diffusing surface instead of looking directly into the optical system. The illumination level was, in fact, some 30 times lower than that available in Stiles' trichromator.

The actual results obtained in these investigations will be described in the next chapter as an introduction to the C.I.E. system. It will be useful, however, to consider first the algebraical manipulations of colorimetric quantities and the geometrical properties of the chromaticity chart, before describing the experimental work.

THE ALGEBRA AND GEOMETRY OF COLORIMETRY

We have already described how a colour C can be matched in a colorimeter by the appropriate mixture of the red, green and blue matching stimuli used in the instrument. The match can be represented by the equation

$$q\,.\,(C) \equiv u\,.\,(R) + v\,.\,(G) + w\,.\,(B)$$

where u, v and w might be the amounts of (R), (G) and (B) as given directly by the scales on the instrument and q is some convenient measure of C. The symbolism of this equation calls for some comment, as we have to recognize that there are two distinct kinds of symbol in this equation; thus (*C*), (*R*), (*G*) and (*B*) have no numerical significance but are introduced to identify the lights that are being mixed or measured, whereas q, u, v and w represent the amounts of the lights involved in the observation.

The equation is, in fact, merely a short-hand way of recording a colour match, and the use of brackets round *C*, *R*, *G* and *B* is a device that has been adopted (and formally approved in 1955 by the C.I.E.) to distinguish a symbol describing the kind of stimulus involved, from that representing its amount. Thus just as we might speak of 10 lb of apples, so we can speak of u units of the radiation (*R*). The symbols *R*, *G*, and *B* without the brackets are then conveniently available to represent what are called the tristimulus values of (*R*), (*G*) and (*B*), as described presently. Where a bracketed symbol, for example (*R*), appears by itself in an equation, then it is understood that unit quantity of (*R*) is implied, the unit being based on whatever scale of (*R*) has been defined. Generally, this will be on some defined tristimulus system of units.

As an added distinction between a colour-matching equation and a normal type of algebraic equation in which all the symbols can be multiplied, divided, etc., in the usual way, it has been agreed by the C.I.E. (1955) that colour-matching equations can be written using the equivalent, or identity, sign $\equiv$. This is a logical and convenient practice in so far as a colour-matching equation implies that the two sides of the equation represent stimuli which are identical in colour, while it can hardly lead to confusion with the wider significance of algebraic identity. As a further reminder of the meaning of a colour equation, it is useful to adopt the custom of reading the equivalent sign as 'matches' or 'is matched by'.*

In the equation for *C* above, it has been suggested that u, v and w might represent the amounts of the instrumental or matching stimuli as given on the scales of the instrument. Alternatively, the

* *Note.* There is also the question how far the brackets should be used in the text and diagrams, as well as in equations. Any decision must be based in the first place on the need to avoid ambiguity and uncertainty, but since the symbolism is somewhat clumsy and involves extra typing and printing, the use of the brackets should be reduced to a minimum. In this book the following procedure has been adopted: in all equations, the brackets will be used round symbols describing the kind of stimuli involved. In the text and diagrams, the brackets will also be retained where the symbols refer to the matching stimuli, or reference stimuli, such as (*R*), (*G*), and (*B*), or (*X*), (*Y*), and (*Z*); this will be done because the unbracketed symbols *R*, *G* and *B*, or *X*, *Y* and *Z* are

quantities might be put on a less arbitrary basis by converting *u*, *v* and *w* into physical units of power or energy, or photometric units of luminous flux or luminance, but this rarely turns out to be very useful or to have much meaning. An energy scale, for example, would not be particularly easy to calibrate and might be quite misleading if the red filter of a filter colorimeter transmitted a considerable amount of infra-red radiation. Also, a considerable simplification of colour calculations is often effected by separating the total amount of light from its colour quality.

It has therefore become the practice to base the units of (*R*), (*G*) and (*B*) on a match on a white of some defined quality. Since white may be regarded as a colour in which neither red nor green nor blue predominates, the arbitrary but reasonable convention that equal amounts of (*R*), (*G*) and (*B*) are required to match white may be adopted. It should be noted that this is not a statement of the kind which future investigation might subsequently show to be slightly inaccurate; it does not depend on any assumption that white stimulates the three response mechanisms in the retina by equal amounts, although it is conceivable that this does occur. It is merely a convention which has been adopted because it is convenient and because it fits in with our conception of white as a neutral colour possessing no attribute of hue.

We can, then, visualize a colour-measuring system in terms of the red, green and blue stimuli of some trichromatic colorimeter with the units of the stimuli adjusted to be equal in a match on a standard white. If the matching stimuli of the colorimeter were monochromatic radiations, their quality could be defined by their wavelengths, while the white could be defined by its colour temperature, if it were a Planckian radiator. On these units, the colour equation for *C* is then given by

$$c\,.\,(C) \equiv R\,.\,(R) + G\,.\,(G) + B\,.\,(B)$$

also used in the text to describe the tristimulus values of the stimuli. Where other colours are being described and are given symbols such as (*C*), (*W*), (S_A) (λ), etc., in colour equations, the brackets will in general be dropped in the text and diagrams, since the unbracketed symbols have no other connotation. Where a particular trichromatic system is being described, this will be referred to as the (*RGB*) system or the (*XYZ*) system to avoid the clumsiness of individual brackets round the three symbols. Sometimes, as will appear later, the units of the stimuli are changed, while their quality remains the same, and this distinction is marked by the use of square, instead of round, brackets. On occasion, therefore, it may be necessary to include brackets in the text to distinguish whether *C*, say, is being specified in one unit system or the other.

where R, G and B are numerical quantities and are called tristimulus values. The amount of C represented by this equation is given by c, where on this system

$$c = R + G + B$$

(It should be noted that this is an ordinary algebraic equation.)

In many problems it is convenient to separate the colour quality of C, as represented by the proportions of R, G, B in the match, from the total amount of light being matched. This can be done, for example, by dividing R, G and B by $(R + G + B)$ to give

$$1{\cdot}0(C) \equiv r\,.\,(R) + g\,.\,(G) + b\,.\,(B)$$

where

$$r = \frac{R}{R+G+B}, \quad g = \frac{G}{R+G+B}, \quad b = \frac{B}{R+G+B}$$

and hence

$$r + g + b = 1$$

This equation for C is known as a unit trichromatic equation, and the amount of C represented in such an equation as 1 trichromatic unit, or 1 T-unit for brevity; c above is also measured in T-units. Since the quality of C is now a function of only two variables, it can be represented graphically on a two-dimensional diagram known as a chromaticity chart; r, g and b are therefore called the chromaticity co-ordinates of C.

When r is plotted as abscissa and g as ordinate, we have a chromaticity chart of the type illustrated in Plate 7. A colour C is located by its co-ordinates (r, g), the value of b being given by

$$b = 1 - (r + g)$$

The red matching stimulus (R) has the co-ordinates $(1{\cdot}0, 0{\cdot}0)$; (G) is located at $(0{\cdot}0, 1{\cdot}0)$; and (B) is located at $(0{\cdot}0, 0{\cdot}0)$. Since the units of (R), (G) and (B) have been adjusted to be equal in the match on the standard white, this particular white, to which we can give the symbol W, will be located at $(0{\cdot}333, 0{\cdot}333)$.

The line $(R)(G)$ is the locus of points in which $(r + g) = 1$ and for which the value of b is consequently zero. From our knowledge of the results of additive colour mixture outlined earlier, we can deduce that $(R)(G)$ is the locus of saturated orange, yellow and yellow-green colours. Similarly, $(G)(B)$ is the locus of colours in which r is zero, and includes blue-greens of varying predominance of green or blue, while $(R)(B)$ is the locus of purples in which g is zero. Between the boundaries of the triangle and the white point, the desaturated colours are located. This distribution is illustrated in Plate 7 but the location of the colours cannot be accurate because of the impossibility of representing spectral colours by printing inks. In any case, while it is sometimes convenient to use the chromaticity chart as if it were a colour map of the trichromatic system, we must always do so with caution because its true function is to express the results of colour mixture andcolour matching.

The trichromatic specification (i.e. the tristimulus values and chromaticity co-ordinates) of a colour C_3 obtained by the additive mixture of two other colours C_1 and C_2 can be readily found as follows: Suppose the unit equation for C_1 (Fig. 3.4) is

$$(C_1) \equiv r_1 \,.\, (R) + g_1 \,.\, (G) + b_1 \,.\, (B) \qquad (r_1 + g_1 + b_1 = 1)$$

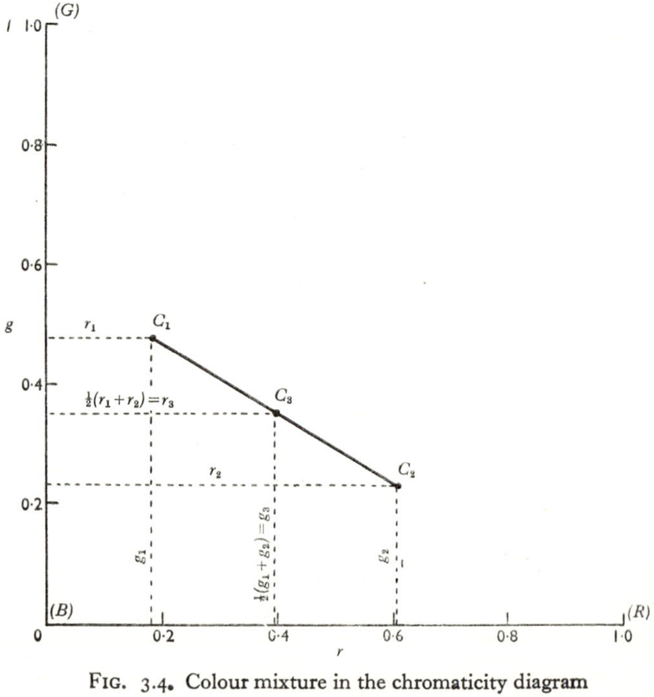

FIG. 3.4. Colour mixture in the chromaticity diagram

and that for C_2 is

$$(C_2) \equiv r_2 \,.\, (R) + g_2 \,.\, (G) + b_2 \,.\, (B) \qquad (r_2 + g_2 + b_2 = 1)$$

If one T-unit of C_1 is to be mixed with one T-unit of C_2, then on the additivity principle, the new colour C_3 will be represented by

$$\begin{aligned} c_3 \,.\, (C_3) &\equiv (C_1) + (C_2) \\ &\equiv (r_1 + r_2) \,.\, (R) + (g_1 + g_2) \,.\, (G) + (b_1 + b_2) \,.\, (B) \end{aligned}$$

c_3 represents the number of T-units of C_3 in the mixture and is given by the sum of the tristimulus values of (R), (G) and (B) in the above equation, i.e.

$$c_3 = r_1 + r_2 + g_1 + g_2 + b_1 + b_2 = 2$$

There are therefore two T-units of C_3 in the mixture, and the unit equation for C_3 will be given by

$$(C_3) \equiv r_3 \,.\, (R) + g_3 \,.\, (G) + b_3(B)$$

where

$$\begin{aligned} r_3 &= \tfrac{1}{2}(r_1 + r_2) \\ g_3 &= \tfrac{1}{2}(g_1 + g_2) \\ b_3 &= \tfrac{1}{2}(b_1 + b_2) \end{aligned}$$

From simple geometrical considerations, it follows that in the chromaticity chart C_3 must lie on the straight line joining C_1 and C_2 at a point midway between them, as shown in Fig. 3.4.

Suppose, now, m units of C_1 are mixed with n units of C_2. Repetition of algebra similar to the above will show that $(m + n)$ units of a new colour C_3 will be formed, and that C_3 will be situated at a distance along the line C_1C_2 such that

$$\frac{C_1C_3}{C_3C_2} = \frac{n}{m}$$

Hence the T-unit enables colour-mixture problems to be very readily envisaged and evaluated on the chromaticity chart and is the basic unit for conducting such colour-mixture calculations.

For three or more colours, the same principle applies, and it follows that the colour reproduced by the mixture of positive amounts of three radiations will always be represented by a point lying within the triangle formed by points representing the radiations in the chromaticity diagram. Any positive mixture of (R), (G) and (B), for example, will produce a colour lying within the triangle $(R)(G)(B)$, and the location of W at (0·333, 0·333) is, of course, the centroid of the triangle $(R)(G)(B)$.

A particularly important application of these colour-mixture relations is to the derivation of the colour specification of a stimulus C when its spectral composition $P(\lambda)$ is known. Before this can be done, we need to know the amounts of (R), (G) and (B) that would be required to match each wavelength in turn in the equal-energy spectrum. They can be obtained from appropriate colour-matching experiments as described in the next chapter, and because they are basic data for colorimetry, they have been given the special symbols $\bar{r}_\lambda$, $\bar{g}_\lambda$ and $\bar{b}_\lambda$. These distribution coefficients, as they have been called, are now more frequently referred to as the spectral tristimulus values of the equal-energy spectrum.

The specification of C is then derived by the repeated summation of the tristimulus values of its component wavelengths exactly as the equations for C_1 and C_2 were summed to give the specification of C_3. Expressed algebraically, this summation can be written

$$R = \Sigma(P_\lambda \, . \, \bar{r}_\lambda)$$
$$G = \Sigma(P_\lambda \cdot \bar{g}_\lambda)$$

and

$$B = \Sigma(P_\lambda \, . \, \bar{b}_\lambda)$$

The unit equation for C will then be

$$(C) \equiv r \, . \, (R) + g \, . \, (G) + b(B),$$

where

$$r = \frac{R}{R + G + B}, \text{ etc.}$$

It has to be admitted that the concept of the T-unit in this kind of computation has often been a source of difficulty to colorimetrists, since it has no very obvious correlation with the more orthodox measures of light, such as watts or lumens. Moreover, the chromaticity chart itself has some peculiar properties, since it includes, for example, stimuli which have colour but no light. Colorimetric algebra is extremely simple, but unless the concepts associated with the algebra are understood, colorimetry is always likely to be a baffling subject. The following analogy is put forward with some hesitation as an aid to that understanding.

Fig. 3.5 is intended to represent several glass dishes containing

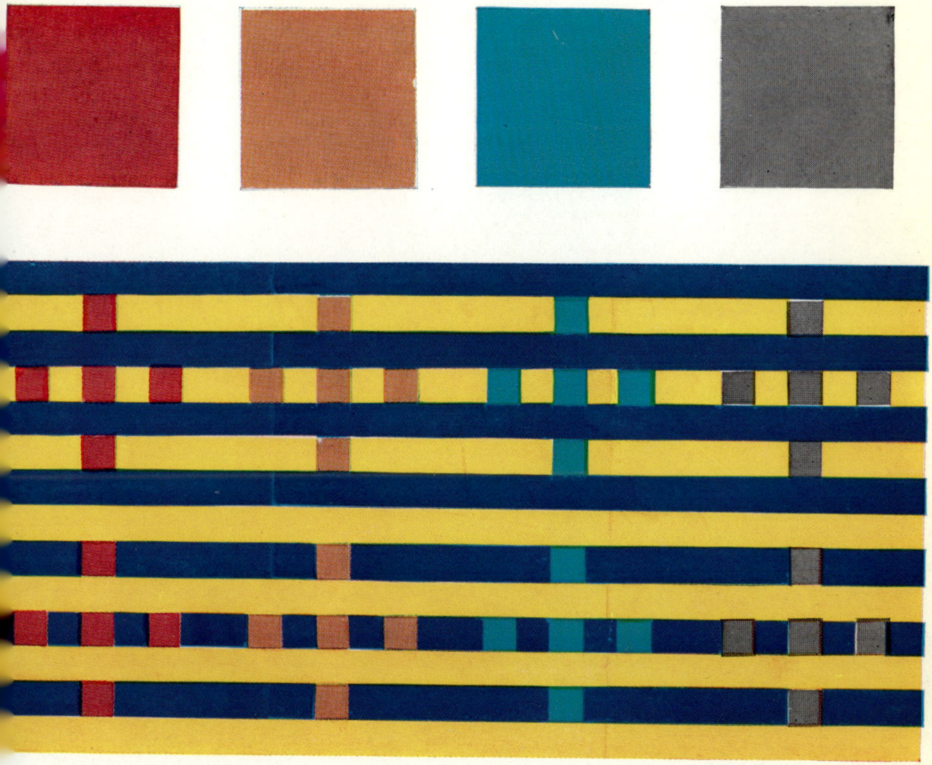

Colour appearance in fine patterns

When the illustration is viewed from a distance of one or two metres and with the pattern of stripes running horizontally, a striking difference in colour develops between the large squares and the small squares printed with the same pigments. This is probably due to a combination of effects, e.g. contrast, light scatter in the eye, chromatic aberration in the eye, eye movements and local adaption. An analogous change is produced if the pattern is seen on edge by tilting the illustration (see page 53).

PLATE 7

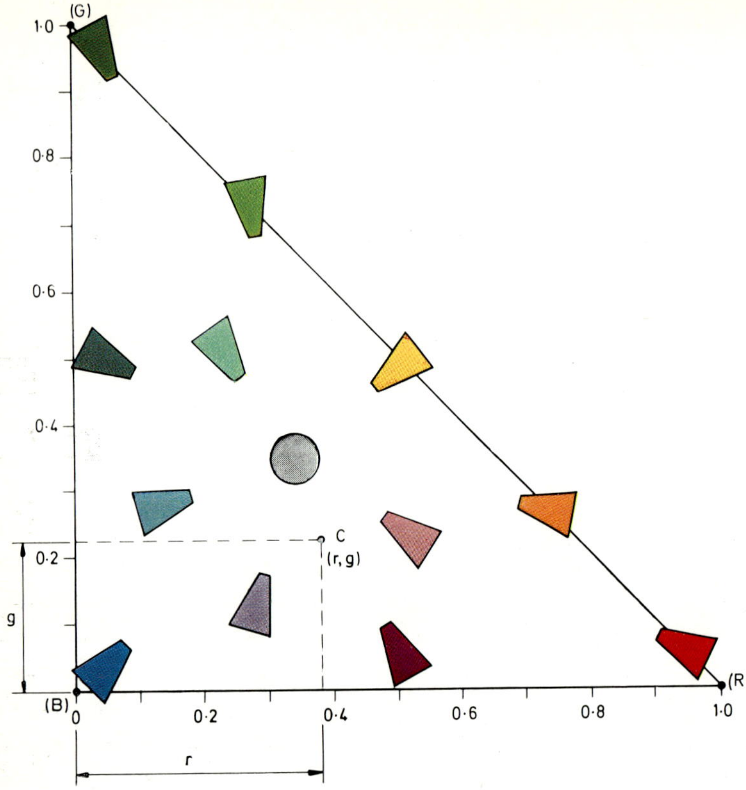

The chromaticity diagram, in which the colour quality of a stimulus is defined in terms of two co-ordinates. The coloured circles are intended to give no more than an indication of the distribution of the colours when the diagram is drawn in terms of some reference stimuli, (*R*), (*G*), and (*B*) (see page 84)

aluminium rivets, lead shot, and copper rivets, and mixtures of all

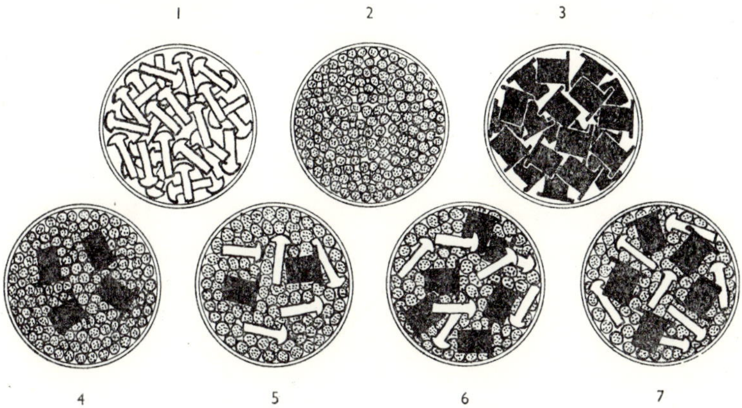

FIG. 3.5. A metallic analogy to colour mixture
Dishes 1, 2, and 3 contain pure aluminium, lead, and copper respectively
The others contain various mixtures of all three

three. If we wish to specify the quantity of rivets or shot in a dish, we can do so in three different units, namely by their number, their weight or their volume. This is analogous to the measurement of light in terms of quanta, luminous flux or T-units, in which:

The number of rivets or shot corresponds to number of quanta (radiometric units)
Their weight corresponds to luminous flux (photometric units)
Their volume corresponds to T-units (colorimetric units)

We can then construct the metal equivalent of a chromaticity chart by using the volumes of each metal as the co-ordinates of the chart. Such a chart is shown in Fig. 3.6, in which the volume of copper is plotted as abscissa, and the volume of lead shot as ordinate. Unit volume could be specified on some absolute basis, or it might be given some arbitrary value, such as the unspecified volume of the dishes illustrated in Fig. 3.5. The latter corresponds to the usual procedure in colorimetry, where it is rarely necessary to define T-units on an absolute scale.

On this basis, a dish of pure copper rivets will be located at (1·0, 0·0), a dish of pure lead shot at (0·0, 1·0) and of pure aluminium rivets at (0·0, 0·0). A mixed sample will be located at *C*,

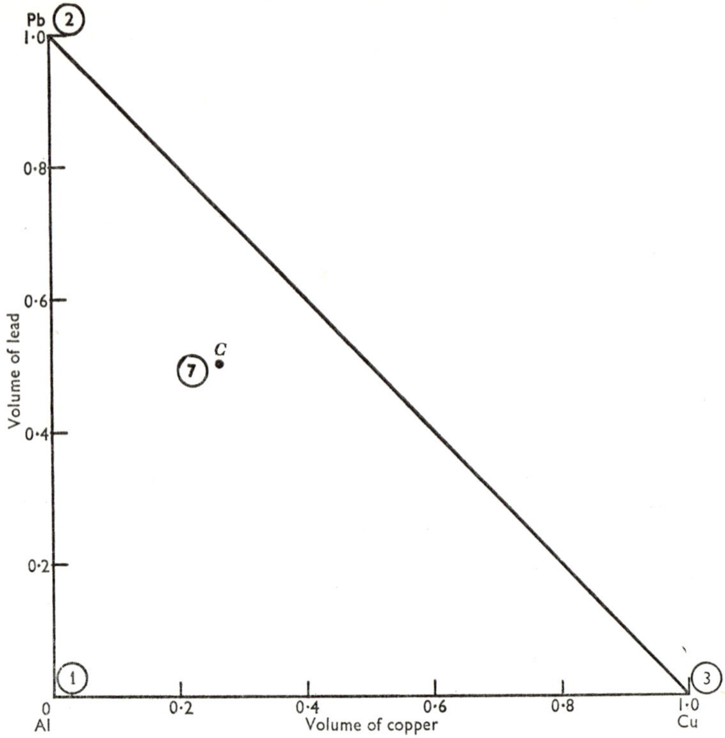

FIG. 3.6. The metallic equivalent of the chromaticity diagram. The encircled numbers refer to the numbered dishes in Fig. 3.5

where the position of C is determined by the mixture by volume of the three pure components so as to give the same composition as the sample. Two samples, C_1 and C_2, can be mixed to give a new composition C_3, whose position can be determined by calculations exactly analogous to those described above for colour mixtures, since the mixture process will obviously be additive; the volumes of C_1 and C_2 will correspond to the number of T-units of colour being mixed, while the sum of their volumes will give the resultant volume of C_3, just as the number of T-units of C_3 was given by the sum of the T-units of C_1 and C_2. Evidently, also, the analogy can be extended to the mixture of three or more samples of the metals. The tristimulus specification of a sample would correspond to the total amounts, by volume, of the three metals in the sample, using the volume of the dish as the unit, while the unit equation would correspond to the amounts, by volume, of the three metals in one dishful of the sample.

In a particular demonstration of the analogy, one dish of copper rivets was found to weigh 400 g, a dish of lead shot 600 g, and a dish of aluminium rivets 150 g. This is similar to the chromaticity chart, where green is the most luminous and blue the least luminous of the matching stimuli; thus, the relative luminous efficiencies of T-units of one particular set of matching stimuli were 1·00 for the red, 4·59 for the green and 0·06 for the blue. Yet in spite of this, T-units can be a valid and useful measure of the amount of the stimulus in colorimetry, just as volume can be a perfectly valid and useful measure of metals.

We must now consider the possibility of transforming a specification from one set of co-ordinates to another. We will do this first for a colour specification and then compare it with the analogous problem of the metals. As we shall see shortly, we shall need to distinguish unit quantities of the same stimuli in two different chromaticity charts, and a convenient way to do this will be to use different types of brackets for the stimuli in the two charts. We shall therefore now employ square brackets for the $[R][G][B]$ stimuli, although when only one system is under discussion, we can always revert to the round brackets.

The only description of $[R]$, $[G]$ and $[B]$ given so far is that they represent red, green and blue radiations. There are many different radiations which might be considered as included by these terms, so that any standard system of colour specification would require that $[R]$, $[G]$ and $[B]$ should be given an exact colorimetric definition. We will assume, without at the moment going into details, that this has been done, and will now consider the possibility of transforming a specification of a colour C defined by the unit equation

$$[C] \equiv r\,.\,[R] + g\,.\,[G] + b\,.\,[B]$$

into a new unit equation of the form

$$(C) \equiv x\,.\,(X) + y\,.\,(Y) + z\,.\,(Z)$$

where (X), (Y) and (Z) are a new set of reference stimuli.

Examination of Fig. 3.7. which shows $[X]$, $[Y]$ and $[Z]$ plotted in the [*RGB*] chromaticity chart, suggests that such a transformation must be possible. We need not at this stage consider what colours [*X*], [*Y*] and [*Z*] represent; our immediate concern is with the geometry of the transformation, and this can be demonstrated as well with one set of points as with another. Referring to the colour *C* in Fig. 3.7. the amounts of [*R*], [*G*] and [*B*] required to match it are given directly by the co-ordinates *r*, *g* and $1 - (r + g)$. But it is also evident that since *C* lies inside the triangle formed by [*X*], [*Y*] and [*Z*], it must be possible to match *C* by a suitable mixture of [*X*], [*Y*] and [*Z*]. If, for instance, a line is drawn from [*X*] through *C* to intersect [*Y*][*Z*] at *D*, then we can conclude that a mixture of [*X*] with *D* in the proportion, in T-units, of the distance *DC* to *C*[*X*], will match the colour

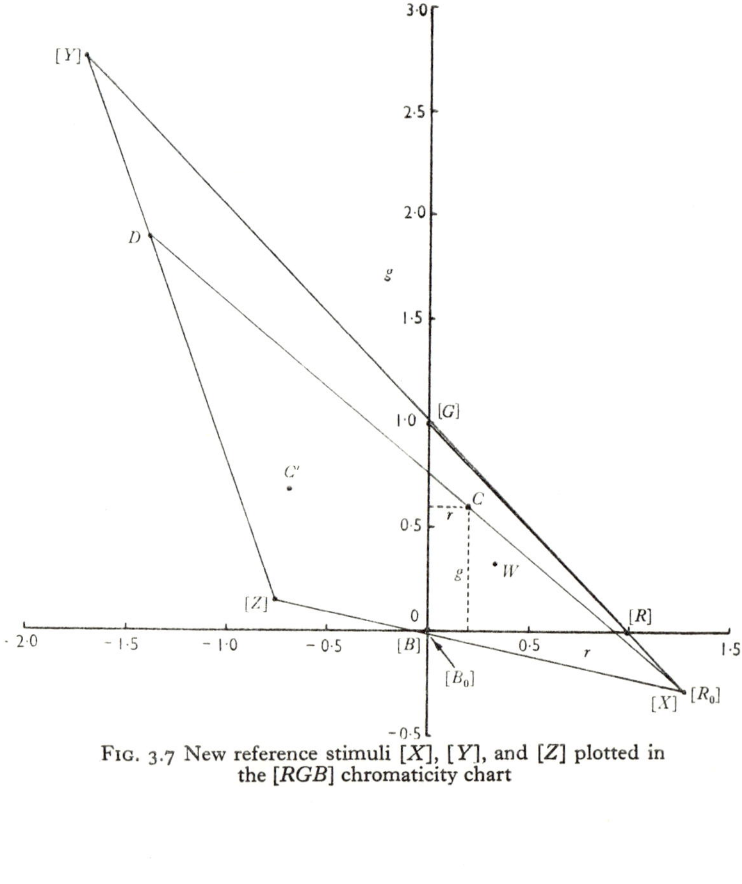

FIG. 3.7 New reference stimuli [*X*], [*Y*], and [*Z*] plotted in the [*RGB*] chromaticity chart

at C; further, D can be obtained from a mixture of $[Y]$ and $[Z]$ in the proportion of $[Z]D$ to $D[Y]$ respectively. Hence from the geometry of the figure, it is possible not only to see that C could be matched by a mixture of $[X]$, $[Y]$ and $[Z]$, but also to deduce the proportions required to give such a match.

We might conclude that the transformation of the specification for C from the $[RGB]$ system to the (XYZ) system would then be complete, but this would be incorrect. If C is to be expressed in terms of (X), (Y) and (Z) without any reference to the $[RGB]$ system, then the units of (X), (Y) and (Z) must be defined in terms of their own system and not, as at present, in terms of T-units of the $[RGB]$ system. If the same criterion is to be used to fix the units of (X), (Y) and (Z) as was used for $[R]$, $[G]$ and $[B]$, namely that equal amounts of the three stimuli are required in the match of W, then we must anticipate that this condition will not in general be satisfied if $[X]$, $[Y]$ and $[Z]$ are still measured in T-units of the $[RGB]$ system. This would occur only if W happened to be located at the centre of gravity of $[X]$, $[Y]$ and $[Z]$ as well as of $[R]$, $[G]$ and $[B]$, which is certainly not so in Fig. 3.7. Unit quantities of (X), (Y) and (Z) in the (XYZ) system will therefore be different from unit quantities of $[X]$, $[Y]$ and $[Z]$ in the $[RGB]$ system, so that some adjustment to their magnitude will be required in the course of the transformation.

The transformation problem and the relevant data may then be expressed as follows: A colour C is specified on the $[RGB]$ system by the equation

$$[C] \equiv r\,.\,[R] + g\,.\,[G] + b\,.\,[B]$$

while the units of $[R]$, $[G]$ and $[B]$ are adjusted to satisfy the condition

$$[W] \equiv 0{\cdot}333[R] + 0{\cdot}333[G] + 0{\cdot}333[B]$$

It is required to find an expression for C in terms of three new reference stimuli (X), (Y) and (Z), whose colour qualities are defined by their positions in the $[RGB]$ chromaticity diagram as given by the equations

$$[X] \equiv r_x\,.\,[R] + g_x\,.\,[G] + b_x\,.\,[B]$$
$$[Y] \equiv r_y\,.\,[R] + g_y\,.\,[G] + b_y\,.\,[B]$$
$$[Z] \equiv r_z\,.\,[R] + g_z\,.\,[G] + b_z\,.\,[B]$$

The units of (X), (Y) and (Z) on the new system are to be adjusted to satisfy the condition that

$$(W) \equiv 0{\cdot}333(X) + 0{\cdot}333(Y) + 0{\cdot}333(Z)$$

The equations for $[X]$, $[Y]$ and $[Z]$ can be regarded as three simultaneous equations in $[R]$, $[G]$ and $[B]$, which can be solved to give

$$[R] \equiv x_r' \,.\, [X] + y_r' \,.\, [Y] + z_r' \,.\, [Z]$$
$$[G] \equiv x_g' \,.\, [X] + y_g' \,.\, [Y] + z_g' \,.\, [Z]$$
$$[B] \equiv x_b' \,.\, [X] + y_b' \,.\, [Y] + z_b' \,.\, [Z]$$

Since $[R]$, $[G]$ and $[B]$ each represent one T-unit of colour on the $[RGB]$ system, and since $[X]$, $[Y]$ and $[Z]$ are still measured in units of that system, it follows that the above equations are unit equations in which

$$x_r' + y_r' + z_r' = 1$$
$$x_g' + y_g' + z_g' = 1$$
$$x_b' + y_b' + z_b' = 1$$

(This can be confirmed algebraically by carrying the transformation through in terms of r_x, g_x, b_x, etc.; when an actual calculation is being made, the summing of the coefficients to unity provides a useful check on the arithmetic at this stage.)

The expressions for $[R]$, $[G]$ and $[B]$ can now be substituted in the equation for $[W]$, to give an equation of the form

$$[W] \equiv x_e' \,.\, [X] + y_e' \,.\, [Y] + z_e' \,.\, [Z]$$

where again $\qquad x_e' + y_e' + z_e' = 1$

In the (XYZ) system the equation for (W) is, by definition,

$$(W) \equiv 0{\cdot}333(X) + 0{\cdot}333(Y) + 0{\cdot}333(Z)$$

To avoid complicating the algebra unnecessarily by the introduction of superfluous constants, we will assume that $(W) \equiv [W]$. Then the magnitudes of unit amounts of (X), (Y) and (Z) relative to unit amounts of $[X]$, $[Y]$ and $[Z]$ respectively, can be derived as follows:

$$x_e' \,.\, [X] \equiv 0{\cdot}333(X), \quad \text{giving} \quad [X] \equiv \frac{0{\cdot}333}{x_e'} \,.\, (X)$$

$$y_e' \,.\, [Y] \equiv 0{\cdot}333(Y), \quad \text{giving} \quad [Y] \equiv \frac{0{\cdot}333}{y_e'} \,.\, (Y)$$

and $$z_e' \,.\, [Z] \equiv 0{\cdot}333(Z), \quad \text{giving} \quad [Z] \equiv \frac{0{\cdot}333}{z_e'} \,.\, (Z)$$

Insertion of these values of $[X]$, $[Y]$ and $[Z]$ in the equations for $[R]$, $[G]$ and $[B]$ then gives

$$[R] \equiv \frac{0{\cdot}333 \cdot x_r'}{x_e'} \,.\, (X) + \frac{0{\cdot}333 \cdot y_r'}{y_e'} \,.\, (Y) + \frac{0{\cdot}333 \cdot z_r'}{z_e'} \,.\, (Z)$$

$$[G] \equiv \frac{0{\cdot}333 \cdot x_g'}{x_e'} \,.\, (X) + \frac{0{\cdot}333 \cdot y_g'}{y_e'} \,.\, (Y) + \frac{0{\cdot}333 \cdot z_g'}{z_e'} \,.\, (Z)$$

$$[B] \equiv \frac{0{\cdot}333 \cdot x_b'}{x_e'} \,.\, (X) + \frac{0{\cdot}333 \cdot y_b'}{y_e'} \,.\, (Y) + \frac{0{\cdot}333 \cdot z_b'}{z_e'} \,.\, (Z)$$

The coefficients of (X), (Y) and (Z) in each of these equations will not now sum up to unity, so that although $[R]$, $[G]$ and $[B]$ still represent unit quantities of colour on the $[RGB]$ system, they do not represent unit quantities on the (XYZ) system. The relation between unit quantities of $[R]$ and (R) is then given by

$$[R] \equiv \left(\frac{0{\cdot}333 \cdot x_r'}{x_e'} + \frac{0{\cdot}333 \cdot y_r'}{y_e'} + \frac{0{\cdot}333 \cdot z_r'}{z_e'} \right) . \, (R)$$

with corresponding equations for $[G]$ and $[B]$. We shall not, in fact, make use of these relations in transforming the equation for C, but it is important to realize that in general the coefficients on the left-hand side of a colour equation are equal to the sum of the coefficients on the right-hand side. The absence of this equality in the equations for $[R]$, $[G]$ and $[B]$ is due solely to the fact that $[R]$, $[G]$ and $[B]$ are still measured in units of the old system, while (X), (Y) and (Z) are measured in the new system.

The advantage of retaining the equations for $[R]$, $[G]$ and $[B]$ is that they can be substituted directly in the equation

$$[C] \equiv r\,.\,[R] + g\,.\,[G] + b\,.\,[B]$$

to give a new equation that can be reduced to the form

$$[C] \equiv x'\,.\,(X) + y'\,.\,(Y) + z'\,.\,(Z)$$

This will again not be a unit equation, since $[C]$ is a unit on the old system, but by writing

$$[C] \equiv c\,.\,(C)$$

where

$$c = x' + y' + z'$$

we obtain the unit equation for (C) as

$$(C) \equiv x\,.\,(X) + y\,.\,(Y) + z\,.\,(Z)$$

where

$$x = \frac{x'}{c}, \quad y = \frac{y'}{c}, \quad \text{and} \quad z = \frac{z'}{c}$$

If calculations represented by the above algebra had to be carried through for each colour transformation, the labour might reasonably be regarded as excessive. However, once the equations for $[R]$, $[G]$ and $[B]$ in terms of (X), (Y) and (Z) have been derived, that part of the computation does not have to be repeated for subsequent calculations, so that the additional work required to transform each new colour specification is not considerable.

Sometimes the transformation problem may arise in a slightly different form. Whereas $[C]$ may be given by the equation

$$[C] \equiv r\,.\,[R] + g\,.\,[G] + b\,.\,[B]$$

and $[W]$ by

$$[W] \equiv 0{\cdot}333[R] + 0{\cdot}333[G] + 0{\cdot}333[B]$$

it may happen that (R), (G) and (B) are already known in the (XYZ) system by the equations

$$(R) \equiv x_r\,.\,(X) + y_r\,.\,(Y) + z_r\,.\,(Z)$$
$$(G) \equiv x_g\,.\,(X) + y_g\,.\,(Y) + z_g\,.\,(Z)$$
$$(B) \equiv x_b\,.\,(X) + y_b\,.\,(Y) + z_b\,.\,(Z)$$

These are now unit equations for (R), (G) and (B) on the (XYZ) system, and cannot therefore be substituted in the equation for $[C]$ to secure the transformation, until the relations between (R) and $[R]$, (G) and $[G]$, (B) and $[B]$ have been determined. Suppose

$$[R] \equiv k_r \,.\, (R), \quad [G] \equiv k_g \,.\, (G), \quad [B] \equiv k_b \,.\, (B)$$

Then

$$[R] \equiv k_r \,.\, x_r \,.\, (X) + k_r \,.\, y_r \,.\, (Y) + k_r \,.\, z_r \,.\, (Z)$$
$$[G] \equiv k_g \,.\, x_g \,.\, (X) + k_g \,.\, y_g \,.\, (Y) + k_g \,.\, z_g \,.\, (Z)$$
$$[B] \equiv k_b \,.\, x_b \,.\, (X) + k_b \,.\, y_b \,.\, (Y) + k_b \,.\, z_b \,.\, (Z)$$

Substitution in the equation for $[W]$ will then give

$$\begin{aligned}[W] &\equiv 0{\cdot}333(k_r \,.\, x_r + k_g \,.\, x_g + k_b \,.\, x_b)(X) \\ &\quad + 0{\cdot}333(k_r \,.\, y_r + k_g \,.\, y_g + k_b \,.\, y_b)(Y) \\ &\quad + 0{\cdot}333(k_r \,.\, z_r + k_g \,.\, z_g + k_b \,.\, z_b)(Z)\end{aligned}$$

But on the (XYZ) system, (W) is also given by

$$(W) \equiv 0{\cdot}333(X) + 0{\cdot}333(Y) + 0{\cdot}333(Z)$$

Hence

$$k_r \,.\, x_r + k_g \,.\, x_g + k_b \,.\, x_b = 1$$
$$k_r \,.\, y_r + k_g \,.\, y_g + k_b \,.\, y_b = 1$$
$$k \,.\, z_r + k_g \,.\, z_g + k_b \,.\, z_b = 1$$

from which k_r, k_g and k_b can be found. The expressions for $[R]$, $[G]$ and $[B]$ in terms of (X), (Y) and (Z) can then be derived and the transformation can proceed as before.

When transformations as described above are carried through for the chromaticity chart illustrated in Fig. 3.7. the new chart shown in Fig. 3.8. is obtained. The elongated triangle $[XYZ]$ of Fig. 3.7. has been compressed into a right-angled triangle, while the right-angled triangle $[RGB]$ has become more nearly equilateral. The position of W has also shifted relative both to $[R]$, $[G]$ and $[B]$ and to $[X]$, $[Y]$ and $[Z]$, although its actual position has remained unchanged at (0·333, 0·333). However, although changes of shape and disposition have occurred, the transformation has been carried through by means of linear equations. This means that lines which are straight in one system remain straight in the other and that one chart is merely an oblique projection of the other.

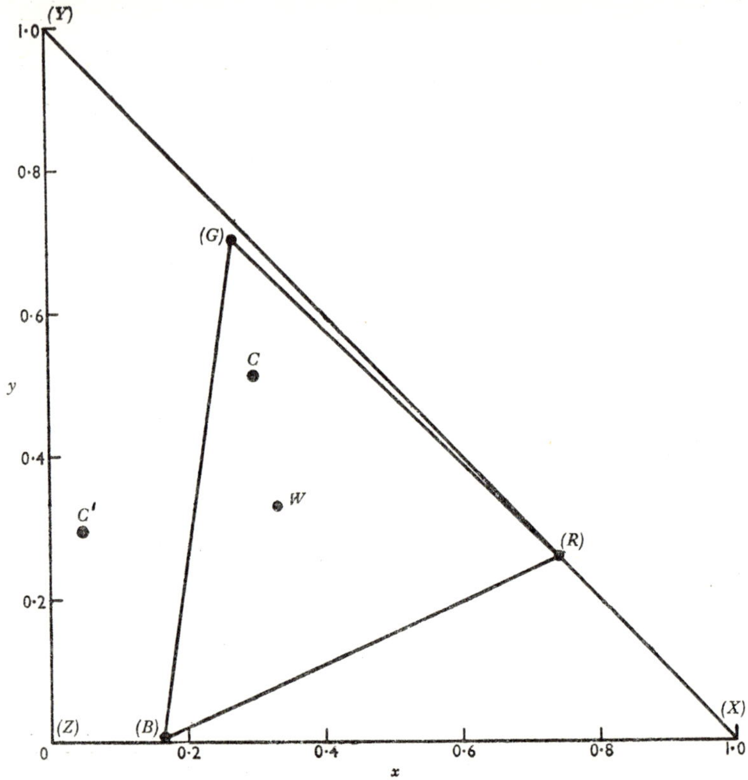

FIG. 3.8. The chromaticity chart of Fig. 3.7. transformed to co-ordinates of (*X*), (*Y*), and (*Z*), in which the locations of the old reference stimuli (*R*), (*G*), and (*B*) are shown

One consequence of this is that a point such as *C* which lies inside the triangle [*RGB*], will always remain inside the triangle in any system of co-ordinates, since a positive mixture of [*R*], [*G*] and [*B*] will still be required to produce a match of *C*, assuming the sign of [*R*], [*G*] or [*B*] has not been changed. What significance, then, should be attached to a colour such as *C'* which lies outside the triangle [*RGB*]? In terms of the chart of Fig. 3.8. *C'* has an all-positive specification and is in fact defined by the equation

$$(C') \equiv 0{\cdot}045(X) + 0{\cdot}295(Y) + 0{\cdot}660(Z)$$

In the co-ordinates of Fig. 3.7. however, it is defined by the equation

$$[C'] \equiv -0{\cdot}715[R] + 0{\cdot}700[G] + 1{\cdot}015[B]$$

The occurrence of the negative sign seems difficult to understand, and although there is nothing abnormal about the algebra, we cannot readily conceive of a negative amount of colour. This difficulty, however, can be easily resolved by re-writing the equation in the form

$$[C'] + 0{\cdot}715[R] \equiv 0{\cdot}700[G] + 1{\cdot}015[B]$$

This is algebraically equivalent to the previous equation, but now has the easily understood significance that if a certain amount of the red stimulus is added to C', the combination can be matched by a mixture of blue and green.

At this stage it may be helpful to return to the three-metals analogy, and to consider the implications of a transformation there from one system to another, and of negative quantities. Fig. 3.9 shows the metal chart again with pure copper, lead and aluminium at the corners of the triangle CuPbAl. Suppose, now, that for some reason it was impossible to obtain pure lead shot, and that the best that could be obtained always included a certain amount of copper impurity. This quality of shot would then be located at some point such as Pb′ in the chart. If we had to specify some mixture C we could do so in the usual way in terms of the proportions of pure Cu, Pb and Al required to match its composition. We could also, however, specify it in terms of some mixture of Cu, Pb′ and Al, since C lies inside the triangle formed by these three points. We need not go through the actual transformations to derive the new specification, since it is of the same form as the colour transformations already considered, but provided the position of Pb′ on the chart is known a new specification can evidently be calculated.

Can a similar specification be derived for a sample such as C' which lies outside the triangle CuPb′Al? Any positive mixture of Cu, Pb′ and Al will contain more copper than C', and the matching of its composition will be impossible because of the postulated inability to obtain pure lead shot. However, its specification would still be possible if a known amount of copper were added to C' to give the mixture a composition corresponding to a point inside the triangle CuPb′Al. This mixture could then be matched, the amount of copper which had been added could be subtracted, and a specification for *C′* could be obtained which was not at all unusual algebraically, but from the metallurgical point of view would have the unexpected property of a negative amount of copper.

This negative quantity has arisen merely because the specification is in terms of an impure sample of lead, and in the same way negative quantities can occur in colour measurements with a colorimeter, because the matching stimuli may not be pure red, green or blue stimuli. From the results that will be described in the next chapter, it appears that even spectral red, green and blue stimuli are not physiologically pure, in the sense that each of them stimulates at least two of the colour processes in the retina. In particular, the green wavelengths appear to stimulate the red process to quite a significant extent, a result due to the breadth of the red spectral sensitivity curve, which extends to the green wavelengths and beyond (see Fig. 2.7, p. 45). In practice, this means that saturated blue-green colours have to be desaturated by the admixture of some of the red stimulus before they can be matched and measured with a colorimeter.

A further extension of the metal analogy is to consider the possibility of referring the specification to hypothetical metals lying outside the triangle Cu Pb Al, and in particular to points lying below the line Cu Al. We can illustrate the unusual properties possessed by such points by reference to the system already quoted on p. 89, in which one dish of lead shot weighed 600 g, one dish of copper rivets 400 g, and one dish of aluminium rivets 150 g. For points along the line, PbAl, the weight of one dish of the sample will decrease steadily from 600 g at Pb to 150 g at Al. Below Al, we shall be dealing with hypothetical mixtures of a positive amount of aluminium and a negative amount of lead; the weight per dish will continue to decrease and at the point Al_0, co-ordinates (0·0, —0·33) in Fig. 3.9. the weight will become zero. Below Al_0, the weight will even become negative. A similar zero-weight position, Cu_0, co-ordinates (3·0, —2·0) can also be found along the extension of the line PbCu (Fig. 3.9.), and in fact the straight line Al_0Cu_0 is the locus of all such zero-weight metals. We might go further and develop a system of specification in terms

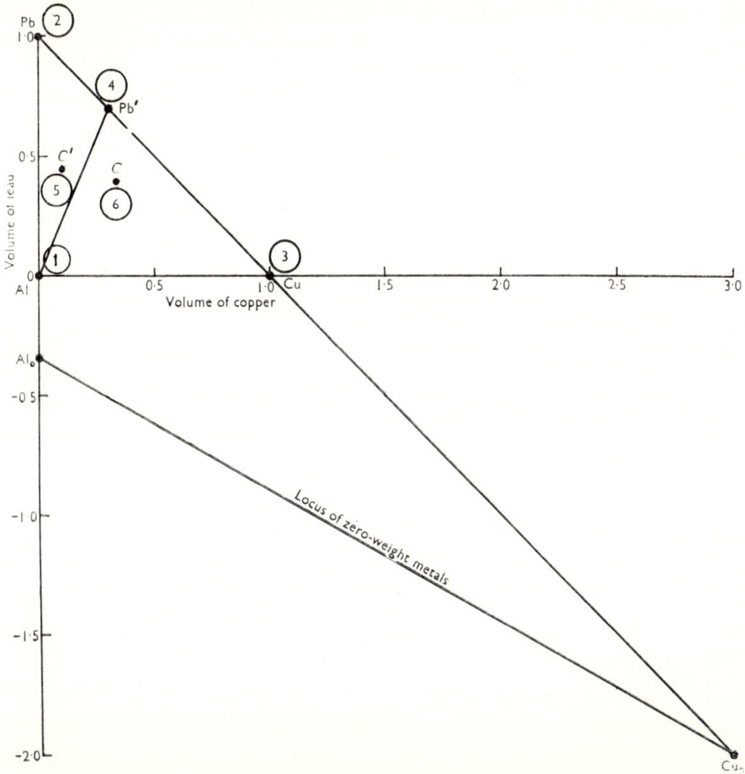

FIG. 3.9. The transformation of a metallic specification from units of Cu, Pb, and Al to units of Cu, Pb′, and Al, and the occurrence of negative quantities in a metal specification and of metal mixtures of zero weight

of Cu_0PbAl_0 which would have the curious, but perhaps useful property that the weight of any sample would be given entirely by the Pb component in the mixture.

Something closely analogous to this is possible with the chromaticity chart. In a trichromatic system it is possible to measure the light flux (or the luminance) of one T-unit of each of the three matching stimuli [R], [G] and [B], just as it was possible to weigh one dishful of each of the three metals. Suppose these photometric values are $V_{[R]}$, $V_{[G]}$ and $V_{[B]}$ respectively. (In practice, relative values are generally all that are required.) Then the flux $V_{[C]}$ in one T-unit of C as defined by the unit equation

$$[C] \equiv r \,.\, [R] + g \,.\, [G] + b \,.\, [B]$$

will be given by

$$V_{[C]} = r \,.\, V_{[R]} + g \,.\, V_{[G]} + b \,.\, V_{[B]}$$

Now the values of $V_{[R]}$, $V_{[G]}$ and $V_{[B]}$ are very far from being equal for the red, green and blue stimuli of a typical colorimeter, and values have already been quoted on p. 89 for one particular system as

$$V_{[R]} = 1{\cdot}00, \quad V_{[G]} = 4{\cdot}59, \quad V_{[B]} = 0{\cdot}06$$

This is similar to the differences in weight between dishfuls of copper, lead and aluminium. It follows that if we consider colours along the line [*G*] [*B*] in the chromaticity chart of Fig. 3.7, the value of $V_{[C]}$ will steadily diminish as we move down from [*G*] towards [*B*]. And just as with the metal analogy, if we go below [*B*] we reach a point $[B_0]$ where $V_{[C]} = 0$, i.e. we have a stimulus of zero flux (or luminance), just as we had a metal of zero weight. There will similarly be a point $[R_0]$ for which $V_{[C]}$ is also zero on the extension of the line [*G*] [*R*]. Further, the line $[R_0]$ $[B_0]$ is the locus of all stimuli for which $V_{[C]} = 0$ and this line has been given the name *alychne* (lightless).

This concept was first enunciated by Schrödinger* and as we shall see in Chapter 4, it was used by the C.I.E. in defining the reference stimuli (*X*), (*Y*) and (*Z*) of the 1931 system. Fig. 3.7 in fact represents these stimuli with [*X*] and [*Z*] located on the alychne. In this way the photometric aspects of colorimetry are referred entirely to the third stimulus [*Y*].

[*X*], [*Y*] and [*Z*] cannot, of course, be reproduced in the laboratory. They are hypothetical stimuli with rather unusual properties but they nevertheless have very real merits as reference co-ordinates in the C.I.E. system.

In the foregoing discussion of the chromaticity chart and colour transformations, the approach which Guild developed of reducing the analysis to a two-dimensional problem and using the concept of the T-unit has been followed. Not everyone finds this method acceptable and the more sophisticated approach operating in three-dimensional space and using vector notation is often preferred. In

* E. SCHRÖDINGER: *Sity. Akad. Wiss., Wien, Abt.* II. a, 1925, **134**, p. 471.

this case the reference stimuli, for example, are defined as the vectors **X**, **Y**, **Z**, in tristimulus space, rather than as points in the chromaticity chart, and the chromaticity chart itself is the unit plane in tristimulus space.*

* See, for example, G. WYSZECKI and W. S. STILES: *Color Science*, (Wiley: New York and London), 1967, p. 232.

CHAPTER 4

THE C.I.E. SYSTEM OF COLOUR MEASUREMENT

THE DEFINITION OF THE 1931 STANDARD OBSERVER

The principles of colorimetry were developed in the previous chapter from the starting point of a colour match made with a visual trichromatic colorimeter. This is historically and academically a justifiable approach, but although much technical colorimetry has been carried out with visual instruments, the precision and accuracy obtainable by these means are limited. The sensitivity of the eye to small colour differences is comparatively low when looking through the eyepiece of an instrument at a small and not very highly illuminated field, while the wide differences in the matches made by different observers are a further source of error and uncertainty.

The alternative is to define a colour physically in terms of its spectral composition. Spectrophotometric measurements may involve elaborate and expensive instruments, but once the data have been recorded, the results are available in a perfectly definite and unambiguous form. Yet while this information is often essential for the solution of a colour problem, the chromaticity of the integrated light can only be calculated by using the spectral distribution coefficients as described in the previous chapter.

We are thus brought back to visual colorimetry and its uncertainties, unless standard tables of distribution coefficients are adopted to represent the colour-matching characteristics of the average eye. But, once a set of coefficients is agreed, colour specifications can be derived which are independent of the colour vision of any particular individual, and which will be comparable between one laboratory and another.

Before such a system of colour measurement can be established, agreement on two separate issues is required. The distribution coefficients give the amounts of three stimuli required to match

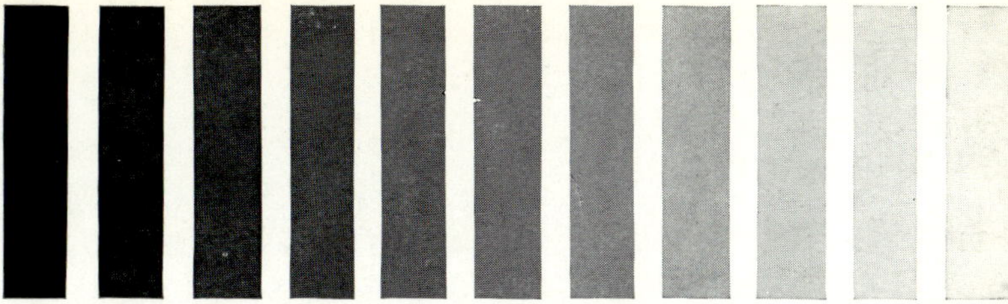

(*a*) Grey series selected to give approximately equal subjective intervals of lightness between neighbouring specimens

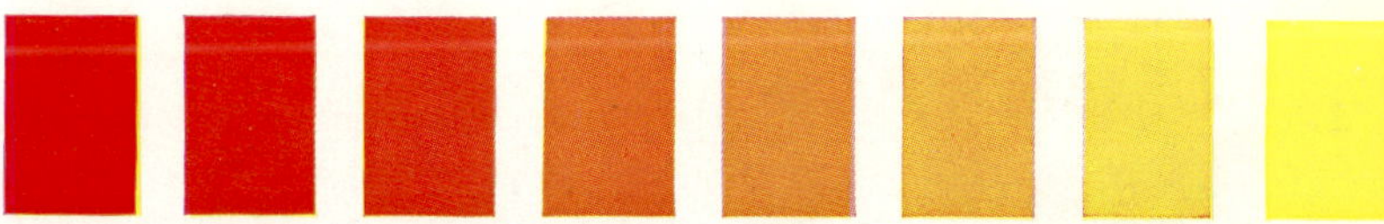

(*b*) Red-to-yellow series giving approximately equal subjective intervals of hue

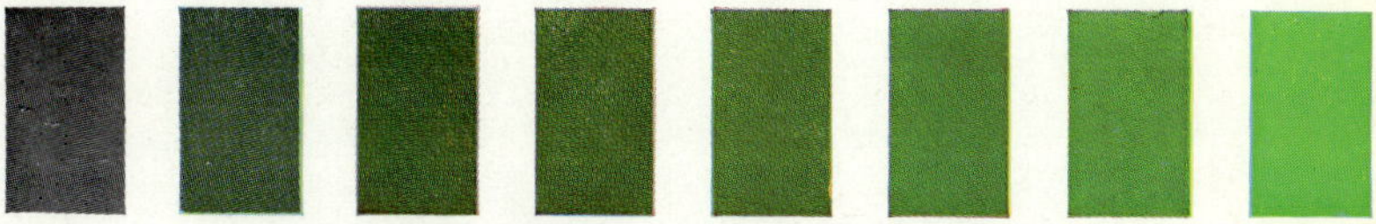

(*c*) Grey-to-green series giving approximately equal subjective intervals of saturation (see page 174)

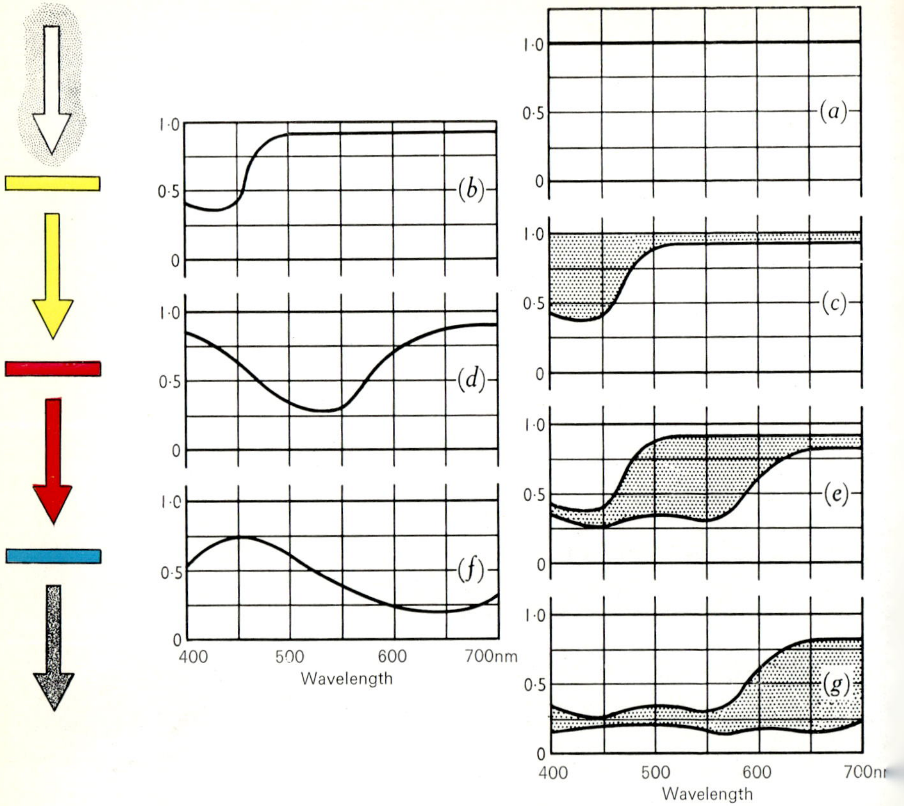

Subtractive colour mixture. The arrows on the left represent the colour of the light as it passes successively through yellow, magenta, and cyan or blue-green filters. The transmission curves of the filters are shown in the middle column (ordinates: transmission factor). The column on the right gives the energy distribution of the light at each stage in its path (ordinates: relative energy). The shaded areas in the graphs show the part of the spectrum that each filter controls by its absorption (see page 199)

the colours through the spectrum and the first stage in the preparation of a standard system of colour measurement is therefore the adoption of colour-mixture data representative of a normal observer. The second stage requires agreement on a set of reference stimuli relative to which the distribution coefficients can be transformed and in terms of which standard colour specifications can be expressed. This second issue can be decided from considerations of practical convenience in technical work without the limitations that are imposed on the choice of stimuli in a visual colorimeter.

Prior to 1931, colour specifications which made any pretence of possessing international significance were based on the colour-mixture data recorded in the pioneer investigations of König and of Abney, but the reliability of their results was limited by the small number of observers used in recording the data and by the lack of refinements in experimental technique compared with more modern researches. The later investigations by Guild and by the author were therefore undertaken independently, using very different instruments, as already described, and with different groups

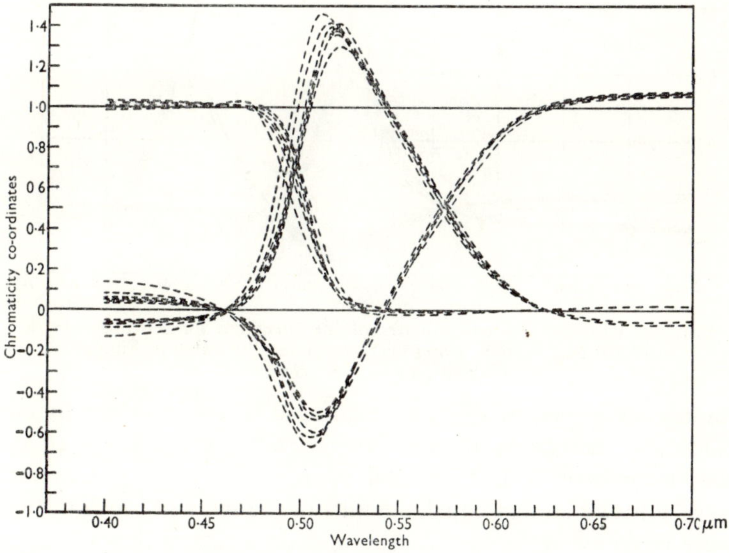

FIG. 4.1. Chromaticity co-ordinates of the spectrum for seven observers as measured by Guild, using his colorimeter illustrated in Fig. 3.1. (Guild, *Phil. Trans.*, 1931).

of observers, but with the common aim of securing a more reliable set of spectral mixture curves.

In Guild's investigation,* seven observers made colour matches through the spectrum and Fig. 4.1 shows the spectral chromaticity co-ordinates which he obtained. They are expressed in terms of his instrumental stimuli after the units had been adjusted to be equal in a match on the N.P.L. white then in use, and are shown here plotted against λ. At each wavelength the co-ordinates add up to unity, but it will be noticed that one or other of the co-ordinates is always negative; this indicates that desaturation of the spectral colours was necessary before they could be matched by a positive mixture of the instrumental stimuli.

Fig. 4.2 shows the corresponding data recorded by the author†

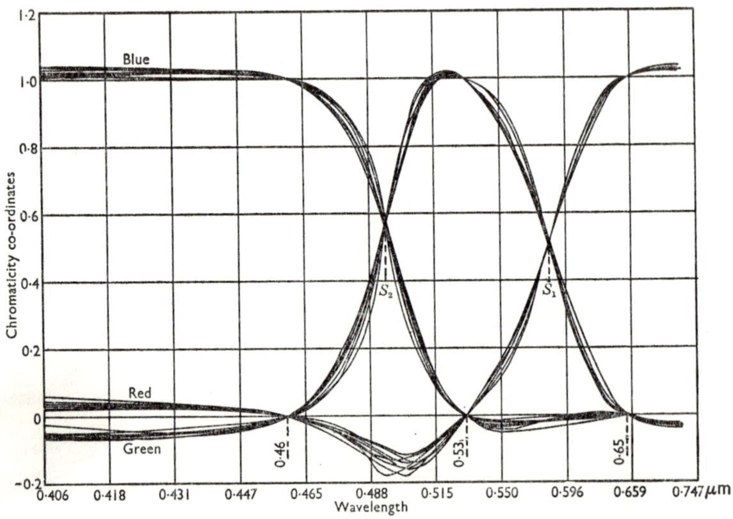

FIG. 4.2. Chromaticity co-ordinates of the spectrum for ten observers as measured by Wright, using his colorimeter illustrated in Fig. 3.2. (Wright, *Trans. Opt. Soc.*, 1928–9)

for ten observers, in this case expressed in terms of the monochromatic radiations of wavelengths 0·65, 0·53 and 0·46 μm that were used as instrumental stimuli. Their units, however, were adjusted so that equal amounts of the red and green stimuli were required in the match of a monochromatic yellow of wave-

* J. GUILD: *Phil. Trans. Roy. Soc., A*, 1931, **230**, p. 149.
† W. D. WRIGHT: *Trans. Opt. Soc.*, 1928-29, **30**, p. 141.

length 0·5825 μm, and equal amounts of the green and blue in the match on 0·4940 μm. This unit system, as explained in the original paper, had the advantage that the spectral co-ordinates were unaffected by pigmentation in the eye and provided an effective method of separating and analysing the variables responsible for the differences between one observer and another. The differences shown in Fig. 4.2, for example, could be due only to differences in the receptor processes, apart of course from experimental inaccuracies. On the other hand, the position of the white point which each observer measured, using the same N.P.L. source as that employed by Guild, was no longer at the fixed point (0·333, 0·333) in the chromaticity chart, but was affected by, and could be used to indicate, the degree of pigmentation of the individual observer.

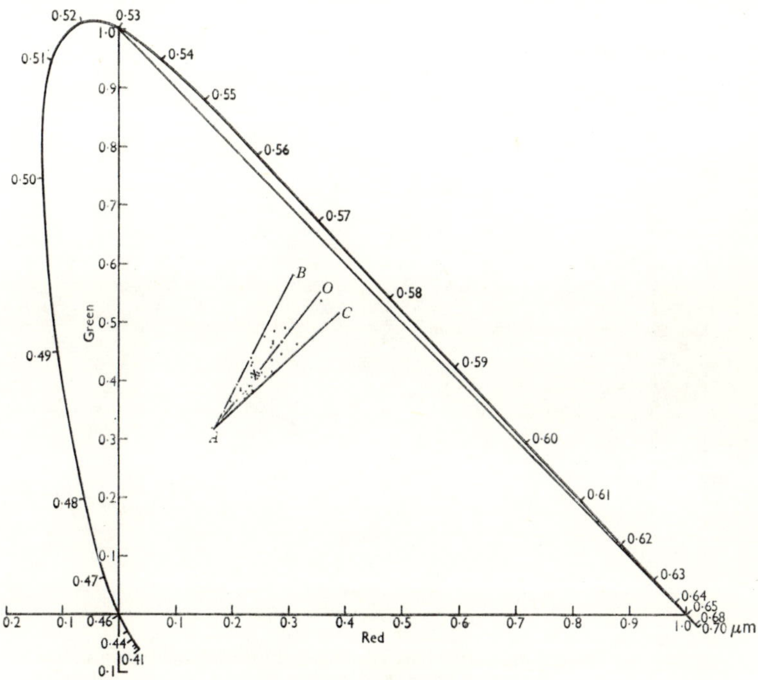

FIG. 4.3. Average data of Fig. 4.2 plotted in the chromaticity chart in terms of spectral matching stimuli of wavelengths 0·65, 0·53, and 0·46μm, with their units based on matches on 0·5825 and 0·4940μm. The white points of thirty-six observers are also shown

(Wright, *Trans. Opt. Soc.*, 1928–9)

Thus Fig. 4.3 shows the chromaticity chart in terms of the author's stimuli and unit system, and includes the spectrum colours plotted at their appropriate chromaticities and connected together to give the spectrum locus in the chart. The curve plotted here is the mean curve for the ten observers represented in Fig. 4.2. In addition, the white points of 36 observers are shown, and their spread across the chart in the blue-yellow direction suggests that the main cause of the differences is probably to be found in the density of yellow macular pigmentation and in the yellowing of the crystalline lens.

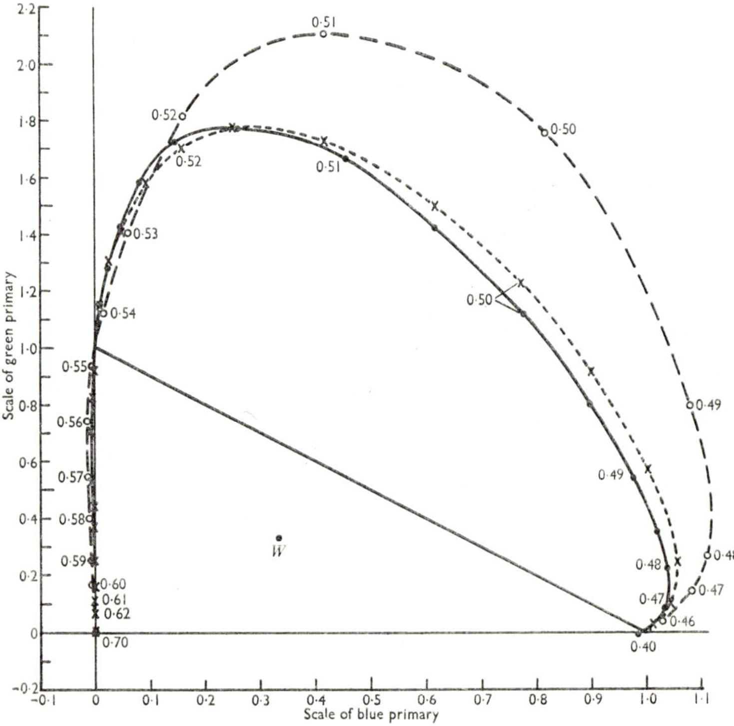

FIG. 4.4. The chromaticity chart plotted in terms of matching stimuli of wavelengths 0·700, 0·5461, and 0·4358 μm. with units based on a defined white. The diagram was prepared by Guild to compare the spectrum loci determined by Guild (continuous line), Wright (short dashes), and Weaver (long dashes). Weaver's results were derived from data obtained by König and Abney

(Guild, *Phil. Trans.*, 1931)

Another advantage of this method of reporting the data was that it enabled at least some observations to be recorded for a fairly large number of observers. A good deal of practice is required before accurate colour matches can be made and, even with practice, many observers are unable to make satisfactory matches unless a great deal of time is spent on the observations. The most reliable observers were therefore used to make a complete set of measurements on both the white point and on the spectral colours, while those who were slower or who had less time to spare confined their measurements to the white point. This enabled the greatest number of observations to be made on the item which showed the greatest spread, namely the white point, and the mean data had a correspondingly greater reliability.

The average results for the two investigations were compared by Guild after both sets of data had been transformed to the same reference stimuli, namely 0·700, 0·5461 and 0·4358 μm, with the units based on the N.P.L. white. Fig 4.2 shows the chromaticity chart on this basis and the two mean spectrum loci, together with a curve that had been derived earlier by Weaver* from the data of Abney and König. It was the close agreement between the Guild-Wright investigations that led to the recommendation and eventual adoption by the C.I.E. of their mean values to define the colour-matching characteristics of the 1931 standard observer. However, it is interesting to include Fig. 4.5, which shows the blue corner of Fig. 4.4 on a greatly enlarged scale, since the two loci did differ significantly in this region.

The mean spectrum locus does not in itself provide a complete specification of the colour-matching characteristics of an average observer. We also need to know the number of T-units at each wavelength in the equal-energy spectrum if we are to calculate the equal-energy distribution coefficients $\bar{r}_\lambda$, $\bar{g}_\lambda$ and $\bar{b}_\lambda$. These coefficients could be obtained from direct colour matches on an equal-energy spectrum, or on a spectrum of known energy content, but since the standard V_λ curve adopted by the C.I.E. in 1924 was already in existence, an alternative route was necessary.

We have already seen on p. 100 that the flux in one T-unit of a colour C is given by

* Report of Committee on Colorimetry, *J. Opt. Soc. Am.*, 1922, **6**, p. 527.

$$V = r_{[C]} \cdot V_{[R]} + g \cdot V_{[G]} + b \cdot V_{[B]}$$

where $V_{[R]}$ is the flux in one T-unit of [R], etc. If the unit equation for λ is

$$[\lambda] \equiv r_\lambda \cdot [R] + g_\lambda \cdot [G] + b_\lambda . [B]$$

the flux in one T-unit of λ will be given by

$$V_{[\lambda]} = r_\lambda \cdot V_{[R]} + g_\lambda \cdot V_{[G]} + b_\lambda \cdot V_{[B]}.$$

Since the flux at wavelength λ in the equal-energy spectrum is proportional to V_λ, the number of T-units at λ in the equal energy spectrum will be proportional to

$$\frac{V_\lambda}{V_{[\lambda]}} \text{ or } \frac{V_\lambda}{r_\lambda \cdot V_{[R]} + g_\lambda \cdot V_{[G]} + b_\lambda \cdot V_{[B]}} = m_\lambda$$

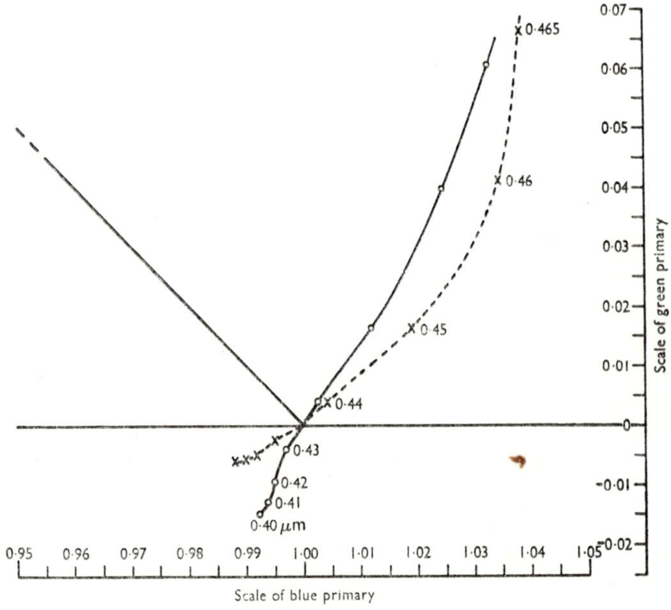

FIG. 4.5. The blue corner of Fig. 4.4. enlarged to show the differences between the Guild (continuous line) and Wright (dashes) determinations of the spectrum locus

(Guild, *Phil. Trans.*, 1931)

Hence with V_λ given, and r_λ, g_λ and b_λ already measured, the only additional information needed to derive m_λ are the values $V_{[R]}$, $V_{[G]}$ and $V_{[B]}$. These can be obtained experimentally by applying one of the techniques of heterochromatic photometry to the three instrument stimuli used in the measurements or, if [R], [G] and [B] are not the actual instrumental stimuli, $V_{[R]}$, $V_{[G]}$ and $V_{[B]}$ can be calculated by transformation from the instrument values. The distribution coefficients are then given by

$$\bar{r}_\lambda = m_\lambda \, . \, r_\lambda$$
$$\bar{g}_\lambda = m_\lambda \, . \, g_\lambda$$
$$\bar{b}_\lambda = m_\lambda \, . \, b_\lambda$$

but since we are generally only interested in relative values, the coefficients for all wavelengths may be multiplied by some common factor if desired.

For the reference stimuli of Fig. 4.4, Guild obtained the values

$$V_{[R]} = 1{\cdot}00, \quad V_{[G]} = 4{\cdot}390, \quad V_{[B]} = 0{\cdot}048$$

based on the values for his own colorimeter, but modified slightly to ensure internal consistency of the data. These results were then used by the C.I.E. to derive the distribution coefficients and so complete the definition of the standard observer. One final modification to the data was made, namely that the units of [R], [G] and [B] were adjusted to be equal in a match on an 'equal-energy' white, i.e. a white in which the energy per unit waveband is constant through the visible spectrum. Such a white is not reproducible in the laboratory from any known lamp-and-filter combination, but as it is basic to all colorimetric calculations, it provides the obvious centroid to any standard chromaticity chart. The curves for $\bar{r}_\lambda$, $\bar{g}_\lambda$ and $\bar{b}_\lambda$ as calculated on this basis are shown in Fig. 4.6.

The observer data adopted in 1931 by the C.I.E. consisted* then of: (i) tables of the spectral chromaticity co-ordinates (at that time called trichromatic coefficients) r_λ, g_λ and b_λ, giving the chromaticities of the spectral colours at every 5 nm from 380 nm to 780 nm in terms of the stimuli 700 nm [R], 546·1 nm [G] and 435·8 nm [B], with units adjusted to be equal in a match on an equal-energy white; (ii) tables of the tristimulus values, or dis-

* *Comm. Int. de l'Éclairage, Compte Rendu*, 1931, p. 19.

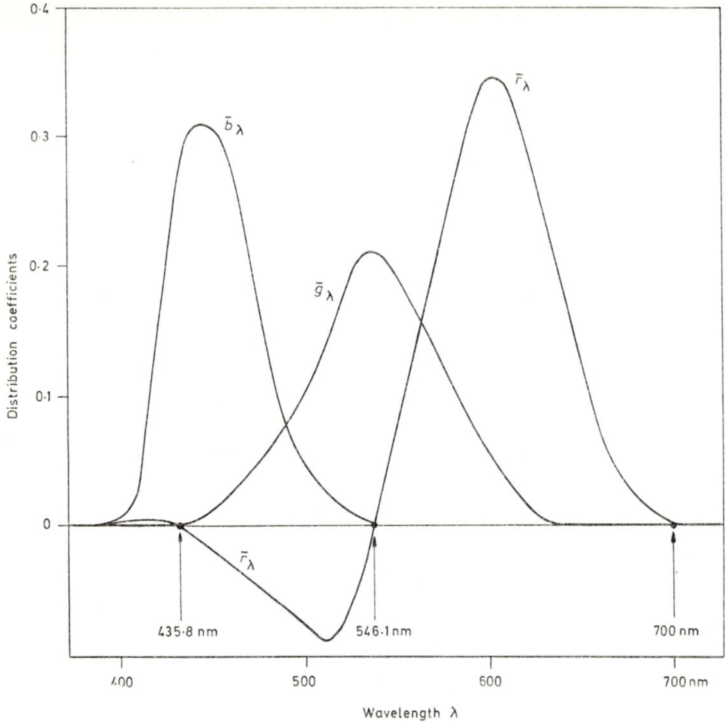

FIG. 4.6. The spectral distribution curves for the equal-energy spectrum for the 1931 C.I.E. standard observer, expressed in terms of the matching stimuli 700 nm, 546·1 nm and 435·8 nm and with their units adjusted to be equal in a colour match on an equal-energy white

tribution coefficients, $\bar{r}_\lambda$, $\bar{g}_\lambda$ and $\bar{b}_\lambda$ for the equal-energy spectrum at the same wavelength intervals and for the same stimuli and units; and (iii) the relative luminous efficiencies (then called relative luminosities) of unit quantities of [*R*], [*G*] and [*B*], which had the values $V_{[R]} = 1{\cdot}00$, $V_{[G]} = 4{\cdot}5907$ and $V_{[B]} = 0{\cdot}0601$.

Since the distribution coefficients were derived to satisfy the equation

$$V_\lambda = (\bar{r}_\lambda \,.\, V_{[R]} + \bar{g}_\lambda \,.\, V_{[G]} + \bar{b}_\lambda \,.\, V_{[B]}) \times (\text{a constant})$$

the 1924 V_λ curve formed an essential element in the 1931 standard observer data for colorimetry. This was technically a great convenience since photometry and colorimetry are so closely related, but it did have the disadvantage that two entirely different sets of results relating to two entirely different groups of observers had to be merged in such a way as to form an internally consistent system. Unfortunately, as has already been mentioned, the 1924 V_λ curve has been shown to be too low in the violet end of the spectrum and although those wavelengths usually make only a very small contribution to the light flux, so that the error from the photometric point of view is unimportant, the effect on the colorimetric data might have been significant, at least in some applications. It was partly for this reason that the C.I.E. recommended that a redetermination of the colour matching functions should be undertaken and for which the Stiles trichromator described on p. 78 was constructed.

Fig. 4.7 shows how close the new 2° measurements made by

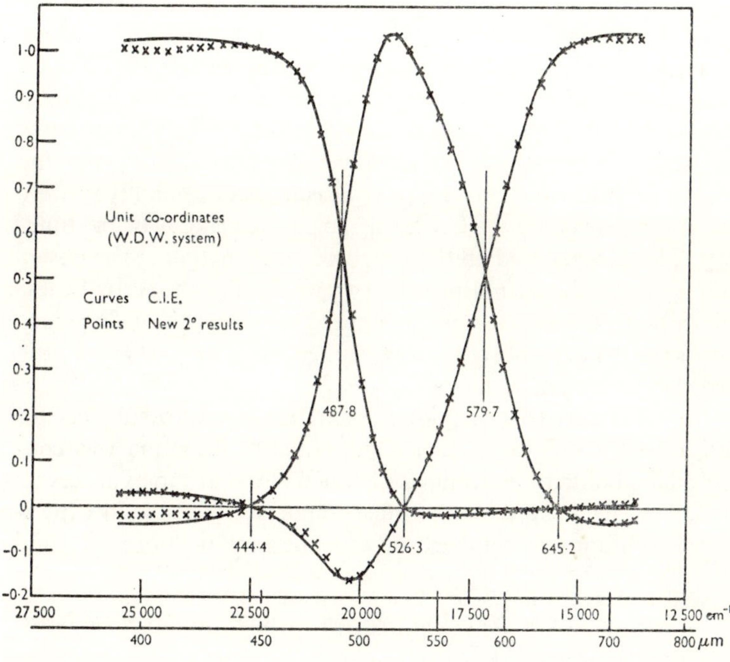

FIG. 4.7. Chromaticity co-ordinates of the spectrum as recorded by Stiles in a pilot investigation using a 2° field, compared with the 1931 C.I.E. data (*Physical Society Year Book*, 1955)

Stiles* were to the 1931 data, at least so far as the chromaticity co-ordinates of the spectrum are concerned, and it is evident that no important errors were present in this part of the definition of the 1931 observer. In fact the varied field tests that have been undertaken recently on the initiative of the C.I.E. have now confirmed that in general the 1931 data give satisfactory colour match predictions up to field sizes of 4°. No revision of the 1931 standard observer data is therefore to be expected in the near future.

THE DEFINITION OF THE 1964 STANDARD OBSERVER

The demand for large-field colour matching data mentioned on p. 77 was met in 1963 when the Colorimetry Committee of the C.I.E. recommended the adoption of an additional set of colour matching functions for use as a supplement to the 1931 C.I.E. colour matching functions whenever more accurate correlation with visual colour matching of fields of large angular subtense was desired.† These functions had already been prepared in 1959 on the basis of the Stiles and Speranskaya measurements,‡ and they were formally approved by the C.I.E. in 1964 as official C.I.E. 10° functions. They were based on the weighted combination of the observations of 49 subjects (34 men and 15 women) recorded by Stiles§ and 18 by Speranskaya, and the subsequent extension of Speranskaya's results to 27 observers (23 women and 4 men) showed little change in her mean set of curves.‖

Stiles first carried out pilot measurements on 10 subjects for both 2° and 10° field sizes, but subsequently attention was concentrated on obtaining results for a much larger group of observers for the 10° field. Perhaps the most important advance in experimental technique in comparison with the investigations of Guild

* W. S. Stiles: *Physical Society Year Book* (London), 1955, p. 44.

† D. B. Judd: *J. Opt. Soc. Am.*, 1963, **53**, p. 1012. *C.I.E. Compte Rendu*, 1964.

‡ *C.I.E. Compte Rendu*, 1959, Vol. A, p. 95.

§ W. S. Stiles and J. M. Burch: *Optica Acta*, 1959, **6**, p. 1.

‖ N. I. Speranskaya: *Optics and Spectroscopy*, (Trans. by Opt. Soc. Am.) 1959, **7**, 424.

and the author, was in the direct calibration of the energy of the monochromatic test and matching stimuli. This meant that instead of deriving the distribution coefficients indirectly from the chromaticity co-ordinates, the V_λ curve and the V-values of the matching stimuli, the colour matching functions could be calculated directly in terms of the energies of the matching stimuli required to match each wavelength in the equal-energy spectrum. The data were subsequently weighted by the V-values of the matching stimuli and summed to give the V_λ curve, although Stiles regarded the colour mixture curves as his main concern.

In the course of the work Stiles made certain tests on additivity under 10° viewing conditions and although some evidence of a failure of additivity was found, he considered that the departures from additivity were not large enough relative to differences of colour matching to prove a valid objection to the adoption of a linear system of colour-matching functions.* He also found that rod vision was interfering with the data to some extent, but its effect was reduced by using a yellow matching stimulus instead of a green for some of the matches and also by carrying out some of the observations at a higher illumination level.

Before presenting the data for consideration by the C.I.E. a further extrapolation for rod intrusion was made by Stiles and a similar correction was subsequently made on the Speranskaya data. Both sets of colour matching functions were then transformed to spectral matching stimuli of wavelengths 645·2 nm, 526·3 nm and 444.4 nm (corresponding to wave-numbers 15500, 19000 and 22500 respectively). The weighted mean curves then gave amounts, in energy units, of these matching stimuli required to match the spectral colours in the equal-energy spectrum. As described later in this chapter, the mean data were transformed to new reference stimuli giving all-positive distribution curves and having similar photometric properties as the 1931 C.I.E. system. This meant that the relative luminous efficiencies of the matching stimuli were also required and the values used were those reported by Stiles and

* W. S. STILES: *N.P.L. Symposium on Visual Problems of Colour*, (H.M. Stationery Office, London), 1958, p. 211; *Optica Acta*, 1963, **10**, p. 257.

R. D. LOZANO and D. A. PALMER: *Vision Research*, 1967, 7, p. 929; *J. Opt. Soc. Am.*, 1968, **58**, 1653.

Burch on the basis of observations by flicker photometry as follows: $V_{6452}=0{\cdot}166$, $V_{5263}=1{\cdot}000$, $V_{4444}=0{\cdot}088$. These were for unit energies of the stimuli and not for T-units. The values obtained by Stiles using a direct-comparison method of heterochromatic photometry instead of a flicker method showed major differences from these values, a further example of the ambiguity associated with the photometric evaluation of stimuli of different colours.

Stiles has recently published tables giving the individual colour matching functions of 20 of his 49 observers.* A selection of these functions has been plotted in Fig. 4.8 to help visualize the spread that can occur in colour matching among a group of observers. These curves refer to colour matching with a 10° annular field, the central, foveal area being ignored. It would have been useful to make a comparison with the corresponding spread in the individual colour mixture curves from which the 1931 2° distribution curves of Fig. 4.6 were derived, but unfortunately their method of derivation involving the 1924 V_λ curve makes this impossible. However, the spread in the 2° chromaticity co-ordinates of the spectrum and of the white points has been illustrated in Figs. 4.1, 4.2 and 4.3 and is certainly comparable to the spread in the 10° functions of Fig. 4.8. Indeed, it is likely to be greater because of the greater influence of the macular pigmentation on the small field observations.

As significant differences in colour matching occur among a group of observers, it might be suggested that there is little point in striving after very great accuracy in colour measurement and specification. This is certainly a factor which has to be remembered when fixing permissible tolerances in a colour specification, but, as we shall see in the next chapter, the size of a tolerance is also related to the degree of metamerism between the samples being compared. When colours in the spectrum are being matched with a trichromatic colorimeter, there is a very high degree of metamerism and observer differences are exaggerated. To that extent, the observer differences implied by the spread among the curves in Fig. 4.8 is of less significance for commercial colour matching than might at first be thought.

* G. Wyszecki and W. S. Stiles: *Color Science* (Wiley: New York and London), 1967.

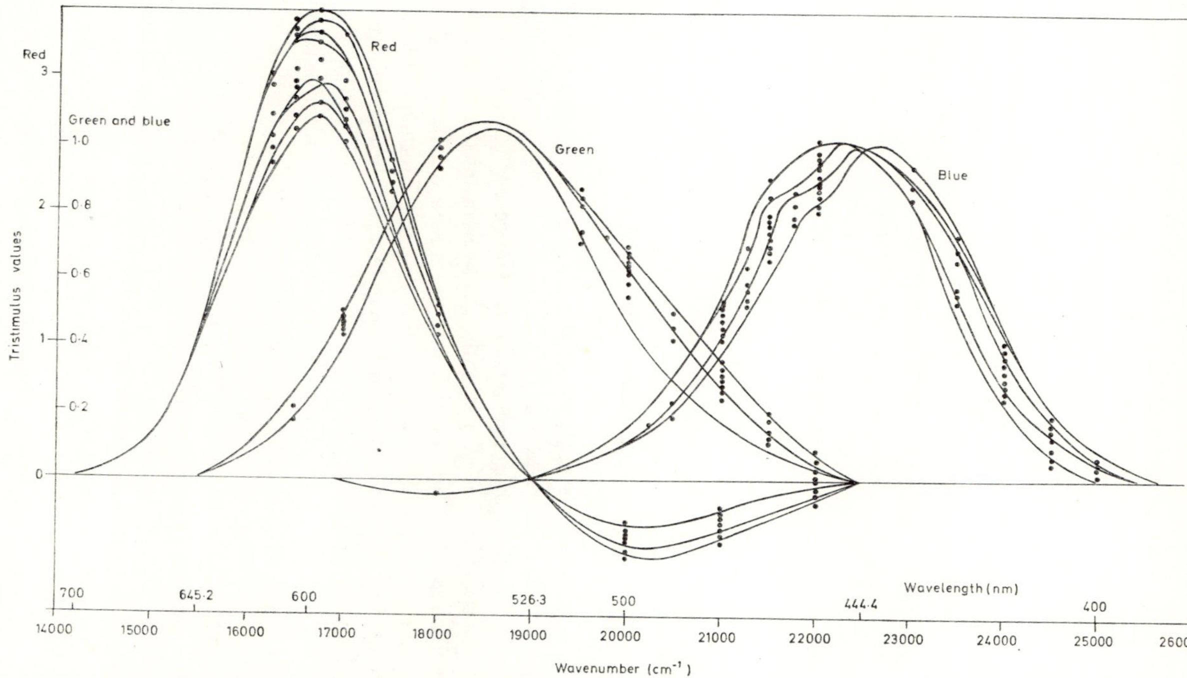

FIG. 4.8. A selection of colour matching functions of individual observers as recorded by Stiles using a 10° matching field. To avoid confusion, only a small number of continuous curves have been drawn. A further indication of the spread has been given by plotting individual points at certain wavelengths. (From data published by Wyszecki and Stiles: *Color Science*)

THE 1931 CO-ORDINATES, CHROMATICITY CHART AND DISTRIBUTION CURVES

Once the colour-matching characteristics of the standard observer were defined, the C.I.E. had to consider the second stage of the standardization, i.e. agreement on the reference stimuli. It would have been quite feasible to use the [*RGB*] system as the standard system for colour specification, but it was obviously desirable to consider whether a different choice of co-ordinates would not be more convenient for technical purposes. Reference has already been made on p. 104 to the fact that the colour equations for the spectral radiations in terms of [*R*], [*G*] and [*B*] involve negative quantities, and there was considerable discussion preceding the adoption of the C.I.E. system about the advisability of using a system in which negative values might occur. It was argued with some force that the meaning of a negative amount of a colour was unlikely to be readily appreciated by many of those anxious to use the system, and the inadvertent omission of the negative sign might be an occasional source of error. Moreover, the negative sign occurs not only in the chromaticity co-ordinates but also in the distribution coefficients. Hence evaluation of the tristimulus specification of a stimulus by weighting the distribution coefficients by the spectral energy values of the stimulus would have involved both positive and negative products. These would then have had to be summed separately and the one summation subtracted from the other. Apart from the extra computing operations, the design of mechanical integrators would have been more difficult and the development of photoelectric tristimulus colorimeters would also have been greatly hampered.

For reasons such as these, it was eventually agreed that while the [*RGB*] system was an appropriate framework within which to define the colour-mixing characteristics of the standard observer, an all-positive system should be adopted as the standard reference system for colour specification, using the reference stimuli [*X*], [*Y*] and [*Z*] shown in the [*RGB*] chromaticity chart in Fig 4.9.

It is necessary to consider why these particular stimuli should have been chosen out of all the possible points in the chroma-

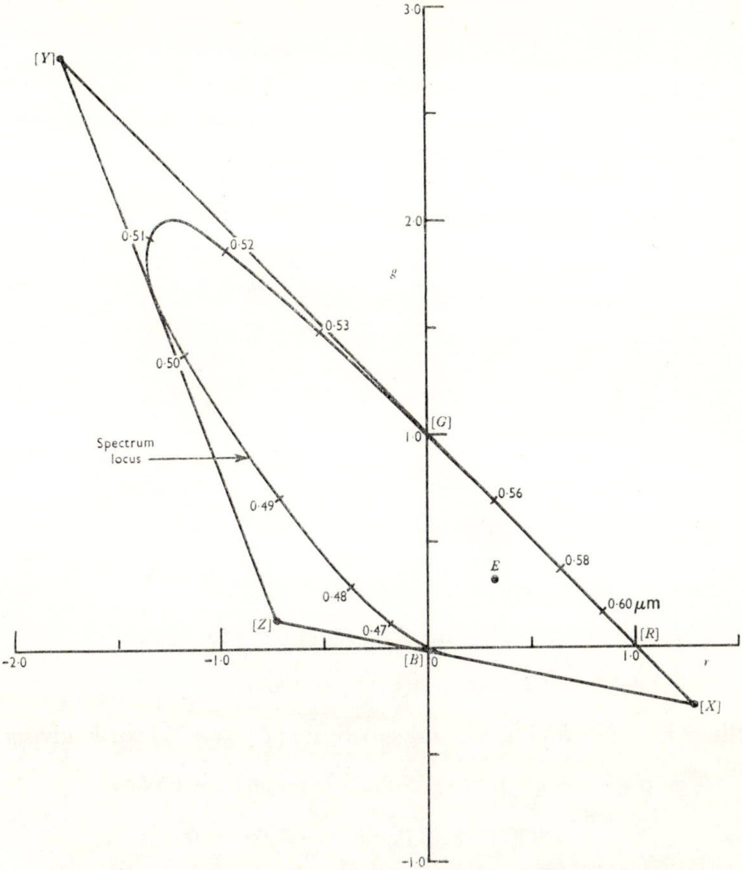

FIG. 4.9. The spectrum locus as defined by the C.I.E. in 1931 for the standard observer in terms of the stimuli 700, 546·1, and 435·8 nm, with the units based on an equal-energy white E. The locations of the stimuli [X], [Y], and [Z] selected by the C.I.E. for use as reference stimuli in standard colour specifications are shown relative to the spectrum locus

ticity chart. In the first place, if the system is to be all-positive, all real physical stimuli must lie inside the triangle [XYZ] and in particular the spectrum locus must lie entirely on or within the triangle. Secondly, it was decided that the line [X][Y] should lie along the spectrum locus at the red end of the spectrum where the locus is straight. In this way, the blue co-ordinates could be made zero for an appreciable range of wavelengths in the red, a condition likely to save a useful amount of computational labour in certain calculations. Thirdly, use was made of the device suggested by Schrödinger and elaborated by Judd* in which $[X]$ and $[Z]$ are chosen so that $V_{[X]}$ and $V_{[Z]}$ are zero and therefore fall on the alychne, as described on p. 100.

The equation for the alychne in the $[RGB]$ system can be derived quite simply as follows: If we have a colour C defined by the unit equation

$$[C] \equiv r\,.\,[R] + g\,.\,[G] + b\,.\,[B]$$

then the relative luminous efficiency of 1 T-unit will be

$$V_{[C]} = r\,.\,V_{[R]} + g\,.\,V_{[G]} + b\,.\,V_{[B]}$$

In the $[RGB]$ system

$$V_{[R]} = 1{\cdot}00, \quad V_{[G]} = 4{\cdot}5907, \quad V_{[B]} = 0{\cdot}0601$$

as already quoted, so that

$$V_{[C]} = r + 4{\cdot}5907g + 0{\cdot}0601b$$

If C is to lie on the alychne, $V_{[C]}$ must be zero, so that

$$0 = r + 4{\cdot}5907g + 0{\cdot}0601b$$

Since $r + g + b = 1$, we can substitute $(1 - r - g)$ for b, giving

$$0 = r + 4{\cdot}5907g + 0{\cdot}0601 - 0{\cdot}0601r - 0{\cdot}0601g$$

or

$$0{\cdot}9399r + 4{\cdot}5306g + 0{\cdot}0601 = 0$$

This is the equation for the alychne and it will be seen to be a straight line. The co-ordinates of $[X]$ and $[Z]$, as located by the C.I.E. in the $[RGB]$ chromaticity chart, are

$$r_X = 1{\cdot}275, \qquad r = -0{\cdot}743$$
$$g_X = -0{\cdot}278, \quad g_Z = 0{\cdot}141$$

These values satisfy the equation for the alychne and therefore prove that $V_{[X]}$ and $V_{[Z]}$ are zero. The values of $V_{(X)}$ and $V_{(Z)}$ in the (XYZ) system will also necessarily be zero from the relation between unit quantities of (X) and $[X]$, etc., as given on p. 93.

* D. B. Judd: *Bur. Stand. J. Res.*, 1930, **4**, p. 515.

The immense advantage of employing a system in which two of the reference stimuli have zero luminous efficiencies, lies in the fact that the photometric value of the light represented by a colour equation is given solely by the third tristimulus value. Thus in a colour C expressed by the tristimulus equation

$$c\,.\,(C) \equiv X\,.\,(X) + Y\,.\,(Y) + Z\,.\,(Z)$$

the luminous flux F will be proportional to

$$X\,.\,V_{(X)} + Y\,.\,V_{(Y)} + Z\,.\,V_{(Z)}$$

Since $V_{(X)}$ and $V_{(Y)}$ are zero this reduces to $Y\,.\,V_{(Z)}$, so that

$$F = Y \times \text{constant}$$

Moreover, this choice of (X) and (Z) means that the $\bar{y}_\lambda$ distribution curve becomes the same curve as the V_λ curve, as can be shown by repeating the derivation of the distribution curves as given on p. 108, but now for the (XYZ) system. Thus the luminous efficiency $V_{(\lambda)}$ of one T-unit of wavelength λ is

$$\begin{aligned} V_{(\lambda)} &= x_\lambda\,.\,V_{(X)} + y_\lambda\,.\,V_{(Y)} + z_\lambda\,.\,V_{(Z)} \\ &= y_\lambda\,.\,V_{(Y)} \end{aligned}$$

since $V_{(X)}$ and $V_{(Z)}$ are zero. The number of T-units in the equal-energy spectrum at each wavelength λ is therefore proportional to

$$\frac{V_\lambda}{V_{(\lambda)}} = \frac{V_\lambda}{y_\lambda\,.\,V_{(Y)}}$$

The $\bar{y}_\lambda$ distribution coefficients are given by the product of the number of T-units in the equal-energy spectrum at wavelength λ and the co-ordinate y_λ in the unit equation for (λ); hence

$$\begin{aligned} \bar{y}_\lambda &= \frac{V_\lambda}{y_\lambda\,.\,V_{(Y)}}\,.\,y_\lambda \times \text{constant} \\ &= V_\lambda \times \text{constant} \end{aligned}$$

The constant can be given any convenient value such as unity.

This identity between V_λ and $\bar{y}_\lambda$ is particularly useful when the colour of a surface or filter is being calculated from its spectral

I

reflection or transmission curve, since the calculation to determine the Y tristimulus value is the same as that for the luminance factor or the transmission factor, which need not, therefore, be calculated separately. This type of calculation will be elaborated in the next section.

The basic 1931 C.I.E. tables for evaluating a colour are given in Tables 3 and 4 (Appendix II). In Table 3, the chromaticity co-ordinates x_λ, y_λ and z_λ are given for wavelengths from 380 nm to 780 nm at intervals of 5 nm. Table 4 gives the distribution coefficients $\bar{x}_\lambda$, $\bar{y}_\lambda$, $\bar{z}_\lambda$ for the equal-energy spectrum, again at wave-length intervals of 5 nm; as explained above, the $\bar{y}_\lambda$ coefficients are identical with the V_λ values adopted by the C.I.E. in 1924. It will be noticed that the sums of the three sets of coefficients are given, and that they are equal to one another, with a value of 21·371. This equality is evidence of the internal consistency of the system, and of the assumption of additivity, since the units of (X), (Y) and (Z) were adjusted to be equal in a match on an equal-energy white. The value of the sum, 21·371, is of no significance at all; it could without prejudice to the system be adjusted to any arbitrary value by multiplying all the distribution coefficients by a constant factor. This particular value occurs merely because the V_λ values were employed directly as the $\bar{y}_\lambda$ coefficients, and they in their turn had been arbitrarily adjusted in 1924 to bring the maximum of the curve to a value of 1·000.*

The data of Tables 3 and 4 can be plotted to give the diagrams in Figs. 4.10 and 4.11 respectively. Fig. 4.11 shows the (XYZ) chromaticity chart with (x_λ, y_λ) plotted to give the spectrum locus. The equal-energy white E is shown located at the centroid of the triangle (0·333, 0·333) while the three illuminants S_A, S_B and S_C

* More extensive tables have recently been produced by Mahr (K. Mahr: *Die Farbe*, 1961, 7, p. 323) in which the 1931 colour matching functions and spectral chromaticity co-ordinates have been extrapolated to cover the wave-length range from 360 to 830 nm. They have also been interpolated to 1 nm intervals. Special attention has also been paid to the smoothing of the data and at some wavelengths the values are given to as many as 11 or 12 decimal places. In exceptional cases such refinement might be advantageous to avoid dis-crepancies in computations through rounding-off errors, but would not be justified in normal colorimetric calculations. Only 4 decimal places are therefore given in Tables 3 and 4, and indeed in the other tables in this book. Wyszecki and Stiles [G. Wyszecki and W. S. Stiles: *Color Science* (Wiley: New York and London), 1967] have reproduced Mahr's tables in full but with minor modifica-tions.

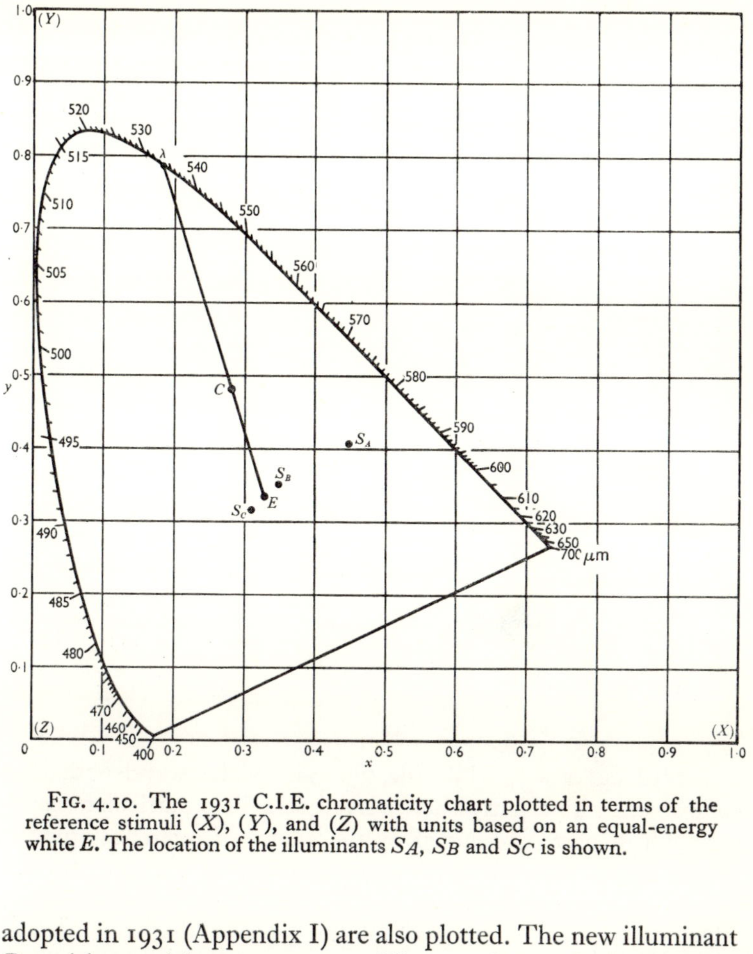

FIG. 4.10. The 1931 C.I.E. chromaticity chart plotted in terms of the reference stimuli (*X*), (*Y*), and (*Z*) with units based on an equal-energy white *E*. The location of the illuminants S_A, S_B and S_C is shown.

adopted in 1931 (Appendix I) are also plotted. The new illuminant D_{6500} (chromaticity 0·3127, 0·3290) lies too close to S_C to be shown separately in this diagram.

There are a number of features to be noted about the spectrum locus. First, it will be seen that from the extreme red end of the spectrum to the green at a wavelength of about 530 nm, the locus is very nearly straight and part of it lies along the line (*X*)(*Y*). Thereafter the curve bends sharply round as the colour changes from green to blue-green; the blue-green wavelengths from 510 nm to 480 nm are spread out over a stretch of the locus which retains

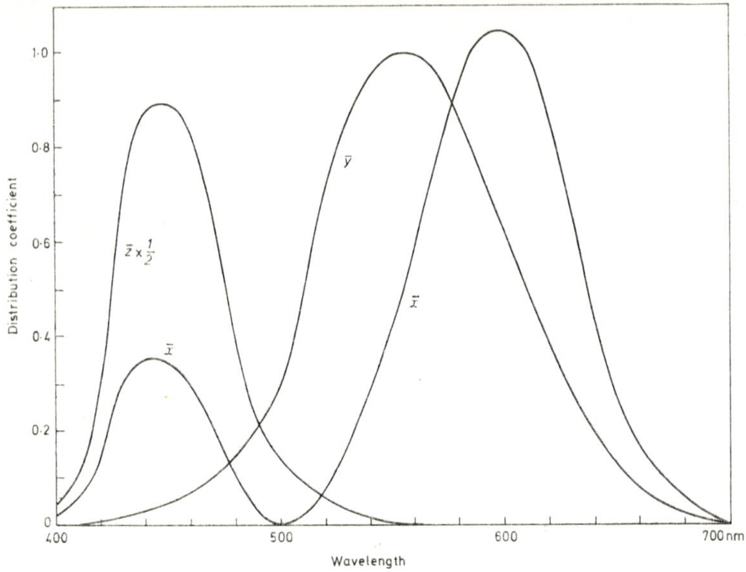

FIG. 4.11. The distribution coefficients $\bar{x}$, $\bar{y}$, and $\bar{z}$ plotted against wavelength to give the equal-energy distribution curves of the 1931 C.I.E. system

a marked curvature, while the blue and violet wavelengths are compressed into a fairly short curved tail. This particular shape, although in part determined by the choice of reference stimuli, is primarily a function of the spectral sensitivity curves of the receptor processes in the retina.

An important characteristic of the locus is that it is always either straight or convex, but never concave. From this it follows that the colour resulting from the mixture of any two wavelengths must lie either on the locus or within the area bounded by the locus, but never outside it. This is even more true when the stimulus falling on the retina is composed of a number of wavelengths. Hence the area bounded by the spectrum locus and the straight line joining its red and violet extremities defines the region of the chromaticity chart outside which no physical stimulus will be located. The significance of points representing nonphysical stimuli has already been discussed in the previous chapter.

In addition to defining the area over which physical stimuli must be located, the spectrum locus gives greater meaning to the colour quality associated with any point within that area, such as *C* in Fig. 4.10. Thus if a straight line is drawn from *E* through *C* to intersect the spectrum locus at λ, we know from the geometry of the chart that a suitable mixture of λ and *E* will match *C*. If the hues of the spectrum are familiar to the colour worker, the value of λ will give an idea of the hue of *C*, while the distance of *C* along the line from *E* towards λ will give an indication of its saturation. The advantage and limitations of this method of locating *C* in the chromaticity chart will be considered in the final section of this chapter dealing with dominant wavelength and purity.

A study of the shape of the spectrum locus will also provide a general explanation of why saturated yellow, orange and red surfaces can be produced which reflect a large percentage of the incident light, whereas saturated green and blue surfaces are much darker. Because of the straightness of the red-yellow-green part of the locus, it is possible to combine any or all of these wavelengths to produce a colour which is hardly less saturated than a spectrum colour itself. Since this part of the spectrum also includes the wavelengths having the highest values of V_λ, the luminance factor can also be high. Moreover, many yellow, orange and red dyes and pigments have just the type of absorption curve to take maximum advantage of this state of affairs, with their high reflection at the longer wavelengths and a rapid drop to low reflection at the shorter wavelengths.

If a green surface with a high luminance factor is required, the red and the blue radiations will have to be present in significant amounts in addition to the green, and on account of the curvature of the spectrum locus at the green corner of the chart the integrated colour will be pulled towards the centre of the chart and therefore be desaturated. Increase in saturation can be obtained only at the cost of a reduction in light reflection. For blue surfaces the position is aggravated still more by the small V_λ values at the short wavelengths. Hence, a saturated blue surface which reflected light from, say, only 470 nm to the violet end of the spectrum would be very dark indeed. However, when we come to discuss colour-discrimination data we shall find that the distribution of colours that can just be discriminated is very far from being uniform over the C.I.E. chromaticity chart, since there is a densely populated area in the blue corner and a sparsely populated area in the green. To some extent, then, our assessment of degrees of saturation and desaturation may have to be modified to take account of this non-uniformity.

The distribution curves of Fig. 4.11 call for little comment. They are now all-positive curves, in contrast to the $\bar{r}_\lambda$, $\bar{g}_\lambda$, $\bar{b}_\lambda$ curves of Fig. 4.6 with their negative lobes. Then the $\bar{y}_\lambda$ curve is itself identical to the V_λ curve of Fig. 4.6, whereas in Fig. 2.6 the V_λ curve would be given by multiplying the red curve by 1·000, the green by 4·5907 and the blue by 0·0601 (the values of $C_{[R]}$, $V_{[G]}$ and $V_{[B]}$) and summing the three curves together. The areas under the $\bar{x}$, $\bar{y}$ and $\bar{z}$ curves are equal, remembering that in Fig. 4.11 the $\bar{z}_\lambda$ curve is plotted to half scale. This equality of area is the counterpart to the equality in the sums at the foot of Table 4 and to the fact that the units of (X), (Y) and (Z) are adjusted to be equal in a match of an equal-energy white.

The secondary maximum of the $\bar{x}_\lambda$ curve in the blue and violet is in part due to the redness that reappears in the violet, but it is exaggerated in the C.I.E. system on account of the location of (Z) in the chromaticity chart so far away from the violet end of the spectrum locus. In some respects it would have been advantageous if (Z) had been chosen nearer to the locus, but this would have meant that (Y) would have had to be more remote, since the line $(Y)(Z)$ had to be outside the locus.

The distribution curves are really the basic curves of the C.I.E. system, since all else, including the chromaticity chart, can be derived from them. The colour specification of a stimulus can be calculated from the distribution curves and the spectral composition of the stimulus, or it can be measured directly with a photoelectric tristimulus colorimeter (Fig. 9.7) if the photocells have spectral sensitivities corresponding to the $\bar{x}_\lambda$, $\bar{y}_\lambda$, $\bar{z}_\lambda$ curves. The distribution curves are, in effect, the sensitivity curves of the C.I.E. system and may be likened to the sensitivities of the retinal processes. Indeed, up to a point, the whole of the rather unusual algebra of the trichromatic system might be forgotten once the distribution curves had been derived, and colorimetry could be expounded afresh with the distribution curves as the new starting point.

THE 1964 CO-ORDINATES, CHROMATICITY CHART AND DISTRIBUTION CURVES

The general convenience of the 1931 system led the Colorimetry Committee of the C.I.E. to recommend that the 10° standard observer data should also be transformed to a similar type of co-ordinate system.* Accordingly three reference stimuli (X_{10}) (Y_{10}) (Z_{10}) were selected that would give all-positive co-ordinates for the spectral colours and in which all the luminance information was given by the Y_{10} tristimulus values. The values of the equal-energy distribution coefficients $\bar{x}_{10_\lambda}$, $\bar{y}_{10_\lambda}$ and $\bar{z}_{10_\lambda}$ were then calculated and tabulated over the wavelength range from 360 nm to 830 nm. The corresponding chromaticity co-ordinates $x_{10\lambda}$, $y_{10\lambda}$, z_{10_λ} were also derived and tabulated.

The extent to which the 10° system differs from the 1931 system can be seen in Fig. 4.12. At first sight the changes might seem hardly significant, but this is partly because the distribution curves are plotted on a linear scale. In fact the values of the $\bar{y}_{10\lambda}$ curve in the wavelength range from 400 to about 510 nm are quite significantly higher than the $\bar{y}_\lambda$ values for the 2° system and this is an indication of the greater sensitivity of the extra-fovea to the shorter wavelengths. The $\bar{y}_{10\lambda}$ curve can be used as the $V_{10\lambda}$ curve when large-field photometric calculations are being made.

The 10° chromaticity chart is shown in Fig. 4.13 and the spectrum locus appears to differ little in shape from the 1931 locus. It is a mistake, however, to regard the one chart simply as a slightly distorted form of the other chart and in any case there is a considerable shift in the position of some wavelengths along the locus. They are different charts in the sense that the axes do not refer to exactly the same reference stimuli, and pairs of colours which are metameric in one chart and therefore have the same chromaticity, will in general have different chromaticities in the other chart and vice versa. On the other hand, the spectrum locus can be used as a reference framework of identical stimuli in the two charts, and it should be possible to deduce useful relations about colour appearance under 2° and 10° viewing conditions in terms of dominant wavelength and purity. The one stimulus that, by definition,

* *C.I.E. Compte Rendu*, 1959, Vol. A, p. 95.

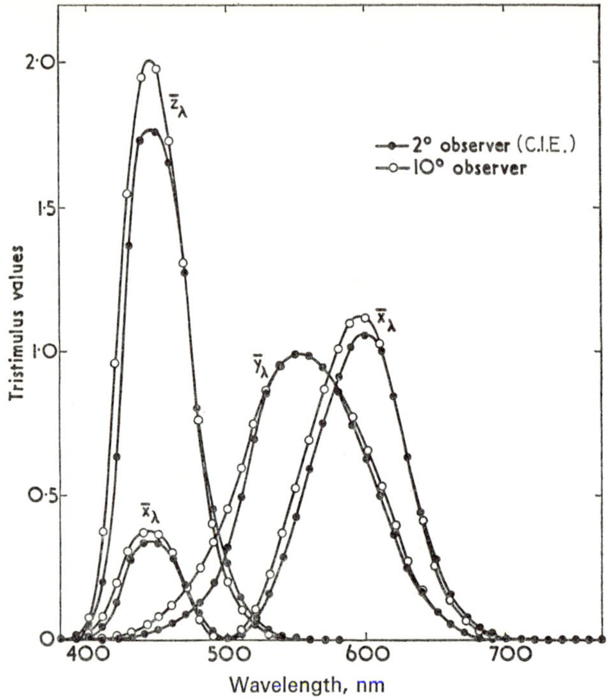

FIG. 4.12. Tristimulus values or distribution coefficients $\bar{x}_\lambda$, $\bar{y}_\lambda$, $\bar{z}_\lambda$ for wavelengths in the equal-energy spectrum for 2° and 10° field sizes. The two sets of distribution curves show the differences between the 2° data adopted by the C.I.E. in 1931 and the 10° data adopted in 1964.
(Diagram from Judd and Wyszecki, *Color in Business, Science and Industry*, 1963)

occupies the same point in the two charts is the equal-energy white

In Table 10 (Appendix V) the 10° chromaticity co-ordinates $x_{10\lambda}$, $y_{10\lambda}$, $z_{10\lambda}$ are given for wavelengths from 380 nm to 780 nm at intervals of 5 nm. The 10° equal-energy distribution coefficients $\bar{x}_{10\lambda}$, $\bar{y}_{10\lambda}$, $\bar{z}_{10\lambda}$ are given in Table 11. As explained on p. 120 these functions have only been tabulated to 4 decimal places here, although the official C.I.E. tables give a greater number of significant figures, are tabulated for closer wavelength intervals and extend over a wider range of wavelengths. This elaboration can be justified as a means of avoiding ambiguity in defining a standard

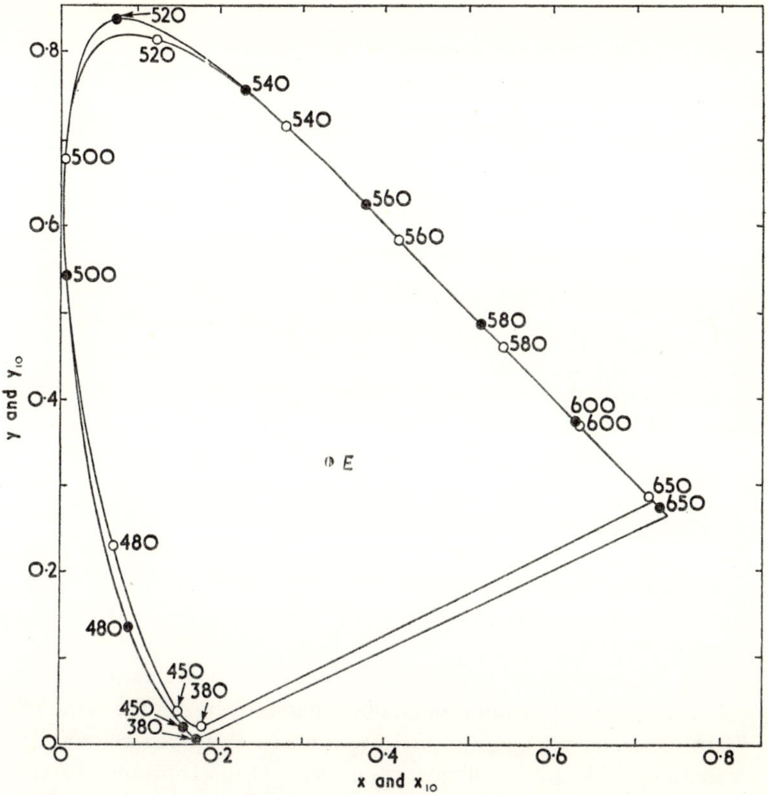

FIG. 4.13. The 1931 C.I.E. chromaticity chart (solid dots) compared with the new 10° chart (open circles)
(Diagram from Judd and Wyszecki, *Color in Business, Science and Industry*, 1963)

system but can rarely be of more than academic interest to a practising colorimetrist.

THE CALCULATION OF TRISTIMULUS SPECIFICATIONS FROM SPECTROPHOTOMETRIC DATA

The procedure for deriving the colour specification of a stimulus C from its spectral composition $P(\lambda)$ has already been established in principle in the previous chapter. Expressed algebraically in terms of the C.I.E. system, the tristimulus values X, Y and Z will be given by

$$X = \Sigma(P_\lambda \, . \, \bar{x}_\lambda)$$
$$Y = \Sigma(P_\lambda \, . \, \bar{y}_\lambda)$$
$$Z = \Sigma(P_\lambda \, . \, \bar{z}_\lambda)$$

and the unit co-ordinates x, y and z will be given by

$$x = \frac{X}{X+Y+Z}, \quad y = \frac{Y}{X+Y+Z}, \quad z = \frac{Z}{X+Y+Z}$$

The tabulated data in Table 4 are frequently used in this way, especially in evaluating the colour of light sources. Strictly, of course, the summation should consist of a process of integration of the areas under the three distribution curves after they have been weighted by the spectral composition of the stimulus, but the wavelength intervals in the tabulated data are sufficiently close to be equivalent to a true integration. Indeed, in many cases computational labour can be saved by taking even fewer wavelengths at wider intervals.

The more common problem, however, is the evaluation of the colour of a surface or filter when illuminated by one of the standard illuminants. If we confine our attention to a surface colour, the essential problem is to calculate the specification from the spectral luminance factors β_λ. One way to do this is to use the data of Table 4 to calculate

$$X = \Sigma(\beta_\lambda \, . \, P_\lambda \, . \, \bar{x}_\lambda)$$
$$Y = \Sigma(\beta_\lambda \, . \, P_\lambda \, . \, \bar{y}_\lambda)$$
$$Z = \Sigma(\beta_\lambda \, . \, P_\lambda \, . \, \bar{z}_\lambda)$$

This involves two multiplications at each wavelength before summing, and a correspondingly greater amount of work. Since the same illuminants are used for many calculations, the labour could evidently be reduced by preparing, once and for all, tables of values of $P_\lambda \, . \, \bar{x}_\lambda$, $P_\lambda \, . \, \bar{y}_\lambda$ and $P_\lambda \, . \, \bar{z}_\lambda$, with which only single products would subsequently have to be made at each wavelength. Such tables have, in fact, been prepared for the standard illuminants, but in doing so advantage has also been taken of the possibility of using the Y value to give the percentage luminance of the surface directly.

Thus for a perfect white diffuser as reference surface, the tristimulus specification will be given by

$$X = \Sigma(P_\lambda \, . \, \bar{x}_\lambda), \quad Y = \Sigma(P_\lambda \, . \, \bar{y}_\lambda), \quad Z = \Sigma(P_\lambda \, . \, \bar{z}_\lambda)$$

since β_λ will be unity throughout the spectrum. As we are at liberty to adjust the distribution coefficients by any factor we please, so long as we apply the same factor to all three columns, we can choose a factor to give

$$\Sigma(P_\lambda \, . \, \bar{y}_\lambda) = 100{\cdot}00$$

This adjustment has been made in Tables 5, 6 and 7, which give the products $P_\lambda \, . \, \bar{x}_\lambda$, $P_\lambda \, . \, \bar{y}_\lambda$, $P_\lambda \, . \, \bar{z}_\lambda$ for the three sources S_A, S_B and S_C. The data as given here were prepared by Smith and Guild* and it will be noted that the sum of each of the $\bar{y}$ products comes to 100·000. The $\bar{x}$ and $\bar{z}$ columns have, of course, different totals, since these must give the tristimulus specification of the white surface under the particular illuminant. Thus for a white magnesium-oxide surface under illuminant S_A, we have

$$X = 109{\cdot}85, \quad Y = 100{\cdot}00, \quad Z = 35{\cdot}58$$

and the corresponding unit co-ordinates

$$x = 0{\cdot}4476, \quad y = 0{\cdot}4074, \quad z = 0{\cdot}1450$$

which are the co-ordinates of S_A in Fig. 4.10.

Suppose, now, these tables were used to calculate the specification of a coloured sample from its β_λ values and we obtained

$$X = 24{\cdot}3, \qquad Y = 35{\cdot}2, \qquad Z = 19{\cdot}4$$

leading to

$$x = 0{\cdot}3080, \quad y = 0{\cdot}4461, \quad z = 0{\cdot}2459$$

We should then know both the chromaticity of the sample and its percentage luminance, 35·2, since the $\bar{y}_\lambda$ values are the same as the V_λ values. This scale is in agreement with the 1959 recommendation of the Colorimetry Committee of the C.I.E. that the tristimulus values of filters and diffusing surfaces should in general be based on a scale having $Y = 100$ for the perfect diffuser. This means that $Y = 100\beta$.

* T. SMITH and J. GUILD: *Trans. Opt. Soc.*, 1931, **33**, p. 73.

The possibility of deriving and including the luminance factor so simply in the colour specification arises directly from the choice of reference stimuli (*X*) and (*Z*) having zero luminous efficiencies. Its derivation with any other set of stimuli, while possible, would require the weighting of all three tristimulus values by their respective luminous-efficiency factors, summing the products and comparing the sum with the corresponding result for the perfect diffuser.

It will be realized that the tristimulus specification of a surface can always be corrected to make the *Y* value correspond to the luminance factor or percentage luminance, whether or not the distribution coefficients have been adjusted to give this result directly, or if the specification has been obtained by observation rather than by calculation. If, for example, the unit equation for a surface under a specified illuminant had been derived from observations with a colorimeter as

$$(C) \equiv x\,.\,(X) + y\,.\,(Y) + z\,.\,(Z)$$

and if, from subsidiary photometric measurements, a luminance factor of β was obtained for the same condition of illumination, then putting $Y = 100\beta$, the tristimulus equation for the surface would be given by

$$\frac{Y}{y}\,.\,(C) \equiv \frac{Y\,.\,x}{y}\,.\,(X) + Y\,.\,(Y) + \frac{Y\,.\,z}{y}\,.\,(Z)$$

In the same way, if the specification for a surface had been calculated using a table of distribution coefficients which had not been adjusted to give a $\bar{y}_\lambda$ sum of 100·0, so that the equation for the perfect diffuser was found as, say

$$m_W\,.\,(W) \equiv X_W\,.\,(X) + Y_W\,.\,(Y) + Z_W\,.\,(Z)$$

while the equation for the sample was calculated as

$$m\,.\,(C) \equiv X\,.\,(X) + Y\,.\,(Y) + Z\,.\,(Z)$$

then the tristimulus equation for the sample, in which the *Y* value gives a direct measure of the percentage luminance of the sample, would be given by

$$\frac{100\,m}{Y_W}\,.\,(C) \equiv \frac{100\,X}{Y_W}\,.\,(X) + \frac{100\,Y}{Y_W}\,.\,(Y) + \frac{100\,Z}{Y_W}\,.\,(Z)$$

This latter case is of some interest when, in order to save computing time, it is decided to use tabulated data such as Tables 5, 6 and 7 at greater wavelength intervals than 5 nm. If, for example, it was considered of sufficient accuracy to work at 10 nm intervals, then the weighted $\bar{y}_\lambda$ coefficients as extracted at alternate wavelengths from Tables 5, 6 and 7 would no longer add up to 100·0. The derivation of the tristimulus values for a sample in the form in which the Y value gives the percentage luminance would therefore require an additional calculation to determine the tristimulus values for the white surface.

In practice the majority of colour calculations are carried out at 10 nm intervals and for this reason condensed tables of weighted distribution coefficients, adjusted to give a weighted $\bar{y}_\lambda$ sum equal to 100·0 have been specially prepared for general colorimetric use. These are given in Table 8, Appendix III, for the 2° system and in Table 12, Appendix V, for the 10° system, in each case for the six illuminants S_A, S_B, S_C, D_{5500}, D_{6500} and D_{7500}. The data given in these Appendices are based on tables prepared by Smith, by Judd and Wyszecki, and by Wyszecki and Stiles.*

These tables may be regarded as the basic working data of colorimetry, certainly for surface colours, but it must be admitted that there are a rather formidable number of them. Sooner or later there must be some rationalization and the C.I.E. Colorimetry Committee anticipate that in the future the vast majority of calculations will be carried out either for illuminant S_A, as a typical tungsten source, or for D_{6500}, as representative of average daylight. Nevertheless, there will be occasions when calculations for other illuminants are required, so that we cannot expect to restrict all problems of colorimetry to just two sources. Similarly, there will be some problems when it will be most appropriate to use the 2° data and others when the 10° colour matching functions correspond more closely to the actual viewing situation. Hence for the moment at least, all the data of Tables 8 and 12 must be provided.

The calculation of the tristimulus specification of a surface can be illustrated graphically, as in Fig. 4.14. Graphical integration is

* T. Smith: *Proc. Phys. Soc.*, 1934, **46**, p. 372.

† D. B. Judd and G. Wyszecki: *Color in Business, Science and Industry* (Wiley: New York and London), 2nd Ed., 1963.

G. Wyszecki and W. S. Styles: *Color Science* (Wiley: New York and London), 1967.

too slow a process for it to be used in preference to calculation from the tabulated data, but illustration of the steps involved is instructive. Thus Fig. 4.14(*a*) shows the spectral distribution $P_B(\lambda)$ for the illuminant S_B. In Fig. 4.14(*b*) the products of curve (*a*) with the three equal-energy distribution curves of Fig. 4.11 are shown. These curves are therefore the distribution curves for the illuminant S_B, and the areas they enclose give the tristimulus values for S_B. In Fig. 4.14(*c*) the spectral luminance factor β_λ is plotted against λ for the surface whose colour specification is required. The final stage is shown in (*d*), where the product of the curves in (*b*) and (*c*) are plotted, i.e. $\beta_\lambda \, . \, P_{B\lambda} \, . \, \bar{x}_\lambda$, $\beta_\lambda \, . \, P_{B\lambda} \, . \, \bar{y}_\lambda$ and $\beta_\lambda \, . \, P_{B\lambda} \, . \, \bar{z}_\lambda$ have been plotted against λ. They are therefore the distribution curves for the surface, and the areas under the three curves will give its

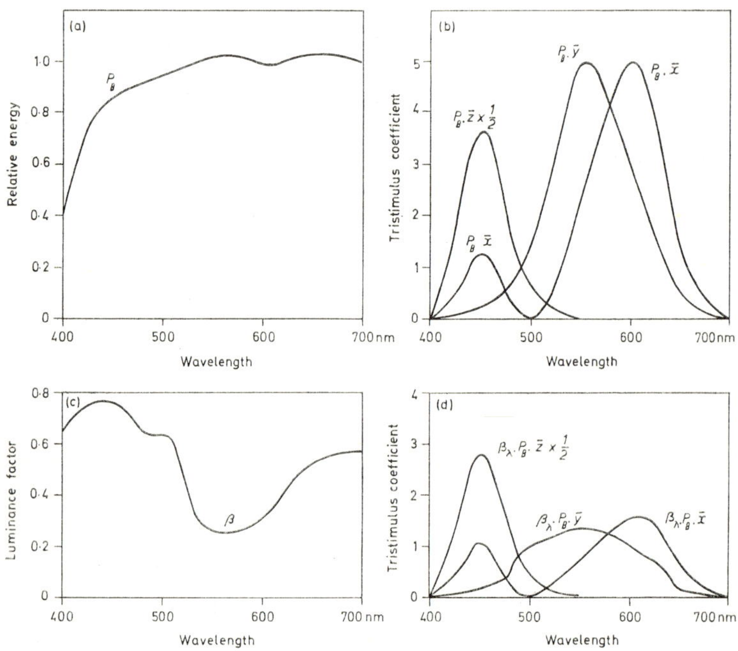

FIG. 4.14. Graphical illustration of the derivation of the tristimulus specification of a coloured surface

(*a*) The energy distribution of the illuminant.
(*b*) The product of (*a*) with the distribution curves of Fig. 4.11.
(*c*) The spectral reflection curve of the surface.
(*d*) The product of (*b*) and (*c*), from which the areas under the three curves will give the tristimulus values, X, Y, and Z, of the surface.

tristimulus specification. The unit equation can then, as usual, be derived by dividing through by the sum of the tristimulus values.

This illustration provides a useful introduction to an alternative method of calculation proposed by Hardy* and described by him as the *selected-ordinate method*, to distinguish it from the *weighted-ordinate method*, the name given to the orthodox procedure already described. In the weighted-ordinate method, the summing of the products β_λ, $P_\lambda \cdot \bar{x}_\lambda$, etc., is merely a numerical method of integrating the areas under the three distribution curves of the sample, as illustrated in Fig. 4.14. The formally correct summation to give the areas under the curves would be to sum the products $\beta_\lambda \cdot P_\lambda \cdot \bar{x}_\lambda \cdot \Delta\lambda$, etc., where $\Delta\lambda$ is the wavelength interval between successive entries in the tables. Since, in the tables given, $\Delta\lambda$ is the same at all wavelengths and since we are interested only in the relative magnitudes of the tristimulus values, there has been no occasion to include the $\Delta\lambda$ in the products. We could, however, draw up a table of distribution coefficients in which $\Delta\lambda$ varied at different points through the spectrum; if we did this we should have to adjust the product $P_\lambda \cdot \bar{x}_\lambda$ in direct proportion to $\Delta\lambda$. If, for example, over a certain region of the spectrum we reduced $\Delta\lambda$ to one-quarter of its normal value, we should have four times as many entries in that region as before, so that in order to keep its contribution to the total sum unchanged, the products $P_\lambda \cdot \bar{x}_\lambda$, etc., would have to be reduced to one-quarter of their previous values. Suppose now, the wavelength intervals $\Delta\lambda_x$ for the $\bar{x}_\lambda$ coefficients are selected to be inversely proportional to $P_\lambda \cdot \bar{x}_\lambda$, so that where the distribution coefficients are large, the wavelength intervals will be small, and *vice versa*; then the product $P_\lambda \cdot \bar{x}_\lambda \cdot \Delta\lambda_x$ will be constant for each of the entries. The X tristimulus value for the sample, which is given by $\Sigma(\beta_\lambda \cdot P_\lambda \cdot \bar{x}_\lambda \cdot \Delta\lambda_x)$, can then be written $P_\lambda \cdot \bar{x}_\lambda \cdot \Delta\lambda_x \cdot \Sigma\beta_\lambda$, where $\Sigma\beta_\lambda$ is the sum of the luminance factors at the mid-points λ_x of the wavebands into which the spectrum has been divided. These wavebands will, of course, be of varying width through the spectrum. In the same way, a different series of wavelengths, λ_y and λ_z, can be found for the $\bar{y}_\lambda$ and $\bar{z}_\lambda$ coefficients respectively, such that $P_\lambda \cdot \bar{y}_\lambda \cdot \Delta\lambda_y$ and $P_\lambda \cdot \bar{z}_\lambda \cdot \Delta\lambda_z$ are each constant.

* A. C. Hardy: *Handbook of Colorimetry* (Massachusetts Institute of Technology: Cambridge, Mass.), 1936.

The tristimulus values of a sample can then be calculated from the spectral reflection curve in the following manner: the X value is given by reading the luminance factors at each of the wavelengths λ_x, adding them together and multiplying the sum by $P_\lambda . \bar{x}_\lambda . \Delta\lambda_x$; the Y value is given by summing the luminance factors at the wavelengths λ_y and multiplying the sum by $P_\lambda . \bar{y}_\lambda . \Delta\lambda_y$; and the Z value is obtained as the sum of the luminance factors at the wavelengths λ_z and multiplying by $P_\lambda . \bar{z}_\lambda . \Delta\lambda_z$.

Hardy has given tables of λ_x, λ_y and λ_z for each of the three sources S_A, S_B and S_C and in each case for 30 and 100 wavelength intervals. The 30-interval tables are reproduced in Appendix IV. The main advantage of the selected-ordinate method is that the greater part of the computation is simple addition, which can be carried through very rapidly with a calculating machine. However with the advent of integrating attachments to recording spectrophotometers, and the increasing use of computer programmes for tristimulus calculations, the selected-ordinate method is likely to fall into disuse. (But see also the reference on p. 254 to the new Japanese integrator which uses the selected-ordinate principle.)

COLOUR SPECIFICATION IN TERMS OF DOMINANT WAVELENGTH AND PURITY

The location of a colour C in the chromaticity chart requires that two co-ordinates be defined, and while the rectangular co-ordinates (x, y) are those most commonly used, there are some merits in the polar co-ordinate system to which a brief reference has already been made on p. 123 and which is illustrated again in Fig. 4.15. Once the chromaticity co-ordinates (x, y) of a colour are known, the colour can be plotted on the chromaticity chart and a line can be drawn from the white point W through C to the wavelength λ_d on the spectrum locus. C can then be specified by the particular value found for λ_d, which is known as the dominant wavelength, and by the distance of C along the line $W\lambda_d$.

This distance is customarily specified by the purity, which is defined as the ratio of the distance CW to the distance λW.* If the

* This concept was known previously as the 'excitation purity' to distinguish it from an alternative and now largely obsolete concept known as the 'colorimetric purity'.

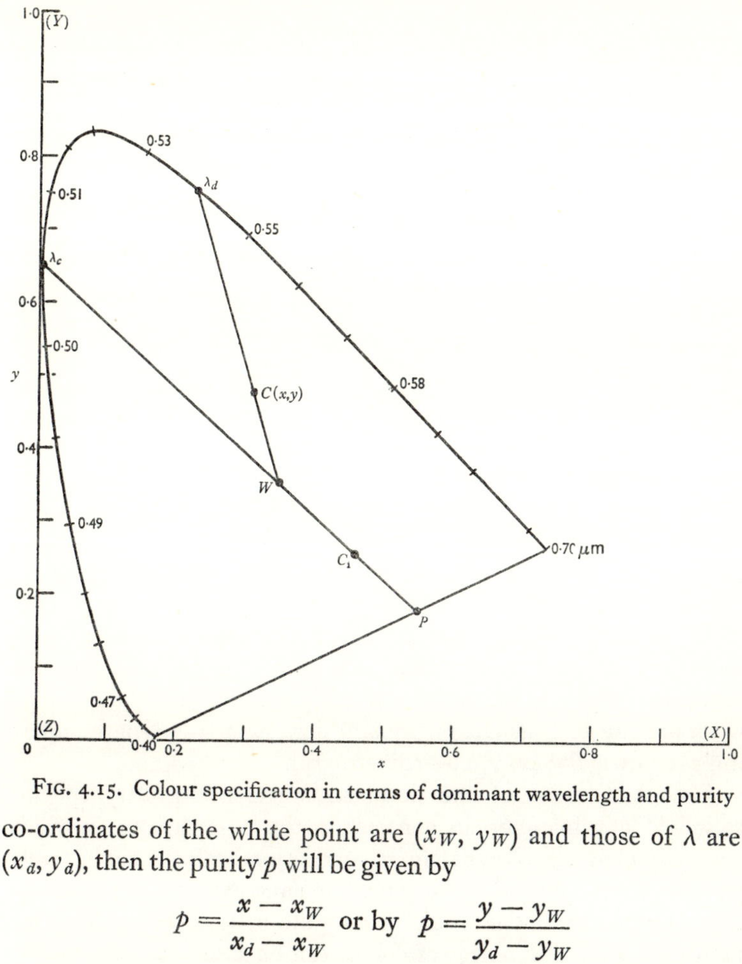

FIG. 4.15. Colour specification in terms of dominant wavelength and purity

co-ordinates of the white point are (x_W, y_W) and those of λ are (x_d, y_d), then the purity p will be given by

$$p = \frac{x - x_W}{x_d - x_W} \quad \text{or by} \quad p = \frac{y - y_W}{y_d - y_W}$$

(These two formulae are equivalent, but when the line $W\lambda_d$ is nearly horizontal the second formula will be inaccurate, and vice versa when the line is near the vertical.)

For colours in the purple segment of the chromaticity chart, a special convention has to be adopted for defining the dominant wavelength and purity, on account of the gap in the spectrum locus between the two extremes of the spectrum. Thus in Fig. 4.15 the dominant wavelength of C_1 is defined by its complementary wavelength λ_c, found by drawing a line from C_1 through W to intersect the spectrum locus at λ_c on the side of W remote from C_1. The purity, on the other hand, is defined as C_1W/PW, where P is the point of intersection between the extension of the line WC_1 and the straight line joining the extremities of the spectrum locus; p will then be given by

$$p = \frac{x - x_W}{x_p - x_W} \quad \text{or by} \quad p = \frac{y - y_W}{y_p - y_W}$$

where (x_p, y_p) are the co-ordinates of P.

The purity varies from zero at the white point to unity at the spectrum locus, but its value, and that of the dominant wavelength also, will depend on the white that is chosen as the reference point. Where self-luminous sources are being considered, the equi-energy point E would normally be taken as the point of origin, but for surface colours the white point should correspond to the colour of the illuminant being used, such as S_A, S_B or S_C. Judd* has described an algebraical method and has prepared tables from which λ_d or λ_c can be calculated for the four points of origin E, S_A, S_B and S_C. Hardy † has also prepared charts from which the dominant wavelength and purity can be found graphically, and although they have been prepared only for the one illuminant S_C, they are in the extremely convenient form of a network of lines of constant purity and dominant wavelength superimposed over the (x, y) co-ordinate grid.

The chief merit of the dominant-wavelength/purity specification of chromaticity, in contrast to the co-ordinates (x, y), is generally claimed to be its closer correlation with the subjectively perceived attributes of the colour. To some extent this advantage may be illusory, but it would seem quite certain that at least the nature of small differences of colour can be assessed more readily from a knowledge of λ_d and p. For example, suppose we consider two samples under illuminant S_C with the following chromaticity co-ordinates

	x	y
C_1	0·546	0·386
C_2	0·526	0·392

* D. B. Judd: *J. Opt. Soc. Am.*, 1933, **23**, p. 359.

† A. C. Hardy: *Handbook of Colorimetry* (Massachusetts Institute of Technology: Cambridge, Mass.), 1936.

Although we can certainly roughly assess the difference in colour by comparing the co-ordinates, a more accurate mental picture can be gained from the corresponding values for the dominant wavelength and purity, namely

	λ_d	p_e
C_1	0·5940	0·82
C_2	0·5920	0·78

Thus we see at once that C_1 is slightly more orange and more saturated than C_2.

One of the most valid reasons for preferring dominant wavelength and purity for certain types of specification is the more direct relation these functions bear to the operations carried out in certain colour industries. When a dyer, for example, is asked to match a given pattern, he chooses the appropriate mixture of dyestuffs to give the correct shade or hue, he adjusts their strengths to control the purity and luminance factor, and he can adjust the luminance factor independently by admixture of a dye of complementary absorption. The dyer's variables are shade, strength and dullness,* in which shade is related to dominant wavelength, dullness is related to luminance factor, and strength is related to both purity and luminance factor. Moreover, dyers are generally concerned with small differences between patterns and would for this reason alone find the specification by dominant wavelength and purity more appropriate.

We shall see, too, in Chapter 8 that the assessment of errors in a colour reproduction can sometimes be more instructive if expressed in terms of dominant wavelength and purity; a case can also be made out for specifying the purity gradients of a system of colour reproduction, as a supplement to the luminance gradients defined by the slope or gamma of the characteristic curve of the process. At least the colour worker should always be ready to convert his data into terms of λ_d and p in case a clearer picture can then be obtained of the problem he is investigating.

* G. S. J. White, T. Vickerstaff, and E. Waters: *Proc. Phys. Soc.*, 1943 **55**, p. 1.

CHAPTER 5

METAMERISM

METAMERIC COLOUR MATCHES

In introducing the subject of trichromatic colorimetry on p. 69, we saw that the two fundamental observations on which colorimetry depends are, first, that a colour can be matched by a suitable mixture of three selected radiations and, second, that two colours matched in turn by the appropriate mixtures of the three radiations will, when combined additively, be matched by the additive combination of the two mixtures. Hence metamerism, which may be defined as the phenomenon of identity of colour appearance between stimuli of different spectral compositions, lies at the heart of colorimetry.

In some colour matches the difference in spectral composition may be quite small, in other cases it may be very considerable. Evidently, therefore, there are varying degrees of metamerism. With a colorimeter such as that illustrated in Fig. 3.2, the matches are likely to be highly metameric because of the big difference in spectral composition that is almost certain to exist between the mixture of the three spectral matching stimuli in one half of the field and the colour under test in the other. An extreme example is illustrated in Fig. 5.1, which represents a match between the standard illuminant S_B and a mixture of the three narrow-band radiations 460, 530 and 650 nm.

The metamerism that can occur in matches between pairs of surface colours is much less marked and it has to be kept to a minimum in commercial matching. Spectral reflection curves of typical fabrics dyed with different combinations of dyes, which are metameric under daylight-illumination, are illustrated in Fig. 5.2. These curves are of commercial importance because metameric fabrics which match under one illuminant are likely to be a mismatch under another illuminant, and the higher the degree of

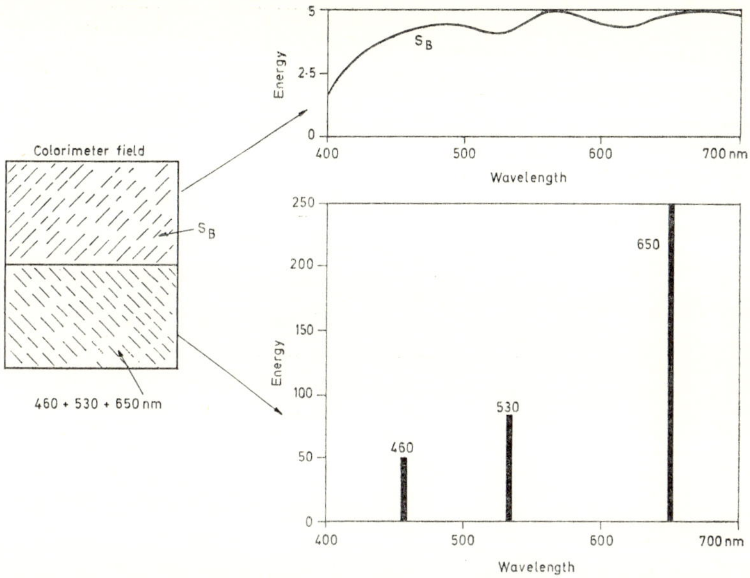

FIG. 5.1. The spectral composition of the two halves of the field in the Wright colorimeter (Fig. 3.2) when light from the illuminant S_B is matched by a mixture of the monochromatic stimuli 460 nm, 530 nm, and 650 nm. (Note different energy scales)

metamerism the greater the mis-match. Similarly, the greater the metamerism the more sensitive will the match be to differences in colour vision between different colour matchers. There is therefore a strong commercial incentive to reduce the metamerism to zero, since the more nearly the spectral reflection curves of two samples match each other, the wider the range of conditions under which the samples will remain a close match and the fewer the complaints from the customer.*

A rather different example of metamerism occurs in three-colour reproduction. In an additive system of colour reproduction such as we have in colour television, the amounts of red, green and blue

* Note: Samples which have identical spectral reflection curves are sometimes described as isomeric pairs. According to Billmeyer and Saltzman [F. W. BILLMEYER and M. SALTZMAN: *Principles of Color Technology* (Wiley: New York and London), 1966], however, this is incorrect terminology, since isomerism and metamerism mean the same, not opposite, things. There is, indeed, no need for any special term, since matches between stimuli with identical spectral compositions can be described simply as spectral matches.

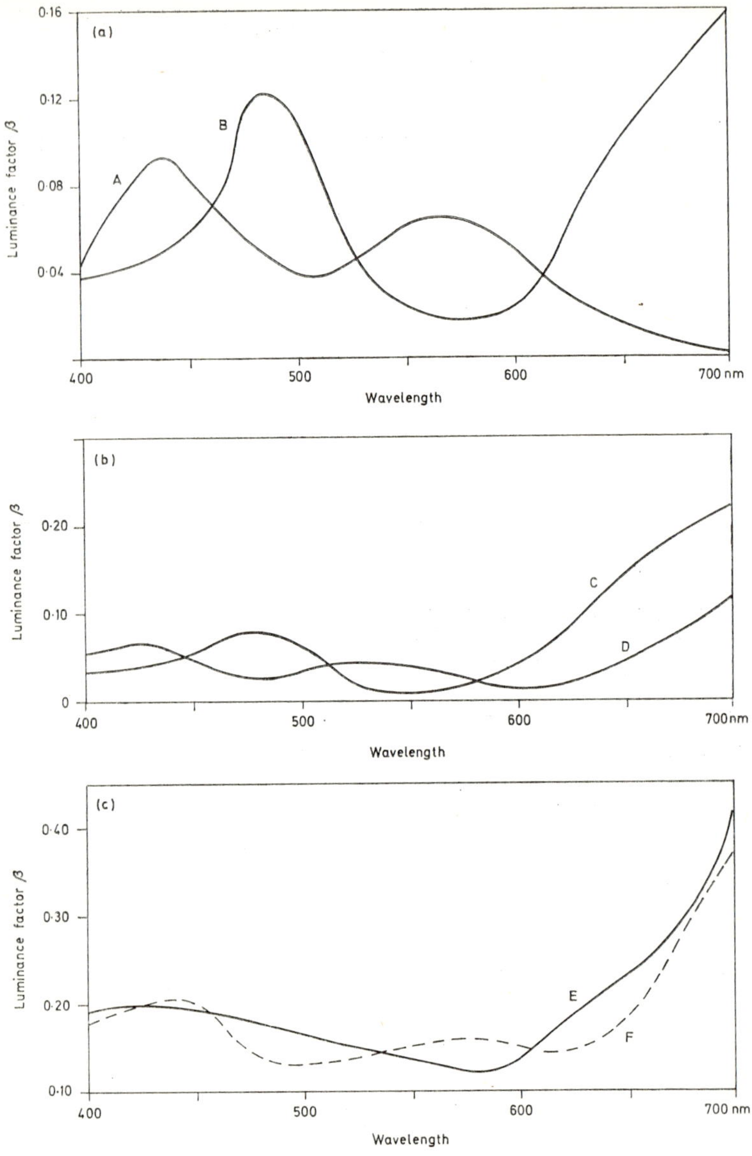

FIG. 5.2. Spectal reflection curves of pairs of dyed fabrics which are metameric under daylight illumination

light are controlled in the receiver to give point-by-point colour matches with the objects in the original scene. In principle, the same situation also exists in subtractive systems such as colour photography, as we shall see in Chapter 9. Hence although we are not often in a position to make a direct comparison of a television picture with the scene being transmitted, there is in effect the equivalent of a metameric match between corresponding areas of the two. (For the present argument we can ignore the departures from an exact match that may be deliberately introduced to compensate for differences in observing conditions.) In televising a cricket match, for example, the metamerism may be quite high between the light from a pattern of the red, green and blue phosphor dots which make up a cricketer's white flannels, compared with the light actually reflected from the flannels under daylight illumination on the cricket field itself. This example is comparable to the metamerism illustrated in Fig. 5.1 except that the phosphors radiate over fairly broad bands of the spectrum and will therefore in combination give a less metameric match with the white than the highly metameric combination of the colorimeter stimuli. In colour photography, on the other hand, the white areas of a photograph are produced by very low densities of the yellow, magenta and cyan dyes, and the metamerism with the white areas of the scene will in consequence be very much less than in colour television.

An analogous, but less deliberate, metamerism also exists between an artist's painting of a subject and the subject itself. Of course, the artist may not be aiming to produce a facsimile of the subject and there will certainly be no conscious attempt to construct his reproduction on any trichromatic principle. Yet as he mixes his pigments to produce the colours he wants, their spectral absorption will in general differ significantly from those of the subject.

Evidently, then, metamerism is a very widespread phenomenon and is currently a subject of very active study among colorimetrists.* There is, however, some confusion and difference of view on how it should be evaluated. To clarify the situation it may

* See, for example, a series of articles on Metamerism in *Color Engineering*, 1967, **5**, No. 3. Also sections 3.9 to 3.13 in *Color Science* by G. Wyszecki and W. S. Stiles (Wiley: New York and London), 1967.

be helpful to review the underlying physiological process on which metamerism depends.

A PHYSIOLOGICAL ASSESSMENT OF METAMERISM

Metamerism is a unique sensory phenomenon and has its origin in the perception of colour through three retinal processes with broad overlapping spectral sensitivity curves such as those illustrated in Fig. 2.7. Once the light incident on the retina has been absorbed in the three groups of receptors, all record of the spectral composition of the incident light is lost. We are left instead with three levels of activity in the red, green and blue processes. Two stimuli can therefore give rise to the same colour sensation, even though they are of quite different spectral compositions, provided they arouse the same levels of activity in the three retinal processes. The actual absorption of the light quanta in the chromophores of the photochemical substances in the receptors will no doubt produce different patterns of molecular disruption depending on the spectral composition of the incident light, but the signal that is emitted by the receptors into the bipolars and ganglions is presumably determined by the total energy absorption. If the curves of Fig. 2.7 are accepted as the sensitivity curves of the actual physiological processes, two light stimuli will therefore be metameric if they react equally with these sensitivity curves. Expressed algebraically, this means that if two stimuli which are metameric have spectral distributions $P(\lambda)$ and $P'(\lambda)$ and if the sensitivity curves of the three processes are $\bar{s}_1(\lambda)$, $\bar{s}_2(\lambda)$ and $\bar{s}_3(\lambda)$, then

$$\Sigma(P_\lambda \cdot \bar{s}_{1\lambda}) = \Sigma(P'_\lambda \cdot \bar{s}_{1\lambda})$$
$$\Sigma(P_\lambda \cdot \bar{s}_{2\lambda}) = \Sigma(P'_\lambda \cdot \bar{s}_{2\lambda})$$

and

$$\Sigma(P_\lambda \cdot \bar{s}_{3\lambda}) = \Sigma(P'_\lambda \cdot \bar{s}_{3\lambda})$$

While these equations lay down the stimulus relations that must be satisfied for metamerism, they give no indication at all of its degree. To assess this we have to know the spectral distribution of the components in each of these summations. The problem can best be understood by studying the curves in Fig. 5.3 for the metameric pair A and B of Fig. 5.2(*a*), and the curves in Fig. 5.4

for the metameric pair E and F of Fig. 5.2(*c*), as seen under illuminant C. Thus the two diagrams in Fig. 5.3(*a*) give the products $(P \,.\, \beta_A \,.\, \bar{s}_1)_\lambda$ and $(P \,.\, \beta_B \,.\, \bar{s}_1)_\lambda$ plotted against λ. In Figs. 5.3(*b*) and (*c*) the corresponding products for $\bar{s}_{2\lambda}$ and $\bar{s}_{3\lambda}$ are given; similar pairs of diagrams are shown in (*a*), (*b*) and (*c*) of Fig. 5.4. The curves in Figs. 5.3 and 5.4 are in effect the tristimulus distribution curves for the metameric stimuli and the condition for metamerism is that the areas under each pair of curves should be equal.

The degree of metamerism, on the other hand, is given by the difference in shape of these distribution curves. Clearly in Fig. 5.3 there is a major difference in shape for each pair of diagrams, and these patterns are therefore highly metameric. On the other hand, the difference in shape between each pair of curves in Fig. 5.3 is relatively slight and the degree of metamerism is correspondingly small.

It would in principle be possible to adopt some formula by which the difference in shape of the metameric distribution curves could be evaluated numerically and from which a scale of metamerism could be constructed. This would lead to a general index of metamerism independent of any specific manifestation of metamerism such as change of illuminant or observer, and it is important to recognize that this general concept of metamerism is valid. In passing it should be noted that although metamerism is related to the physical composition of the matching stimuli, a straight comparison between their spectral distributions would take no account of the varying sensitivity of the eye to different wavelengths. Thus if a metameric pair possessed a major difference in spectral composition at the far red or far violet ends of the spectrum, this might suggest a high degree of metamerism, yet because the eye has only a limited sensitivity to these wavelengths, this conclusion might be quite misleading. It is only after the energy distributions have been weighted by the sensitivity curves of the retinal receptors that a true assessment of the metamerism is possible. We should also note that the degree of metamerism could well be different for each of the three processes $\bar{s}_1$, $\bar{s}_2$ and $\bar{s}_3$.

No formula defining a general index of metamerism has yet been adopted either nationally or internationally and indeed there has been considerable difference of opinion about the practical value

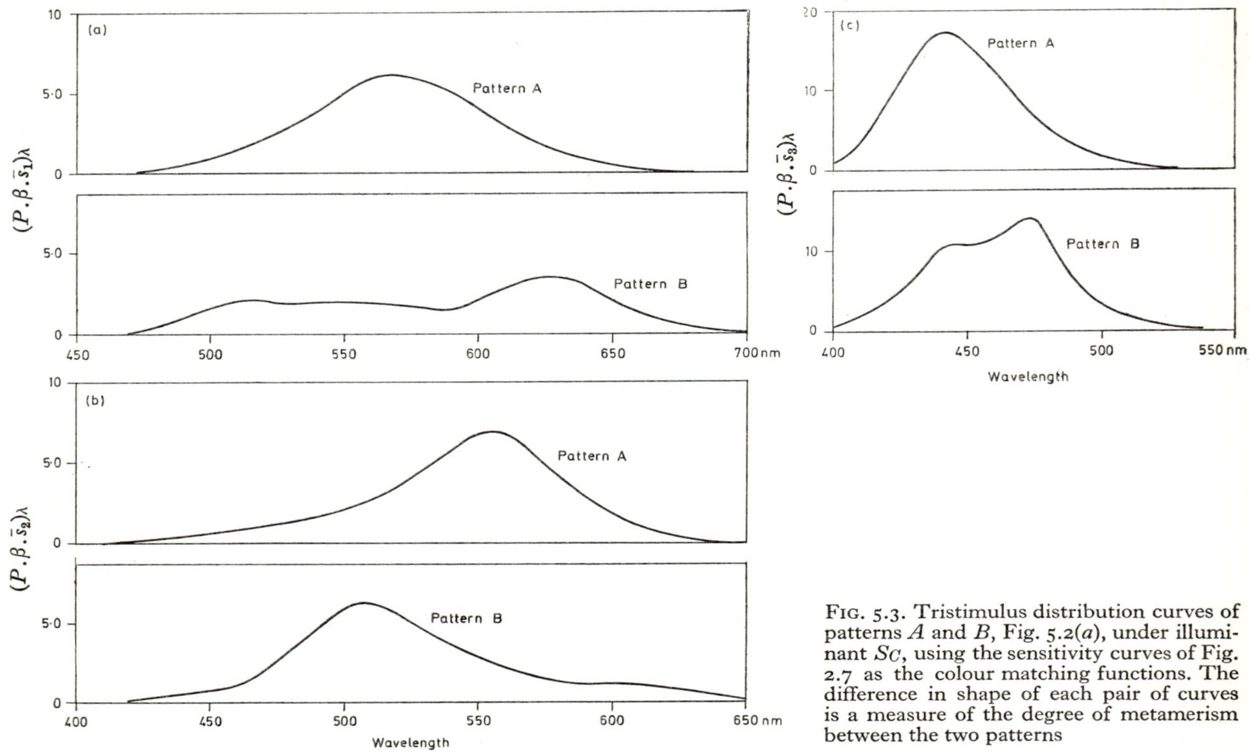

FIG. 5.3. Tristimulus distribution curves of patterns A and B, Fig. 5.2(a), under illuminant S_C, using the sensitivity curves of Fig. 2.7 as the colour matching functions. The difference in shape of each pair of curves is a measure of the degree of metamerism between the two patterns

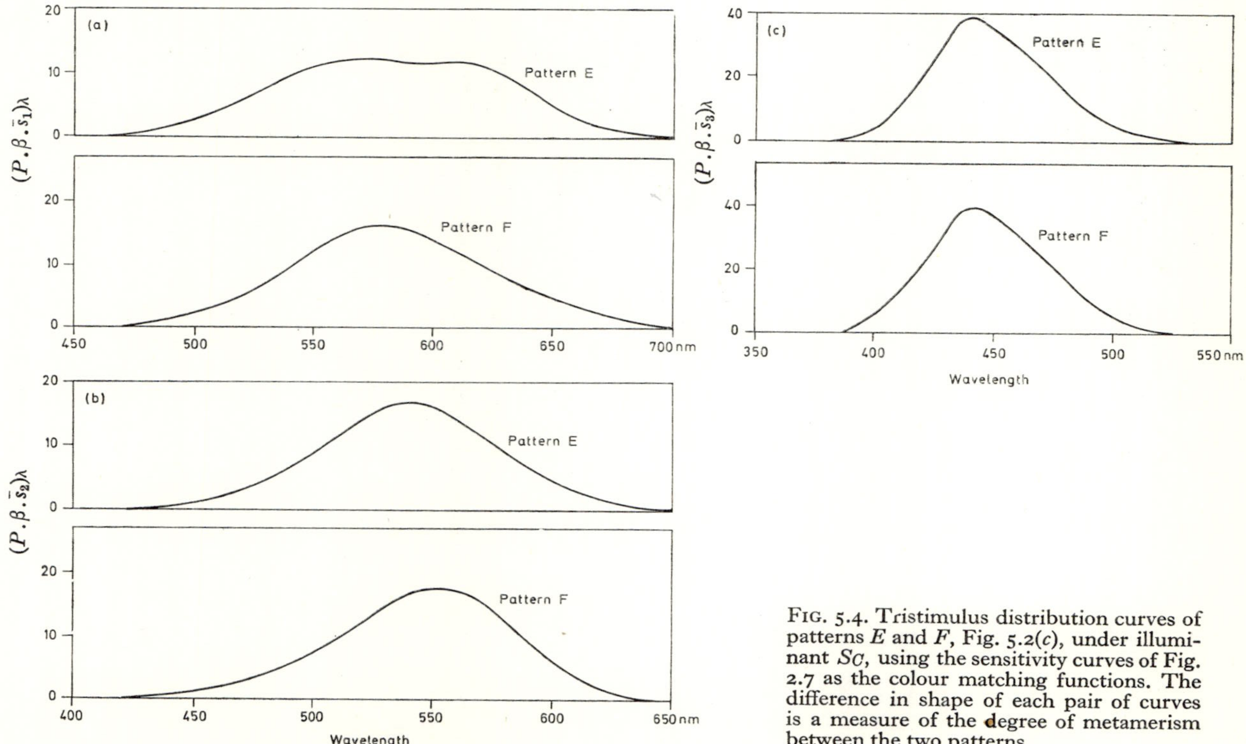

FIG. 5.4. Tristimulus distribution curves of patterns E and F, Fig. 5.2(c), under illuminant S_C, using the sensitivity curves of Fig. 2.7 as the colour matching functions. The difference in shape of each pair of curves is a measure of the degree of metamerism between the two patterns

of adopting such an index. Yet the many different situations in which metamerism occurs and the variety of conditions under which the metamerism of surface colours may manifest itself, certainly suggests that a general index is needed. Indeed, Nimeroff and Yurow* have developed two or three alternative formulae for the calculation of a general index of metamerism based on summations of the squares of the weighted energy differences at wavelengths through the spectrum. This could be regarded as one way of evaluating the difference in shape of the metameric distribution curves and one of the indices they proposed correlated well with the visual determination of metamerism as judged by observer differences in colour matching.

METAMERIC PAIRS OF SURFACE COLOURS

The remainder of this chapter will be devoted to a discussion of metamerism occurring between surface colours, since this is by far the most important industrial problem, particularly the manifestation of metamerism which results in a mismatch between two dyed fabrics when the illuminant is changed. Examples of such fabrics have been illustrated in Fig. 5.2 and their spectral reflection curves are characterized by the common feature that the pairs of curves cross each other at least three times within the visible spectrum. Moreover, each crossing point tends to be located near to the three maxima of the sensitivity curves of Fig. 2.7. Only if this occurs will it be possible for the higher reflection which one of the fabrics possesses at wavelengths on one side of a cross-over point to be balanced against the lower reflections at wavelengths on the other side of a cross-over point for each of the three processes. The precise points of intersection will, of course, depend on the particular dyes and on the illuminant for which they are metameric, but the existence of at least three crossing points is mandatory by virtue of the existence of the three retinal colour processes. There can, of course, be more than three.

We have so far related the phenomenon of metamerism to the sensitivity curves of the actual retinal processes themselves and from the physiological point of view this is the correct approach. From the colorimetric point of view, however, it is more con-

* I. Nimeroff and J. A. Yurow: *J. Opt. Soc. Am.*, 1965, 55, p. 185.

venient to express the metamerism in terms of the C.I.E. distribution curves $\bar{x}_\lambda$, $\bar{y}_\lambda$, $\bar{z}_\lambda$. The condition for metamerism for two surface colours A and B then becomes

$$\Sigma(\beta_{\lambda A} . P_\lambda . \bar{x}_\lambda) = \Sigma(\beta_{\lambda B} . P_\lambda . \bar{x}_\lambda)$$
$$\Sigma(\beta_{\lambda A} . P_\lambda . \bar{y}_\lambda) = \Sigma(\beta_{\lambda B} . P_\lambda . \bar{y}_\lambda)$$
$$\Sigma(\beta_{\lambda A} . P_\lambda . \bar{z}_\lambda) = \Sigma(\beta_{\lambda B} . P_\lambda . \bar{z}_\lambda)$$

where $\beta_{\lambda A}$ and $\beta_{\lambda B}$ are the spectral luminance factors of the two surfaces and P_λ is the spectral distribution of the illuminant.

This condition may be expressed slightly differently as follows:

$$\Sigma(\beta_{\lambda A} - \beta_{\lambda B}) . P_\lambda . \bar{x}_\lambda = 0$$

or, for brevity,

$$\Sigma(\beta'_\lambda . P_\lambda . \bar{x}_\lambda) = 0,$$

where

$$\beta'_\lambda = \beta_{\lambda A} - \beta_{\lambda B},$$

and similarly

$$\Sigma(\beta'_\lambda . P_\lambda . \bar{y}_\lambda) = 0$$

and

$$\Sigma(\beta'_\lambda . P_\lambda . \bar{z}_\lambda) = 0$$

This is the basis of a method of describing metamerism first proposed by Wyszecki* and since the difference function $(\beta_{\lambda A} - \beta_{\lambda B})$ defines the spectral reflection curve of a hypothetical surface which would be black under the particular illuminant P, he referred to such a surface as a 'metameric black'. The surface will have values of β'_λ which are positive for some wavelengths and negative for others, and a typical reflection curve for a black metamer, given by the difference between the curves A and B in Fig. 5.2(*a*), is shown in Fig. 5.5. There must be at least three wavelengths for which β'_λ is zero and again these will in general be located near the maxima of the $\bar{x}_\lambda$, $\bar{y}_\lambda$ and $\bar{z}_\lambda$ curves.

The merit of the black metamer concept is that it provides a possible basis for the classification of at least the more orthodox types of surface metamerism. Thus if a particular surface colour A were defined by its spectral reflection curve ($\beta_{\lambda A}$ plotted against λ), a metameric surface B could be derived by impressing on curve A the curve of a metameric black (β'_λ plotted against λ).

* G. Wyszecki: *J. Opt. Soc. Am.*, 1958, **48**, 451.

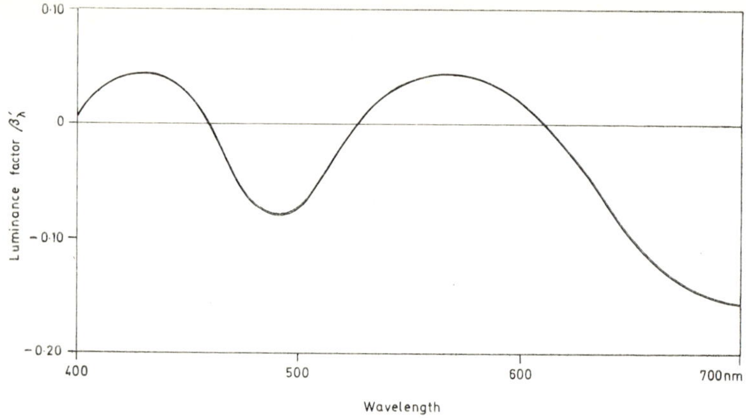

FIG. 5.5. Spectral reflection curve for a typical black metamer, given by the difference between the curves A and B in Fig. 5.2(*a*)

That is to say, the values of $\beta_{\lambda B}$ would be given by

$$\beta_{\lambda B} = \beta_{\lambda A} - \beta'_\lambda$$

Now while there can be many different metameric blacks and therefore many different functions of β'_λ, they are capable of classification in terms of (*a*) the wavelengths for which β'_λ is zero, (*b*) the amplitudes and signs of the lobes on either side of the zero values, and (*c*) the wavelengths at which the lobes have their maximum positive and negative values. Quite a small number of metameric blacks characteristic of the main varieties of surface metamerism might therefore be sufficient to classify the types of metamerism (based, for example, on the wavelengths for which $\beta'_\lambda = 0$), while within each class a scale or index of black metamerism could be established in terms of the amplitudes of the lobes.

ILLUMINANT MANIFESTATION OF METAMERISM

One of the curious features of metamerism is that when we are looking at a pair of metameric samples which are a match to us under a given illuminant, we are unaware of the metamerism unless one of the samples is sufficiently dichroic to exhibit Maxwell's spot. It is only when some change is made in the conditions of viewing, usually when the quality of the lighting is changed, that we realize the match under the first illuminant was metameric. This is revealed because under the new illuminant the match breaks down. Oddly enough, because the match no longer holds, the samples cannot then be described as a metameric pair. We might perhaps call them 'broken metamers'.

The cause of the colour-match breakdown is easy to understand, since the metamerism under the first illuminant depends on a critical balance in tristimulus values as expressed algebraically by the equations on p. 147. Once this balance is upset by a change in the P_λ values, a breakdown in the match must be expected. The magnitude of the mis-match will be a function both of the spectral reflection characteristics of the two samples and of the spectral distributions of the two illuminants. Hence if an 'illuminant' index of metamerism is required, it will have to be specified in terms of two defined illuminants. For general purposes these two illuminants would normally be a standard daylight illuminant such as D_{6500} and a standard tungsten source such as S_A. The index of metamerism would then be given by the colour difference ΔE between the samples under the second illuminant, assuming the samples were a perfect match under the first illuminant. Methods of evaluating ΔE are described in the next chapter. For some applications it might be more appropriate for one of the two illuminants to be a particular fluorescent lamp and if the index of metamerism were required, say, in terms of D_{6500} and a fluorescent lamp of the same correlated colour temperature, then the mis-match ΔE would be a function both of the index of metamerism of the two samples and of the colour-rendering index of the fluorescent lamp. There is, indeed, a close relation between the illuminant manifestation of metamerism and the specification of the colour rendering property of light sources. A more detailed discussion of the colour rendering problem is given on p. 270.

In the colour industries the index of metamerism is required not merely for its own sake, but as a means of keeping metamerism down to a minimum. When a dye formulation is required, one of the conditions laid down by the customer may be that the index of metamerism relative to a reference sample shall be within a certain specified tolerance. Or a dress designer may require that the metamerism between the material of the dress, the buttons and belt to be used on the dress, and the hat, gloves, handbag and shoes to be worn to match the dress, shall have a minimum defined index of metamerism between each other. The index of metamerism may, indeed, be one of the items to be fed into the computer programme used to derive the dye formulation, but in the present state of the art the colourist may also use his experience to select dyes which, from his knowledge of their spectral characteristics, will keep the metamerism to a minimum. Inspection of Figs. 1.2 and 1.3, for example, shows that the tungsten lamp has a much higher ratio of red/blue energy content than daylight has. It follows that differences in reflection for wavelengths at either end of the spectrum, and more especially in the red, should be minimized if illuminant metamerism is to be kept within acceptable limits.

OBSERVER MANIFESTATION OF METAMERISM

We have seen in Chapter 2 that the main sources of observer difference are the variations in density of the yellow macular pigment and increasing yellowing of the crystalline lens with age. In addition, Figs. 4.2 and 4.8 imply variations in receptor sensitivity but these are somewhat indeterminate in character and not easily classified. If we confine our attention to the effects of the yellow coloration of the ocular media, then we have to consider how the absorption illustrated, for example, in Fig. 2.9 for the macular pigment will affect metameric matches.

The main difference between a lightly pigmented observer and one with heavy pigmentation is the reduction in energy of the blue and violet wavelengths in the light reaching the retinal receptors. This is by no means the same as the illuminant manifestation of metamerism, since this owes much of its effect to the excess red energy of tungsten light compared with daylight. Nevertheless, it would be possible to evaluate a comparable observer index of metamerism if standard absorptions were adopted for typical observers with light and heavy pigmentations. Thus if a metameric pair of samples were a match under illuminant D_{6500} for an observer with the standard degree of light pigmentation, the index of metamerism would be given by calculating the colour difference ΔE between the samples when seen through a yellow filter with spectral transmission values τ_λ, where τ_λ is the ratio of the standard heavy-pigment transmission to the standard light-pigment transmission at each wavelength λ. No standards of observer pigmentation have so far been adopted, and for the moment we must be content to point the way towards the establishment of these special indices of metamerism.

FIELD-SIZE MANIFESTATION OF METAMERISM

As mentioned earlier, the presence of metamerism can sometimes be detected by the appearance of Maxwell's spot as the eye scans across the boundary between two matched samples. This has its origin in the absorption of the yellow macular pigment which occupies a retinal area centred on the visual axis and subtending some 4° or so. It was for this reason, as explained in Chapter 3, that the large-field colour matching functions of Stiles and of Speranskaya were recorded for an annular zone of the retina by ignoring or obscuring the central area of the field. An alternative way of describing this effect would be to say that a metameric match observed foveally will become a broken metamer when observed in the extra-fovea and vice versa.

This field-size or retinal-area manifestation of metamerism can also be given an index of metamerism, and in this case the standard C.I.E. 1931 and 1964 colour matching functions for 2° and 10° fields are on hand to provide the basis for such an index. Thus suppose two samples A and B, with spectral luminance functions $\beta(\lambda)_A$ and $\beta(\lambda)_B$ are a match under illuminant D_{6500} when their colour specifications are evaluated according to the 2° functions of the 1931 C.I.E. system. Then in general they will be a mismatch when evaluated using the 10° functions of the 1964 C.I.E. system and the index of metamerism will be given by calculating the colour difference ΔE between the samples when specified on the 1964 system.

COLOUR-MATCH BREAKDOWN DUE TO SURFACE TEXTURE

It is sometimes required that materials which are very different in surface texture should nevertheless be similar in colour or at least have the same hue. This could apply to the example already quoted in this chapter where the colour of a dress had to be matched by the buttons, belt, hat, gloves, handbag and shoes. The surface finish of these different articles is likely to range from the almost completely matt to the highly glossy. Now a match between a matt and a glossy surface will often, but not always, be possible, but where it is possible it will only hold for some given condition of illumination and viewing. Thus when the illumination is highly directional, the top surface reflection from a glossy surface is only seen in the highlights and does not contribute to the main body colour of the sample. With diffuse illumination, on the other hand, there are no highlights and the top surface reflection intermingles with and desaturates the 'true' colour of the material.

It follows that a match between a glossy and a matt surface made under one condition of illumination is likely to break down when the conditions of illumination are changed. This is sometimes referred to as 'geometry metamerism', but while the effect may be of considerable commercial importance it is quite different in origin from the metamerism we have been discussing earlier in this chapter. Nevertheless, it may still be desirable to have some basis on which the colour-match breakdown can be specified. A possible index would be the colour difference ΔE between two samples when seen under diffuse illumination, assuming that they match under directional illumination, or vice versa.

METAMERISM AND COLOUR CONSTANCY

Returning to the illuminant manifestation of metamerism, we have seen that two samples with different spectral reflection characteristics which match under one illuminant will mis-match under another. This means that at least one of the samples, and possibly both, will change in colour appearance with the change in illuminant, although in many cases the alteration in appearance may be so slight that it would pass undetected if the second sample were not available for comparison. We saw in Chapter 2 that the process of visual adaptation could largely compensate for changes in level and colour of the illumination and hence produce a considerable measure of colour constancy in the appearance of many ordinary objects. The compensation is rarely quite complete and with certain types of light source, such as gas discharge lamps, there may be gross colour distortion. Also with certain types of dye, notably those which give a main reflection band in the green and a secondary one in the far red, curve A, Fig. 5.6 the colour may

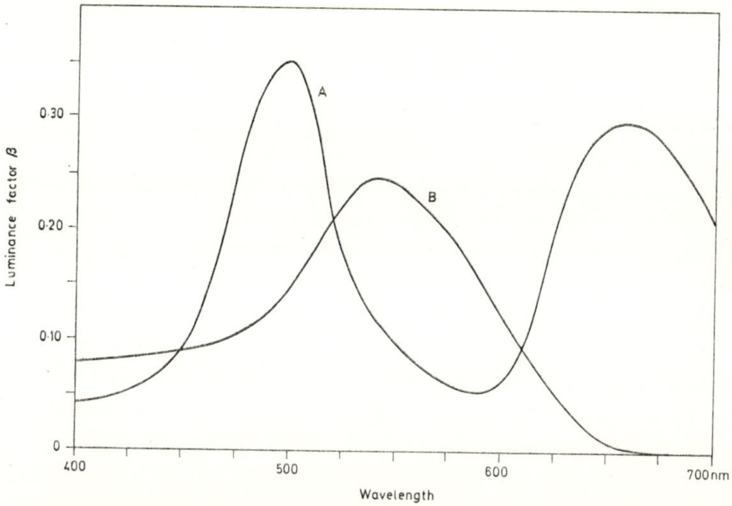

FIG. 5.6. Spectral reflection curves of two patterns which appear green, and are a metameric pair in daylight. Pattern A changes to a brown under tungsten light, but pattern B remains green and shows a high degree of colour constancy

change from green to brown in passing from daylight to tungsten light. A fabric dyed in this way would therefore be highly metameric relative to a more normal green dye, such as curve B, Fig. 5.6.

It is instructive to recognize this relationship between colour constancy and metamerism, for if the colour constancy were always perfect it would be impossible to demonstrate any illuminant manifestation of metamerism. Indeed, metamerism, colour constancy and colour rendering are all inter-related problems. Metamerism, however, is the easiest to handle since it can be analysed in normal colorimetric terms and does not involve subjective measurements of colour appearance.

CHAPTER 6

COLOUR DIFFERENCE METROLOGY

THE LIMITED PRECISION OF COLOUR MATCHING AND COLOUR MEASUREMENT

The colour matching functions of the C.I.E. standard observer specify the mixtures of the red, green and blue matching stimuli which produce perfect matches of the colours in the spectrum for that observer. In practice, however, there is a limit to the precision within which a real observer can make a match. With a visual colorimeter, for example, the instrument controls cannot be re-set to exactly the same readings in successive matches. In dyeing, there are limits, admittedly very fine ones, within which the colourist can judge the correctness of the match between his dyed patterns.

Similarly, there are limits to the precision and accuracy within which colours can be measured with purely physical instruments such as recording spectrophotometers or tristimulus colorimeters. Yet in industry the colour of a product may have to be maintained the same over periods of months or years. This means very close control of the manufacturing process and frequent checking of colour standards. If the colour limits are very tight, the measuring and control techniques will be correspondingly exacting and the manufacturing costs will increase accordingly.

Although it is still comparatively rare for a commercial transaction to include a statement defining a colour tolerance ΔE to which a product has to conform, the concept of a tolerance is in reality a central element in deciding the financial terms of a contract, even though it may not always be recognized as such. Indeed, so important has this concept become, that a new metrology of differential colorimetry is beginning to emerge. It is the aim of this chapter to examine the visual basis on which colour differences can be assessed and the attempts that have been made to develop new co-ordinate systems and formulae with which to express a colour difference ΔE in visually meaningful terms.

THE JUST NOTICEABLE DIFFERENCE

One way to express the ability of an observer to discriminate colours is to measure what is known as the just noticeable difference (j.n.d.) between two colours. The j.n.d. has had a long history as a measure of the perceptual discrimination of an observer, and although other, and no doubt statistically sounder, measures of sensitivity are frequently used, the j.n.d. has the merit of simplicity and is generally found to give essentially similar information to that derived from, say, the standard deviation of a series of matches or from a forced-choice technique in which the observer may have to say whether a stimulus looks greener or bluer than another, or lighter or darker, or whatever it may be.

There are several possible reasons why two stimuli have to differ by a finite amount before a difference in sensation can be detected, but it is by no means easy to decide which factors are the most significant in any particular case. Thus the stimulus itself may not be constant with time, or uniform over its area; the sensitivity of the receptors may vary over the retina or fluctuate from moment to moment; there may be spontaneous discharge in the retina and other sources of physiological 'noise', and the perceptual processes themselves must almost certainly have some discrimination limit, which is likely to be affected not only by the degree of attention that the observer is prepared and able to devote to the task but also by the inherent interest and significance of the stimulus changes.

While we can thus understand that the j.n.d. cannot be vanishingly small, its actual size will also be governed by the psychophysical relation between stimulus and sensation as established through the physiological mechanism of the eye and brain. We can hardly hope to know the exact form of this relationship, but as we saw in Chapter 2 the ability to distinguish between two wavelengths appears to originate in the existence of three types of retinal receptor with different spectral sensitivities. Yet unless there are also different qualities of sensation associated with these different types of receptor, there can be no colour discrimination. Hence, for example, defective colour vision could have its origin either in the retina or in the brain.

The j.n.d. is often regarded as a sensory unit of contrast, and the number of j.n.d.'s between two stimuli A and B as a measure of the contrast between A and B. While this may be a useful assumption, it is not necessarily true, especially if we accept the possibility that physiological 'noise' may affect the size of the j.n.d. to an important extent. For noise may have an influence on, say, luminance discrimination entirely different from that on wavelength discrimination, so that it would be most unwise to assume that the contrast between two luminances L_1 and L_2 separated by, say, 5 j.n.d.'s was the same as the contrast between two wavelengths λ_1 and λ_2 with the same number of j.n.d.'s between them.

Before reviewing the data themselves, we must note the effect of the observing conditions on the size of the j.n.d. Our powers of discrimination are at their best when we are looking at two adjacent areas of colour with a sharp and almost invisible boundary line between them, when the areas are large, when the illumination is good and when we are using unrestricted binocular vision free from any optical encumbrances such as eyepieces or exit-pupils. Moreover, when a trained colourist is matching two pieces of cloth he will transpose them and view them at various angles to the light and to himself, and generally introduce a dynamic element into the viewing situation.

We can understand the reasons for these conditions, since we know from electro-physiological studies of the retina that the receptors are most sensitive to sudden changes of stimulation, and that they react more rapidly when the retinal illumination is high.* When comparing or matching two areas of colour, the observer will naturally look from one area to the other, and this scanning across a sharp boundary provides just the conditions likely to produce the maximum change in physiological activity for any given difference of stimulus between the two halves of the field.

It is perhaps less easy to explain the need for a large area of colour, except that the larger the area the more information available for analysis by the retina and brain. It would seem also as if stimulation of a large number of receptors, accompanied presumably by a correspondingly extensive activity in the brain, is an

* R. Granit: *Receptors and Sensory Perception* (Yale University Press: New Haven), 1955.

essential element in the evocation of a full colour sensation. That the influence of the area is by no means small is evident from the very rapid decrease in uncertainty of a colour match with increase in field size as recorded by Judd* and illustrated in Fig. 6.1. We

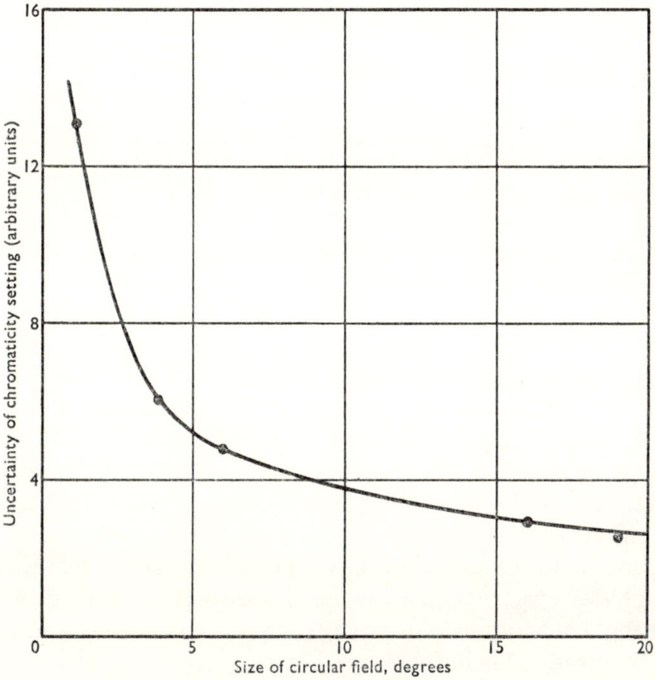

FIG. 6.1 Influence of area on discrimination (Judd)

have also to remember the specific loss of blue sensitivity, leading to small-field tritanopia, which occurs when very small areas are being viewed, to which reference has already been made on p. 55.

From all this it is evident that there is nothing absolute about the magnitude of the j.n.d. Moreover, in addition to the effect of the observing conditions, there will be considerable differences between the sensitivity of one observer and another; training and experience, too, can enhance an observer's ability to detect small colour differences, quite apart from his innate colour sensitivity. Hence, when a statement is made about the size of the j.n.d. or

* D. B. JUDD: *Bur. Stds. J. Research*, 1930, **5**, p. 1161.

about the precision of matching, it will have little significance unless it is accompanied by information of the conditions under which the observation was made; also, when we consider data on colour discrimination as a guide to the spacing of colours in the colour solid or the chromaticity chart, we must ensure that the observing conditions have remained constant throughout the experiments.

COLOUR DISCRIMINATION THROUGH THE SPECTRUM AND IN THE CHROMATICITY CHART

The spectrum is so fundamental to our understanding of colour perception that observations on the ability to discriminate one wavelength from another have always been regarded as of special interest. The basic experiment is as follows: The two halves of a photometric field are illuminated by light of the same wavelength λ and their luminances are adjusted to be equal. The wavelength of one half of the field is then gradually changed to $(\lambda \pm \Delta\lambda)$ until a difference in colour between the two halves of the field can just be detected, the luminances meanwhile being maintained equal. The difference that is detected is then solely due to the change in wavelength. If the experiment is repeated for other values of λ through the spectrum, $\Delta\lambda$ can be plotted against λ to give the wavelength discrimination curve, a typical curve as recorded in one investigation* being shown in Fig. 6.2.

These observations were made on the author's colorimeter with a 2° field, and under those conditions a wavelength difference of about 1 nm will give rise to a just detectable colour difference in the yellow-orange part of the spectrum and again in the blue-green. In view of the effect illustrated in Fig. 6.1, we can deduce that under large-field conditions a difference of only 0·3 nm might be detectable in these regions of the spectrum, and that colour matches could be repeated to an even higher precision. It is this remarkable capacity for colour discrimination that makes it so difficult for a photoelectric colorimeter to compete with the eye. On the basis of Fig. 6.2, some 150 different colours can be dis-

* W. D. Wright and F. H. G. Pitt: *Proc. Phys. Soc.*, 1934, **46**, p. 459.

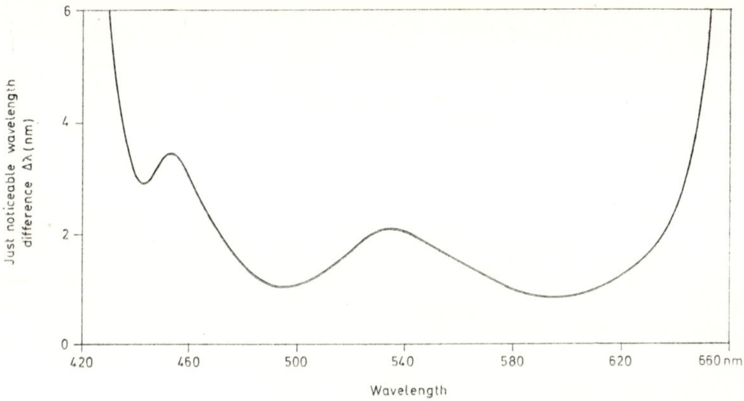

FIG. 6.2 Typical wavelength discrimination for a normal observer using a 2° field (Wright and Pitt)

tinguished in the spectrum, but under large field conditions the number would be many more.

An alternative method of representing the data is to convert the values of $\Delta\lambda$ into the equivalent distances in the C.I.E. chromaticity chart and to represent them by short lines at the appropriate wavelengths round the spectrum locus. When this is done, it is found that the lines are very long in the green parts of the locus, very short in the blue and violet, and of intermediate length in the yellow and red. This non-uniform spacing of the colours in the chart was anticipated by Judd* from data on purity discrimination and on the sensitivity to changes in colour temperature. When the whole chart came to be tested by MacAdam and by the author, the full extent of the non-uniformity was revealed.†

For the results illustrated in Fig. 6.3, the colour steps were measured by the author by varying the mixture of two monochromatic radiations taken from opposite sides of the spectrum locus, and of a monochromatic radiation with one of the purple colours, P_1, P_2 or P_3. The proportion of one constituent was increased relative to the other, until a noticeable change in colour

* D. B. JUDD: *J. Opt. Soc. Am.*, 1935, **25**, p. 24.
† D. L. MACADAM: *J. Opt. Soc. Am.*, 1942, **32**, p. 247.
W. D. WRIGHT: *Proc. Phys. Soc.*, 1941, **53**, p. 93.

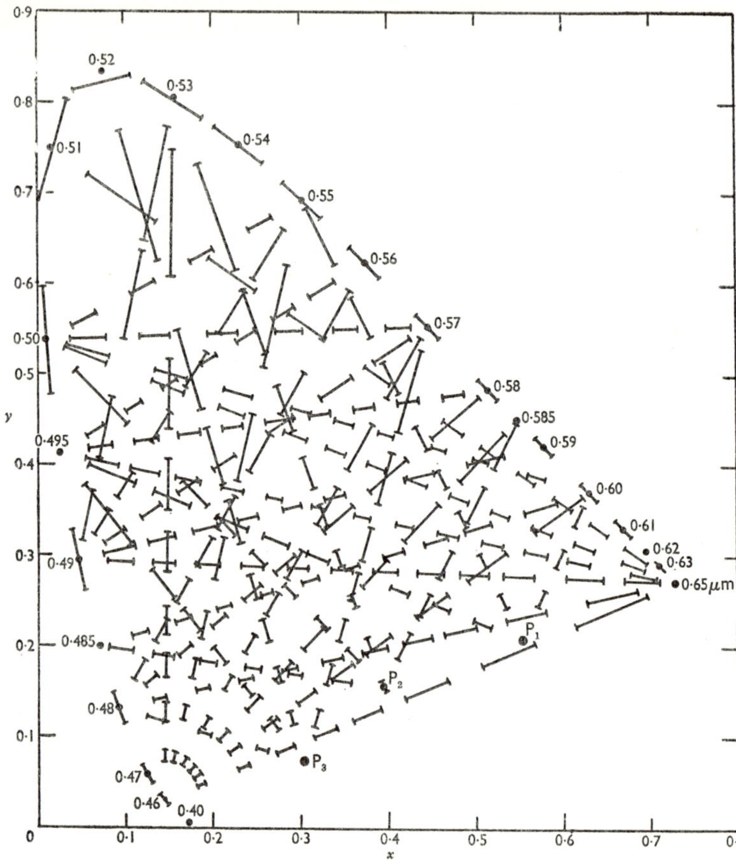

FIG. 6.3 Subjectively equal colour steps plotted in the C.I.E. chromaticity chart. The lines represent a colour difference about three times the just noticeable difference for a 2° field (Wright)

quality was detected by comparison with a similar mixture seen in the other half of a photometric field, the luminances of the two halves being maintained equal during the observation. The change in the mixture was then converted into C.I.E. co-ordinates and the size of step expressed in terms of a short length in the chromaticity chart. The observations were repeated for various mixtures of the two radiations, and for different pairs of radiations. The author's results given in Fig. 6.3 were generally confirmed for a selection of the lines by three other observers.

In MacAdam's investigation, the observer made a large number of matches on a series of test colours located at various points in

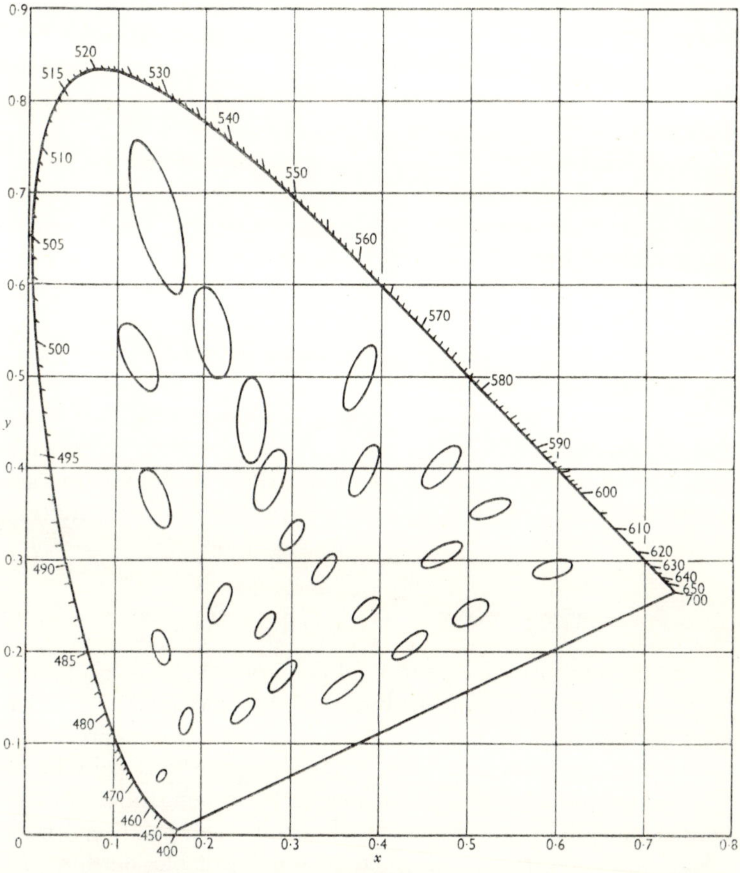

FIG. 6.4 Sensitivity to chromaticity differences expressed in terms of the standard deviations of colour matching. The ellipses represent the standard deviations multiplied by a factor of 10. Results are for one observer (After MacAdam)

the chromaticity chart. The spread in the settings was used as the criterion of sensitivity to colour differences and plotted on the C.I.E. chart in the form of ellipses of varying size (Fig. 6.4), or as steps along a series of lines across the chart. Both investigations gave essentially similar results* and revealed large variations in step size across the chart. Thus, in Fig. 6.3, the ratio of the shortest to the longest step is at least 20 : 1. It should be noted that the size of step represented by each line in this diagram is larger than the j.n.d. for a 2° field by a factor of the order of 3, while measurements made with a visual colorimeter with the same size of field should have a precision of about $\frac{1}{20}$ of the size of step illustrated in the diagram.

An alternative way of describing the non-uniform colour distribution in the C.I.E. chart is to consider the density of population of colours separated from each other by small, equally perceptible, differences of chromaticity. On this basis, the population is some three or four hundred times more dense in the blue corner of the chart than it is in the green.

The experiments described here have been supplemented by further major investigations by MacAdam and Brown.† In particular, the problem has been tackled as a three-dimensional one, and ellipsoids have been derived to represent colour discrimination when luminance as well as chromaticity is allowed to vary. These workers have also studied the effect of surround conditions on discrimination.

UNIFORM COLOUR SCALES

The non-uniform distribution of colours in the C.I.E. chromaticity chart is disappointing but hardly surprising. The C.I.E. co-ordinates were chosen for their convenience in representing colour mixture data and little or no attention was paid to the representation of discrimination data in the system. Indeed in 1931 no general studies of chromaticity discrimination had been undertaken, although some information was available on hue and saturation discrimination. Nevertheless the non-uniformity of the normal C.I.E. chart can be a handicap and various proposals have been made for alternative projections or transformations of the C.I.E. co-ordinates to secure greater uniformity.

* W. D. Wright: *J. Opt. Soc. Am.*, 1943, **33**, p. 632.

† W. R. J. Brown and D. L. MacAdam: *J. Opt. Soc. Am.*, 1949, **38**, p. 808. W. R. J. Brown: *J. Opt. Soc. Am.*, 1951, **41**, p. 684, 1952, **42**, p. 837, and 1957, **47**, p. 137.

One reason in favour of a more uniform scale is that the specification of colour tolerances might be simplified. For example, in the ideal case of a perfectly uniform chart, colours could be specified by a point C in the chart, all acceptable colours having to fall within the area of a circle described with C as centre and radius t. The permissible tolerance is then specified simply by defining the magnitude of t. This would be a valid method of specifying the tolerance, provided that visual discrimination was the predominant factor determining the limit and provided also that errors of hue and errors of saturation were equally important. In industry, manufacturing problems and the high cost of maintaining accurate colour control may determine the tolerance, and it may also happen that errors of hue and saturation are not equally serious. For reasons such as these, MacAdam* has considered that a more uniform chromaticity chart would have little advantage over the C.I.E. chart and that even if the principle of a uniform chromaticity scale were adopted, several different charts would be necessary to meet the needs of various industries and conditions of observation. It is true that, even with a uniform chart, the circular tolerance area might in many cases have to give way to ellipses, triangles, rectangles, etc., and much of the advantage of simplicity of specification would disappear. Nevertheless, the use of a more uniform projection of the C.I.E. chart would avoid the gross discrepancies of tolerance area as exhibited, for example, with signal glasses (see Fig. 10.10, p. 288).

Probably the more serious effect of the lack of uniformity of the C.I.E. chart is the erroneous impression it can give regarding the relation between different colours when plotted on the chart. It is inevitable that the distance apart of two colours in the chart should to some extent be regarded as a measure of their sensory difference; if this is done, the magnitude of the errors in colour matching or of the inaccuracies of colour reproduction may be seriously misjudged. Again, the condemnation of some process of colour reproduction or some type of colorimeter because it is unable to record colours in certain areas of the chromaticity chart may be quite unwarranted if the number of colours that can be discriminated in those areas is relatively small. It is true that once the non-uniformity is appreciated, some allowance can be made in assessing the

* D. L. MacAdam: *Amer. J. Psychol.*, 1939, **52**, p. 414.

significance of given distances and areas in the chart, but on many occasions the chart may be employed by persons who, although they may have a general knowledge of the subject, are not sufficiently conversant with the details of the system to know that any such allowance is necessary.

To produce a more uniform distribution of the colours, a projection of the C.I.E. chart is required that will tend to foreshorten the green corner and elongate and expand the blue corner. Fig. 6.5 is typical of several of the modified chromaticity charts which have

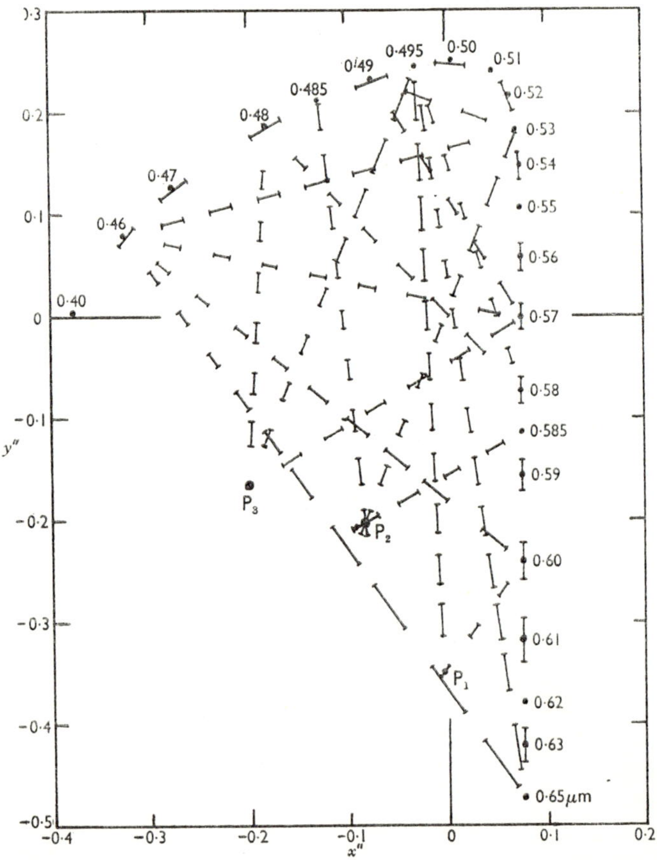

FIG. 6.5 Subjectively equal colour steps plotted in the rectangular uniform chromaticity chart of Breckenridge and Schaub. A selection of the data used for Fig. 6.3 transformed to the co-ordinates of the Breckenridge and Schaub chart

been suggested.* This diagram is plotted in the co-ordinates of the rectangular chromaticity scale of Breckenridge and Schaub, and a selection of the lines given in Fig. 6.3 has been replotted relative to the new co-ordinates. The improvement in distribution is most noticeable in the manner in which the green, and to a less extent the yellow, lines have been shortened and at the same time the short lines in the blue corner have been lengthened. On the other hand, in the red corner the lines have been drawn out in one direction and contracted in the transverse direction, so that, taking the colour chart as a whole, the ratio of the shortest step to the longest is still about 1 : 4.

It is quite certain that no linear projection of the C.I.E. diagram can give exact equality of the colour steps, but it is also true that, considering the limited accuracy to which the steps can be measured, the variation among observers and the varied uses to which a uniform scale would be put, there is no need to demand exact equality over the whole diagram. For practical purposes, a scale might be acceptable if the steps did not differ by a ratio greater than, say, 2 : 1, and on this criterion practically the whole of Fig. 6.5 could be regarded as uniform with the exception of the purple-red corner.

In spite of these arguments for a more uniform chromaticity chart, the type of projection illustrated in Fig. 6.5, which is characteristic of most uniform chromaticity charts including the transformation recommended by the C.I.E. (p. 166), is not in practice a very satisfactory chart for use with surface colours. In the first place, all the reds, oranges, yellows, browns and greens, which probably represent about 75 per cent of all the colours we use, are squeezed into the quite narrow strip running vertically between the zero axis on the x'' scale and the straight part of the spectrum locus on its right. Hence if we wish to illustrate, say, some colour sequence among such hues, any curves we may draw will be very foreshortened and will lose much of their significance. Further, these uniform chromaticity transformations are based on discrimination observations made at constant luminance. Yet for

* D. B. Judd: *J. Opt. Soc. Am.*, 1935, **25**, p. 24.
D. L. MacAdam: *J. Opt. Soc. Am.*, 1937, **27**, p. 294.
F. C. Breckenridge and W. R. Schaub: *J. Opt. Soc. Am.*, 1939, **29**, p. 370.
G. Wyszecki and W. S. Stiles: *Color Science* (Wiley: New York and London), 1967.

surface colours this is an extremely artificial and, indeed, erroneous restriction. When we come to consider the various colour atlases in the next chapter, we shall see that the bright saturated yellows, for example, have a high luminance factor (Munsell value in Fig. 7.3), whereas the saturated blues have a much lower luminance factor. Yet these are the sequences of colours that we may want to compare in the chromaticity chart. Moreover, in Fig. 7.3 for example, there are as many chroma steps between the most saturated yellow, namely sample *Y* 8/12, and the neutral axis as there are between the most saturated blue, sample *PB* 3/12, and the axis. This situation is therefore much more correctly represented on the normal C.I.E. chromaticity chart than on a projection such as Fig. 6.5, and we may well find that it continues to be the most widely used of any of the two-dimensional diagrams, at least for surface colours.

One transformation that has been widely used is the 'alpha-beta' diagram of Hunter in which the co-ordinates of a sample are derived from its C.I.E. specification by the equations

$$\alpha=\frac{2\cdot4266x-1\cdot3631y-0\cdot3214}{1\cdot0000x+2\cdot2633y+1\cdot1054}$$

and

$$\beta=\frac{0\cdot5710x+1\cdot2447y-0\cdot5708}{1\cdot0000x+2\cdot2633y+1\cdot1054}$$

This transformation has found particular application in association with the Hunter Color Difference Meter* with which α and β can be measured directly.

The transformation which the C.I.E. itself recommended in 1959 for use where a more uniform distribution of colours in the chromaticity diagram is required was based on a co-ordinate system recommended by MacAdam in 1937.† The chromaticity co-ordinates (u, v) in this transformation are given by the equations

$$u=\frac{4x}{-2x+12y+3}$$

$$v=\frac{6y}{-2x+12y+3}$$

* R. S. Hunter: *American Society for Testing Materials, Philadelphia Symposium on Color*, 1941, p. 61.

† D. L. MacAdam: *J. Opt. Soc. Am.*, 1937, **27**, p. 294.

leading to the diagram known as the 1960 CIE-UCS diagram.* Judd and Wyszecki† subsequently confirmed that this transformation appeared as uniform as any other chromaticity chart so far suggested, particularly in the success with which new data obtained by an American Committee on Uniform Color Scales could be represented.

Four years later, in 1963, the C.I.E. Colorimetry Committee considered the need to extend the uniformity concept to include differences of luminance or luminance factor. This meant that in effect a uniform colour solid had to be defined by adding a third co-ordinate to the 1960 CIE-UCS diagram. However, more is involved than merely including a vertical axis to represent luminance discrimination because account also has to be taken of the fact that there is a loss of chromaticity discrimination with dark colours. The UCS diagram therefore has to contract as the luminance factor (or Y tristimulus value) gets less.

Pending further experimental data which might lead to the development of an improved co-ordinate system, the C.I.E. recommended the provisional use of a new co-ordinate system defining the C.I.E. 1964 uniform colour space.‡ Three variables $W^\star$, $U^\star$, $V^\star$ were defined as follows:

$$W^\star = 25Y^{\frac{1}{3}} - 17;$$
$$U^\star = 13W^\star(u - u_0);$$
$$V^\star = 13W^\star(v - v_0)$$

where

$$u = \frac{4X}{X + 15Y + 3Z}$$

and

$$v = \frac{6Y}{X + 15Y + 3Z}$$

(These are, in fact, equivalent transformation equations to those given above for u and v in terms of x and y.) u_0 and v_0 are the values of u and v for the nominally achromatic colour in the ($U^\star$, $V^\star$)

* C.I.E. *Compte Rendu*, 1959, Vol. A, p. 111.

† D. B. Judd and G. Wyszecki: *Color in Business, Science and Industry* (Wiley: New York and London), 2nd Ed., 1963.

‡ *C.I.E. Compte Rendu*, 1963.
D. B. Judd: *J. Opt. Soc. Am.*, 1963, **53**, p. 1012.
G. Wyszecki: *J. Opt. Soc. Am.*, 1963, **53**, 1318.

co-ordinate system, which means that for surface colours, u_0 and v_0 are the co-ordinates corresponding to the chromaticity of the illuminant.

Of the various other proposals that have been made to define a more uniform scale, one that has found considerable use among colorimetrists, especially in the United States, is the Adams' Chromatic-Value Diagram.* This is based on the Munsell system and is described in the next chapter. That there is still uncertainty as to which colour transformation provides the most uniform distribution of colours is shown by the fact that in 1967 the C.I.E. Colorimetry Committee recommended National Committees to make further studies on the perceived sizes of colour differences.† It may well emerge that no one transformation can be described as the 'best' or as 'correct', but in the meantime the 1964 C.I.E. recommendation represents an attempt to unify the very diverse practices which are currently in use.

COLOUR DIFFERENCE FORMULAE AND COLOUR TOLERANCES

On the assumption that the 1964 C.I.E. transformation does provide a uniform three-dimensional scale, the magnitude of the perceived colour difference ΔE between two stimuli U_1^*, V_1^*, W_1^* and U_2^*, V_2^*, W_2^* can be calculated from the equation

$$\Delta E = [(U_2^* - U_1^*)^2 + (V_2^* - V_1^*)^2 + (W_2^* - W_1^*)^2]^{\frac{1}{2}}$$

This formula is expected to apply satisfactorily for pairs of colours of the same size and shape viewed in identical white to middle-grey surroundings by an observer adapted to average daylight. For samples subtending between 1° and 4°, U^*, V^* and W^* should be calculated from tristimulus values X, Y and Z, but for fields larger than 4°, the co-ordinates X_{10}, Y_{10} and Z_{10} derived from the new 10° system should be used.

Several other colour difference formulae have been proposed‡ based on various uniform colour systems and an early proposal by Judd§ led to the establishment of a widely used unit known as the

* E. Q. Adams: *J. Opt. Soc. Am.*, 1942, **32**, p. 168.

† *C.I.E. Compte Rendu*, 1967.

‡ For a recent summary, see G. Wyszecki and W. S. Stiles: *Color Science* (Wiley: New York and London), 1967.

§ D. B. Judd: *Am. J. Psychol.*, 1939, **52**, p. 418.

N.B.S. unit of colour difference. In a revised form based on Hunter's 'alpha-beta' diagram, the difference ΔE between two colours having the co-ordinates α_1, β_1, Y_1 and α_2, β_2, Y_2 is given in N.B.S units by the equation*

$$\Delta E = f_g\{[221\,Y^{\frac{1}{4}}((\Delta\alpha^2)+(\Delta\beta)^2)^{\frac{1}{2}}]^2+[k\Delta Y^{\frac{1}{2}}]^2\}^{\frac{1}{2}}$$

where

$$Y=\frac{Y_1+Y_2}{2},\ \Delta Y^{\frac{1}{2}}=Y_1^{\frac{1}{2}}-Y_2^{\frac{1}{2}},$$

$$\Delta\alpha=\alpha_1-\alpha_2,\ \Delta\beta=\beta_1-\beta_2.$$

f_g is a constant which can be adjusted to take account of the masking influence of a glossy surface on the detection of colour differences. For typical conditions of inspection f_g can be found from the equation

$$f_g=\frac{Y}{Y+2{\cdot}5}$$

k is a further constant to allow for the variation in the contribution which differences in lightness make to the total perceived colour difference depending on the viewing conditions. If the two colours being compared are juxtaposed with a sharp dividing line between them, k will have a value of 12, but with a broad band separating the two colours, k has to be reduced to 3 or 4.

Like all colour difference formulae, the inclusion of these rather arbitrary constants suggests that the foundation on which the expression for ΔE is based cannot have any very fundamental visual significance. Nevertheless, it is important to have a general idea of the order of magnitude of the N.B.S. unit, that is, the colour difference when $\Delta E=1$ as calculated from the above equation. The N.B.S. unit was intended to be about 5 times the smallest difference that can be perceived under the best experimental conditions and is equivalent to about 0·1 of a Munsell value step, 0·15 of a Munsell chroma step, and about 2·5 Munsell hue steps at chroma/1 in the Munsell system described in the next chapter. In the centre of the C.I.E. chromaticity chart, it corresponds to a chromaticity step of the order of 0·0015 to 0·0025 in x or y.

Another unit which is also widely used, especially in expressing

* R. S. Hunter: *J. Opt. Soc. Am.*, 1942, 32, p. 509.

permissible tolerances in colorant formulation, is the MacAdam unit of color difference.* Here the colour difference Δs between two stimuli with C.I.E. co-ordinates x_1, y_1, Y_1 and x_2, y_2, Y_2 is calculated from the equation

$$(\Delta s)^2 = g_{11}(\Delta x)^2 + 2g_{12}\Delta x\Delta y + g_{22}(\Delta y)^2 + g_{33}\left(\frac{\Delta Y}{Y}\right)^2,$$

where

$$\Delta x = x_2 - x_1, \Delta y = y_2 - y_1, \ \Delta Y = Y_2 - Y_1$$

and

$$Y = \frac{Y_1 + Y_2}{2}$$

g_{11}, g_{12}, g_{22} are constants which depend on the chromaticity of the samples and are read off specially prepared graphs, while g_{33} is another constant determined by the viewing conditions. If the samples being compared subtend an angle of about 2° and are separated by a sharp dividing line, the value of g_{33} is approximately unity. A colour difference corresponding to $\Delta s = 1$ as calculated by this equation is known as a MacAdam unit.

According to a table given by Wyszecki and Stiles,† a unit of colour difference on the C.I.E. $U^\star$, $V^\star$, $W^\star$ system is equal in magnitude to the N.B.S. unit calculated on the revised 'alpha-beta' formula, while a MacAdam unit is equal to about 1·4 N.B.S. units. Other units which have been suggested need not concern us for the moment since they have not as yet come into wide use and to include them would only confuse still further an already rather complex situation.

We have indeed to recognize that colour difference metrology cannot be expressed in simple algebraical relations. The critical dependence of the size of the j.n.d. on observing conditions is one complicating factor. The change in the relative importance of differences of luminance and differences of chromaticity with change in observing conditions is another. The non-linear relation between colour differences and luminance, is a third. The non-additivity of colour differences is a fourth, and perhaps the most

* D. L. MacAdam: *J. Opt. Soc. Am.*, 1943, **33**, p. 18 and 1957, **47**, p. 268; *Official Digest*, 1965, **37**, p. 1487 and *J. Paint Technol.*, 1966, **38**, p. 70.

† G. Wyszecki and W. S. Stiles: *Color Science* (Wiley: New York and London), 1967, p. 468.

serious, factor, since it means, for example, that the equality between two small differences may not persist if the differences are doubled or quadrupled.*

Clearly any formula that is used for the calculation of colour differences is likely to have an empirical element in it, and the preference for one formula over another is likely to be determined by its practical convenience and ease of application rather than because of some superior visual validity. The practical colourist in the dyehouse might indeed wonder whether he needs any formula at all, since he is likely to feel more at home with his own visual assessment of colour differences in terms of what he calls 'traces', with which he has been familiar for a long time. However, whether he likes it or not, the invasion of the dyehouse by the computer makes it almost essential to adopt some algebraic expression for a colour difference in order that an appropriate tolerance can be fed into the computer programme.

This need was recognized by Alderson, Atherton and Derbyshire† at an early stage in the development of computer techniques for colorant formulation and they derived transformation equations linking differences in the dyer's variables of strength, shade and brightness with differences expressed either in C.I.E. or Munsell co-ordinates. They also studied the size of a 'trace' difference relative to the N.B.S. unit and for a particular standard blue they quote the following relations:

1 trace of strength difference = 0·47 N.B.S. units
1 trace of shade difference = 0·75 N.B.S. units
1 trace of brightness difference = 0·84 N.B.S. units

This clearly implies a discrepancy between the two measures of colour difference when compared in different directions in colour space and is a further illustration of the uncertainty which surrounds the subject at the present time.

The actual size of a colour difference that is acceptable as a tolerance limit will vary widely from industry to industry and it has to be recognized that the philosophy behind the specification of a colour limit is very different from that underlying the limits specified in engineering. When a shaft is being machined to fit a

* D. NICKERSON and S. M. NEWHALL: *J. Opt. Soc. Am.*, 1943, **33**, p. 419.

† J. V. ALDERSON, E. ATHERTON and A. N. DERBYSHIRE: *J. Soc. Dyers and Colourists*, 1961, **77**, p. 657.

bearing, it must neither be left so large that it will not go into the bearing at all nor be made so small as to lead to rapid wear and inefficiency. It is therefore essential that the parts of a machine should be manufactured to dimensions specified within small and precise limits, otherwise the machine will not work at all. With colour, on the other hand, one customer may demand virtually exact identity of colour between two pieces of cloth, while another may accept a quite perceptible difference, particularly if the material can thereby be purchased at a reduced price. Hence the accuracy required in colour matching is relative to the economic level and aesthetic susceptibility of the public for whom the products are intended, although a firm which can guarantee a greater reproducibility in the colour of its goods will have an advantage when in competition with other firms. The economic justification for time and energy spent on manufacturing a product to an exact colour specification clearly calls for a rather special form of cost accounting.

CHAPTER 7

THE COLOUR SOLID AND THE COLOUR ATLAS

SUBJECTIVE JUDGEMENTS OF COLOUR APPEARANCE AND COLOUR INTERVAL

In previous chapters we have traced the development of the C.I.E. system of colorimetry and its international recognition as a valid framework for the measurement of colour. We know, too, from its widespread and increasing use in both industry and research that the numerical specification of colour is of scientific and commercial value. Nevertheless, the primary interest of most people in colour is æsthetic, and he would be a bold man who designed a colour scheme purely on the basis of tristimulus values. Even a person well versed in the details of colorimetry will prefer to choose his colours from a colour chart rather than from a chromaticity diagram, and our concern in this chapter is to examine the important role which the colour atlas has to play as a bridge between the colour scientist and the colour user.

Everyone with normal colour vision has a clear idea of what he means by redness or greenness or blueness, and a colour system could be established on the basis of colour naming alone. As one example of such a classification, we may note the work reported in the United States* on the ISCC–NBS method of designating colours, in which the colour solid is subdivided into about 300 separate compartments, each with its own name such as strong yellowish-pink, brilliant purple, etc. However, a subdivision of this kind cannot be pressed too far, and as the colour differences are made finer, it quickly becomes necessary to specify colour appearance by actual colour chips—a natural and logical thing to do, but equivalent to describing a colour by saying that it has the same appearance as such-and-such a colour chip. Yet

* D. B. Judd and K. L. Kelly: *Nat. Bur. Stds. Circ.*, No. 553, 1955.

the colour chip is a physical object sending a physical stimulus into the eye and this stimulus can be given a C.I.E. specification. Hence, a colour atlas is in effect a physical representation of a variety of colours each of which can be specified colorimetrically; we are therefore deluding ourselves if we think that the colour chips are more subjective than their C.I.E. specifications. Yet the colour perceptions themselves *are* subjective, and the selection and spacing of the chips must be based on subjective observations. In other words, we have to establish subjective relations between physical stimuli.

We can understand how this might be done if we consider the simple case of a series of neutral colours ranging from near-black to near-white, each of them specified by their luminance factor β. The first and most obvious thing we can do is to arrange them in a series with the lightest at one end and the darkest at the other. Next, we can study the intervals between neighbouring samples and we shall probably find that some pairs are nearly the same in lightness while others show big differences. This suggests that not only can we arrange the samples in order of lightness, but we can also grade the lightness intervals between neighbours in the series. We can even go one stage further and select a series having equal intervals between the specimens, interpolating where necessary to find intermediate values of β that will provide as uniform and smooth a series as possible, as illustrated in Plate 8(*a*) (facing p. 102).

There can be no question at all that judgements such as that just described can be made successfully and consistently, and we need not concern ourselves here with the probably insoluble problem whether the subjectively equal steps correspond to equal arithmetical intervals in the magnitude of the sensation, equal geometrical ratios, or some other relation. We should recognize, however, that we are dealing with very imperfectly understood matters and that we may easily fall into error if we try to apply the laws of physics and mathematics to these judgements. For example, if we have four specimens A, B, C and D, which have been chosen such that the intervals A–B, B–C and C–D appear equal, we cannot take it as axiomatic that the interval A–C is subjectively twice as great as the interval C–D. Indeed, while judgements of equality of interval are possible, estimates of the relative magnitudes of different intervals must be accepted with the utmost reserve. Then again, the series in Plate 8(*a*) may appear equally spaced when viewed against a white background, but will tend to appear too compressed at the light end and too extended at the dark end, when seen against a black background.

Colour chips will differ not only in respect of lightness, but also of hue or saturation, and we might arrange a series having equal intervals of hue or equal intervals of saturation. Thus in Plate 8 we have in (*b*) a series from red through orange to yellow in equal steps of hue, and in (*c*) a series from grey to green in equal steps of saturation. If this process were extended to various hues and lightness levels and degrees of saturation, we could hope to produce a three-dimensional colour arrangement with equal steps in all directions, but we can probably already begin to visualize the difficulties that might arise. For one thing, we should have to equate lightness steps with hue and saturation steps, a difficult type of judgement and one dependent very much on the observer and the observing conditions. Then if the colours were equally spaced along three mutually perpendicular directions, would they also be equally spaced in an oblique direction through the solid?

Another type of subjective judgement that has to be made is to establish the identity of one of the attributes of hue or lightness or saturation in the presence of differences in the other two. Thus, from a random collection of colour chips, we may wish to select all those that are, say, of the same red hue, whether they are light or dark, pale or vivid. This can undoubtedly be done with a considerable degree of certainty and of agreement between observers. However, if we wish to pick out chips of the same saturation, we shall find the selection very much more difficult. Indeed, we shall have to be clear in our minds exactly what we mean by saturation. Everyone would agree that in Plate 8(*c*), the colours increase in saturation from left to right, but if we were asked to compare, say, a dark brown and a pale yellow and state which was the more saturated, we might hesitate a long time before we made up our minds. Part of this hesitation might be due to the inherent difficulty of the judgement, but part also to the fact that there are two different concepts of saturation.

This has been recognized by Pope,* who refers to Saturation I

* A. POPE: *The Language of Drawing and Painting* (Harvard University Press; Cambridge, Mass.), 1949.

and Saturation II, but there is also a national divergence of view, at least between the United States and Great Britain. Thus Judd and Wyszecki* quote from a list of terms defined in America by the Inter-Society Color Council and give the definition of saturation as: 'The attribute determining the degree of its difference from the achromatic color perception most resembling it.' On the other hand in British Standard 1611: 1953, saturation is defined as: 'The attribute of a visual sensation which permits a judgement to be made of the proportion of pure chromatic colour in the total sensation.' On the one hand, therefore, there is the idea of a difference from grey and on the other the idea of a ratio, and while perhaps neither concept can be pressed too hard when we are dealing with such nebulous subjective matters, the divergence is real enough when we come to select colour chips of equal saturation according to the two definitions. Thus, according to the American definition, we should tend to regard colours having the same Munsell chroma (see p. 184) as being equally saturated, while according to the British definition, the choice would correspond more to the isochrome colours in the Ostwald system (see p. 182). We can do no more here than recognize that this difference in outlook exists, and emphasize that, where confusion is likely to arise, we should make clear the sense in which we are using the term saturation.

Returning now to our random collection of colour chips, we might attempt, as a third selection, to choose colours having the same lightness but differing in hue and saturation. This is a more difficult task than choosing samples of constant hue, but less difficult than picking out chips with the same saturation. The main uncertainty is with the highly saturated colours, since their vividness heightens their luminosity, so that, in comparing the lightness of, say, a saturated red with that of a grey, we have to decide how far, if at all, we should discount the vividness in assessing the lightness. There may, for example, be legitimate differences of opinion whether the extreme green in Plate 8(*c*) is lighter or darker than the grey at the other end of the series. Not everyone, perhaps, would even agree that the judgement has much significance, and as we shall see, lightness is not universally employed

* D. B. Judd and G. Wyszecki: *Color in Business, Science and Industry* (Wiley: New York and London), 2nd ed., 1963.

as one of the co-ordinates of a colour atlas. Are we wrong then in trying to isolate a simple subjective attribute to correlate with the photometric concept of luminance factor? We have already met this question in our discussion of heterochromatic photometry, but while we may have to recognize that the situation is more complex than it was thought to be at one time, nevertheless the occurrence of lighter and darker colours is universally recognized and it is easy to show that in general the light colours have a high, and the dark colours a low, luminance factor.

THE CONSTRUCTION OF A COLOUR ATLAS

The function of a colour atlas is to assist the colour user in describing, selecting and matching colours and we must now consider how the samples can best be arranged to achieve this aim. A haphazard collection of several hundred colour chips would be virtually useless; instead, they must be classified in terms of easily recognizable attributes, such as those we have just described, and then be arranged in a logical sequence to facilitate the location of any particular colour.

The colours in an atlas are normally mounted in the pages of a book, but in most systems, each page can be regarded as a part or a whole of a section through a colour solid. As a rule, the vertical axis of the solid contains the neutral colours with black at the bottom and white at the top, while vertical planes radiating out from this axis contain colours of a given hue. Each page of the atlas might then correspond to such a hue plane, with successive pages following in the order of the colours in the spectrum from red to orange, yellow, green, blue-green, blue, violet and back again through the purples to red.

Since the number of patterns in any given atlas must be limited by considerations of cost and by the necessity to keep the atlas down to a size which can be handled conveniently, only a small fraction of the several million surface colours that can be discriminated can possibly be represented in a single atlas. This means that the colour intervals between neighbouring patterns will be several times larger than a just noticeable difference. If the atlas is to be satisfying in appearance and provide a reasonably smooth numerical ordering of the colours, the colour steps between patterns should be as nearly equal as possible or, failing that, there should be a smooth gradation in the intervals. However logical the choice of co-ordinates may be, the atlas will tend to be condemned in the eyes of the user if irregularities of spacing can easily be detected.

We thus have the two conditions that the arrangements of the chips must be systematic and the spacing smooth. As a first stage in meeting these conditions, it is the almost universal practice to make the black-white axis vertical with successive hues located in vertical planes around the axis. Then in each plane the saturated colours are the most distant from the axis, although the detailed positioning of chips of a given hue differs from one atlas to another.

In the Munsell system, for example, colours of constant lightness are located in the same horizontal plane, and the series of colours in Plate 8(*c*) are typical of a horizontal row of chips in the green hue chart. As we shall see, the boundary formed by the most saturated colours at each lightness level is irregular in shape and varies from one hue to another. In the Ostwald system, on the other hand, the colour solid is made to take the form of a double cone so that the patterns in each hue are arranged in the form of a triangle (Fig. 7.1). The most intense or full colours are located at

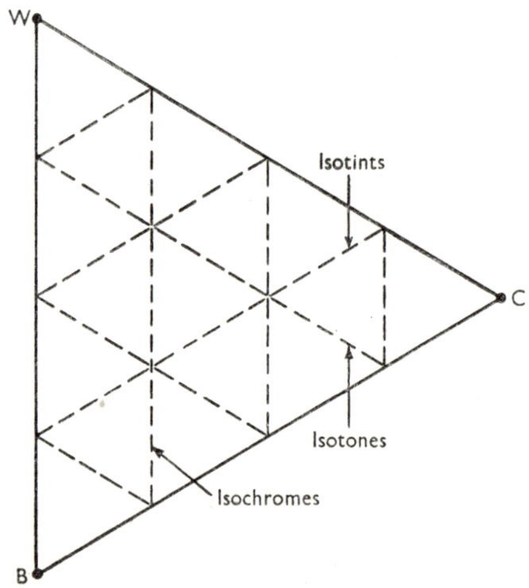

FIG. 7.1. Co-ordinates of the Ostwald system for a given hue plane

the peripheral corners of the triangles away from the neutral axis with black and white at the other two corners, while intermediate colours are located in principle on the basis of the additive mixture of black, white and the full colour. In other systems, such as the Nu-Hue system, the arrangement of colours in a given hue plane is based on colorant mixtures of black pigment, white pigment and the appropriate chromatic pigment.

Some people may feel that a single arrangement should suffice for all purposes, but the multiplicity of atlases has its origin, apart from the desirability of commercial competition, in the different uses to which colour may be put. For the purposes of colour printing, it is not unreasonable to have charts prepared by half-tone colour printing techniques and spaced according to the dot sizes used in the process; for paint mixture, the charts may be more useful if the co-ordinates of the atlas are based on pigment mixtures; for colour design, the arrangement best suited to derive colour-harmony relations may be chosen; for general use in colour description, a system based on subjective appearance and equality of spacing may be the most convenient. A single atlas would also have the disadvantage that it would be representative of only one type of surface texture. Charts for glossy and matt surfaces are desirable, and it has also been found worth while to produce atlases in silk and wool.

If an atlas is to be extensively reproduced and used over a prolonged period of time, we need some assurance that the patterns will be duplicated accurately and be as permanent as possible. This implies that a master atlas must be available against which all other atlases can be checked; the colours in the master atlas should also be standardized on the C.I.E. system to give a permanent record of their colour quality.

COLOUR ATLASES

Many colour atlases have been prepared from time to time, varying widely in the purpose and the public for which they were designed. The simple shade card issued by an individual firm or industry may be regarded as an elementary colour atlas showing the colours produced by a particular firm at a particular time. Each of these is generally of purely local and temporary interest. On a more ambitious scale there are the atlases based on pigment mixtures, such as the Nu-Hue System developed by the Martin-Senour Company, the Colorizer developed by Colorizer Associates, and the Plochere Colour System. Other atlases have been prepared to assist in the description of colours which occur in nature, for example, among flowers or birds, but they have often found considerable use outside the field for which they were originally intended. We may note in this class the *Repertoire des Couleurs* (1905) issued by the Société Française des Chrysanthémistes, *Color Standards and Color Nomenclature* (1912) by R. Ridgway, who was concerned with the plumage of birds, and the Wilson horticultural colour chart (1938 and 1941) prepared by R. F. Wilson initially for use in the description of flowers. The Royal Horticultural Society has recently introduced a revised atlas for this purpose in which the colours are printed four at a time on narrow sheets which can be opened out into a fan of colours.

Colour charts prepared for more general use include *Baumanns neue Farbentonkarte* by P. Baumann, *Code Universal des Couleurs* (1936) by E. Séguy, and the *Dictionary of Color* (1930) by A. Maerz and M. Rea Paul. S. Hesselgren has produced an interesting *Colour Manual* (1953) in which the patterns in each hue are arranged in regard to their lightness, brilliance, saturation, and intensity. The *Colour Harmony Manual*, which is a more recent form of the Ostwald atlas, is described in greater detail below. The British Colour Council has produced a Dictionary of Colour Standards (1934), which is noteworthy for several reasons. The dictionary includes 220 patterns on both smooth and ribbed silk, and each colour is named according to the generally accepted nomenclature among colour users. The historical origin and meaning of each name is given, e.g. beige, Cambridge blue, etc., while each colour has been specified in terms of its C.I.E. co-ordinates from measurements made with a Guild colorimeter, and in terms of the Lovibond red, yellow and blue scales derived from measurements with the Tintometer. The dictionary has been adopted by the British Standards Institution as a method of specifying a colour, and the colours have also been reproduced in wool yarn for use with rougher types of material.

The Munsell Color Company has been perhaps the most active in this field and in addition to the publication of the *Munsell Book of Color*, has sponsored much fundamental research on colour spacing. This is described below. One of the most recent atlases and certainly the most extensive, is the Villalobos colour atlas developed in Argentina. It contains over 7000 samples produced by successive half-tone printing, and very uniform spacing of the colours has been achieved. Each colour patch is 1 cm square in area and has a hole 4 mm in diameter cut out to assist colour matching. This aperture is rather small for accurate comparison, but the atlas is otherwise a very successful publication.

Another recent production is the DIN-Farbenkarte which is the official system of the German Standards Association.* It is based on a colour solid with uniform spacing in the three co-ordinates T, S and D. There are 24 hue planes T in each of which the samples are designed to have the same dominant wavelength rather than exact identity of hue. The saturation scale S is based on subjective estimates of saturation, giving, of course, a non-linear relation between S and the C.I.E. purity. The third co-ordinate D is described as the darkness degree and is calculated from a formula involving the luminance factor of the surface relative to the maximum luminance factor which a surface of the same chromaticity can possess. Initially the system was represented by charts of transparent gelatine samples. More recently an atlas of matt surface colours has been produced, the samples of each hue being mounted in a folder in which much quantitative information about the samples is given, including the equivalent C.I.E., Munsell and Ostwald specifications.

The variables in the Ostwald system are rather different. As shown in Fig. 7.1 for a given hue, the white, black, and full-colour pigments are located at the three corners of the equilateral triangle, while the colours within the triangle are regarded as an additive mixture of white, black and colour, with the sum of the white content W, black content B and colour content C equal to unity, that is

$$W+B+C=1.$$

This corresponds to colour mixture with a rotating disk of three segments of white, black and colour.

Colours located along lines parallel to WC are colours of constant

* M. Richter: *J. Opt. Soc. Am.*, 1955, **45**, p. 223.

black content and are known as *isotones*. The proportion of the white content along these lines has been adjusted to give a uniform spacing of the colours. Colours along lines parallel to *BC* are colours of constant white content and are known as *isotints*. Colours lying on vertical lines parallel to *WB* are colours of equal full-colour content and are known as *isochromes*. Ideally, they also have the same chromaticity and therefore form a shadow series. It should be noted that *B* must actually correspond to a dark grey rather than a true black, for otherwise the line *BC* would become an isochrome instead of an isotint. The colour solid is produced by rotating the triangle about the *WB* axis, and then colours which have constant black, white and colour content but differing hue, are known as *isovalent* colours.

The co-ordinates of the Ostwald system are based in principle on additive colour mixture and can be defined colorimetrically, but their interest and value depend very largely on their close correlation with the subjective assessment of a surface colour in terms of its whiteness, blackness and colourfulness. We continually encounter colour gradations in nature corresponding quite closely to either isotints or isotones, while the different levels of light and shade falling on a coloured surface form a shadow, or isochrome, series. These variables therefore have meaning through association with everyday visual experience, and since any given colour can be associated in four different meaningful ways (through the isotint, isotone, isochrome and isovalent series) with other colours in the solid, the system is of particular value in the aesthetic use of colour.

The practical realization of the system in the form of a colour atlas must inevitably depart to a greater or less extent from the theoretical prescription of the colours. This must be anticipated if for no other reason than that the theory is concerned with additive mixture, whereas the atlas has to be produced from pigment mixtures. There may also be other reasons for deliberately departing from theory, e.g. the desire to include as wide a range of colours as possible.

In the *Color Harmony Manual** produced by the Container Corporation of America and based on the Ostwald system the colour chips consist of a transparent cellulose-acetate base to which

* W. C. GRANVILLE and E. JACOBSON: *J. Opt. Soc. Am.*, 1944, **34**, p. 382. This reference includes the C.I.E. specifications of the colour chips.

a pigment layer has been applied. One side of the chip is therefore glossy and the other matt, which is itself a great advantage. Charts of a given hue number have been designed as far as possible to contain chips of the same dominant wavelength, which means that subjectively the chips may reveal some slight variation of hue. The chips in vertical rows have also, as far as possible, been given the same chromaticity. Initially, the manual contained charts for twenty-four different hues numbered 1 to 24, but a further six intermediate hues have been added to improve the spacing. Each hue chart normally contains twenty-eight chips, but with certain supplementary chips, their total number in the most recent edition of the atlas exceeds 900.

The atlas thus has very great merits, but it does suffer from two disadvantages. First, the lettering system used to locate a colour in a hue chart, as shown in Fig. 7.1, is difficult to remember and does not immediately call to mind the kind of colour being specified. It also does not lend itself readily to interpolation between patterns. Secondly, since the gamut of colours in any given hue is determined by the pigment used to represent the full colour, and since the distribution of colours is by definition bounded by the triangle W B C, there is no room in the system for any brighter pigments that become available through the discoveries of the research chemist. If account were to be taken of such a new pigment, then the new pigment would have to be placed at C and the whole distribution of colours in that hue plane would have to be changed. Admittedly, a situation of this kind is unlikely to arise very often, but the system does possess a somewhat unfortunate rigidity.

The system evolved by A. H. Munsell* was based on the conception of a colour solid similar to that illustrated in Fig. 2.8, but with the co-ordinates shown in Fig. 7.2. The central vertical axis represents the locus of neutral colours, with black at the bottom and white at the top. The lightness of the sample is called the *value*, and the distance of a sample in the solid above the black level is a measure of its value. The value scale, as shown in Fig. 7.2 is divided into eleven subjectively equal steps, with a perfect black as 0 and a perfect white as 10. In practice, the levels from 1 to 9 are those normally employed in the atlas. The distance of the sample

* D. Nickerson: *J. Opt. Soc. Am.*, 1940, **30**, p. 575.

from the central vertical axis is governed by the saturation or intensity of the colour, a quality which is designated as *chroma* on the Munsell system. The chroma is also divided into a series of steps, numbered from the neutral axis outwards, grey having zero chroma.

Differences of hue in the Munsell solid are represented by the different planes around the vertical axis, the whole solid being conveniently divided into ten equal vertical segments. The five principal hues, red R, yellow Y, green G, blue B and purple P, occupy the central planes of alternate segments, while the intermediate hues YR, GY, BG, PB and RP are situated in the mid-planes of the five remaining segments. This arrangement of the hues is shown in the horizontal section of the solid in Fig. 7.2. To provide a closer grading, each segment is divided into ten sections numbered 1 to 10, and arranged so that the main hue in each seg-

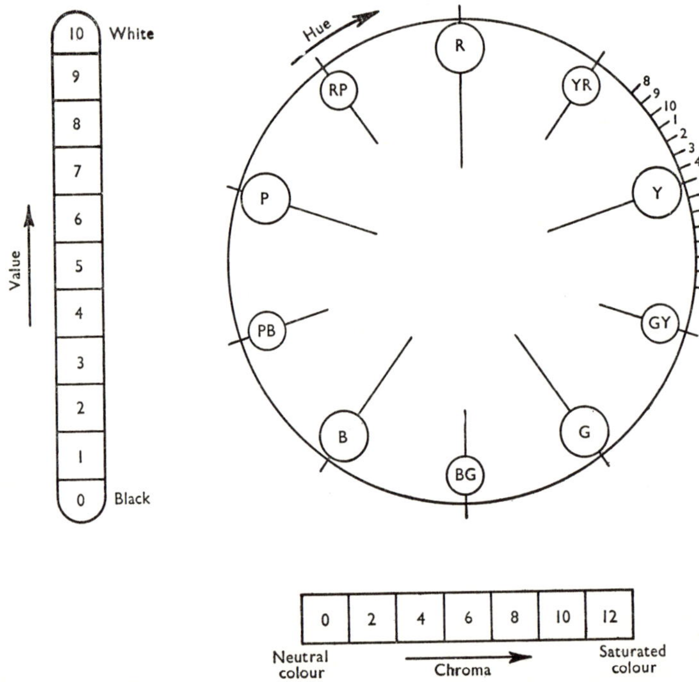

FIG. 7.2. Hue, value, and chroma co-ordinates of the Munsell system

ment is numbered 5, as shown in Fig. 7.2 for the yellow segment. The hue of a sample is therefore designated by a number to indicate the section of the segment in which it occurs, followed by symbols to show the colour of the segment concerned; thus 10P indicates the hue of section 10 of the purple segment.

The complete specification of a sample is expressed by writing the symbols for the hue and value, followed by a stroke and then the chroma. For example, the colour defined as 10P 5/8 has the hue 10P, the value 5 and the chroma 8. We thus have in the system a simple framework within which to arrange the patterns in the atlas and in terms of which a colour may be specified.

The complete atlas comprises some 960 patterns* in 40 hues arranged in a series of charts of constant hue and constant value corresponding to vertical and horizontal sections of the colour solid respectively. Fig. 7.3 represents a vertical section through the

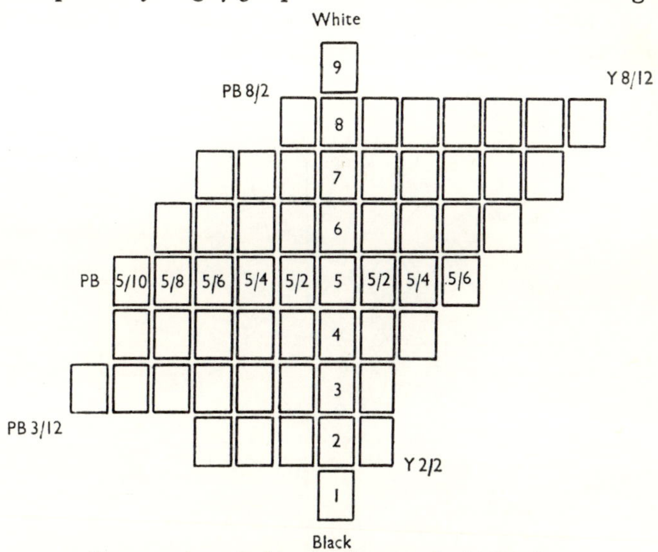

FIG. 7.3. A vertical section through the *Y-PB* plane of the Munsell solid

* Their C.I.E. specifications have been given by:
W. C. GRANVILLE, D. NICKERSON and C. E. FOSS: *J. Opt. Soc. Am.*, 1943, **33**, p. 376.
K. S. GIBSON and D. NICKERSON: *J. Opt. Soc. Am.*, 1940, **30**, p. 591.
J. J. GLENN and J. T. KILLIAN: *J. Opt. Soc. Am.*, 1940, **30**, p. 609.
K. L. KELLY, K. S. GIBSON and D. NICKERSON: *J. Opt. Soc. Am.*, 1943, **33**, p. 455.

Y-PB plane in which the central column of patterns numbered 1 to 9 represents the neutral colours ranging from near black to near white. Patterns to the right of this column are in the Y series, and it will be noticed that the pattern with the highest chroma rating, Y 8/12, is also very light. This is in accordance with the discussion in Chapter 4, in which it was remarked that yellow surfaces with both high saturation and high luminance factor could be produced. On the other hand, to the left of the central column containing the PB series of patterns, the sample at the greatest distance from the centre and therefore with the highest saturation or chroma, is PB 3/12, with a value of only 3.

Fig. 7.4 shows a horizontal section of the solid representing patterns of constant lightness, actually of value 5. At this level, the red samples are the most saturated and the blues, blue-greens and yellow the least.

In the ideal colour-solid the colours should be so spaced that equal distances in any direction and at any position in the solid

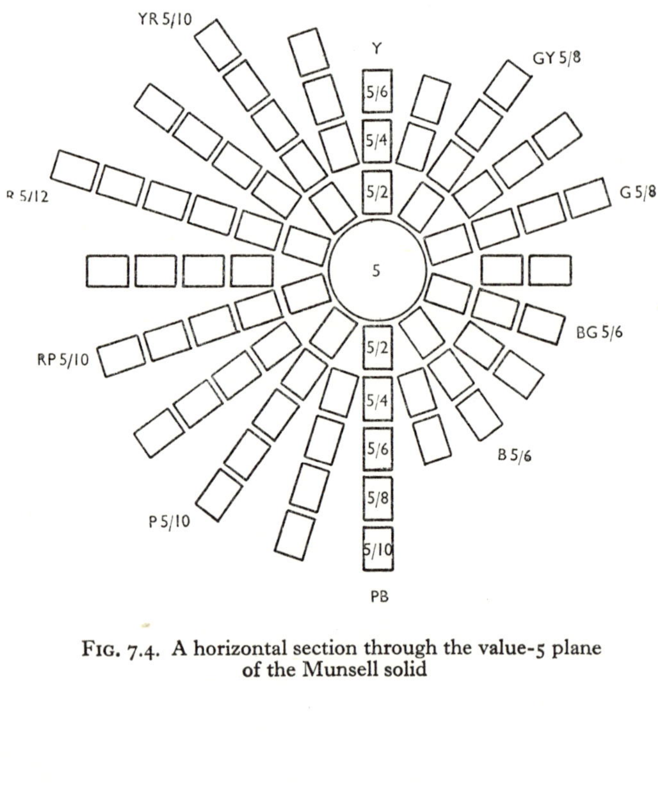

FIG. 7.4. A horizontal section through the value-5 plane of the Munsell solid

should correspond to equal sense differences. This goal has not been reached in the Munsell solid, and it is probably unattainable in any solid except to a very rough approximation; in the Munsell system, 1 unit in value is roughly equivalent to 2 units of chroma and 3 units of hue for samples of chroma 5.* Major alterations in the Munsell scales to bring the three units into closer correspondence are not contemplated at the present time, since this would disturb the notation which is now well established and effective (Newhall). Instead, investigations have been carried out to secure equality in the spacing of the samples in each of the three attributes of hue, value and chroma independently.

As a preliminary, Munsell samples were measured spectrophotometrically and their tristimulus specifications on the C.I.E. system computed. Similar samples were then examined visually and the spacing between neighbouring patterns estimated subjectively in terms of a standard spacing. Thus, the difference between two samples, C_1 and C_2, which differed slightly in one of the three attributes, say chroma, was adopted as a reference standard, and the chroma difference between two other samples, C_3 and C_4, was estimated as a fraction or multiple of the C_1–C_2 difference. An alternative, although less reliable, method was to attempt to give a numerical estimate of, say, the lightness of a pattern C_1 relative to a standard pattern C_2 as it appeared subjectively to the observer.

The difficulties in both recording and interpreting the observations were very considerable. There was, for example, the necessity of securing an immense number of observations from a large and representative group of observers. There were problems of averaging and weighting the results. Tests had to be made on the effect which the background had on the judgement of the differences. Most formidable of all was the question as to what numerical significance could be attached to estimations of sense differences of this kind. In the interim report by Newhall* on the work carried out by a subcommittee of the Optical Society of America, observations were recorded for forty-one observers, from which the average estimates of the hue, value and chroma for the 400 samples were evaluated and tabulated for white, grey and black back-

* S. M. Newhall: *J. Opt. Soc. Am.*, 1940, **30**, p. 617.

grounds. These estimates could be compared with their nominal description according to the existing Munsell classification.

The next stage was to plot the samples in the C.I.E. chromaticity diagram; smooth curves were then drawn through the points for a given chroma to provide a series of iso-chroma contours and through the points for a given hue to give loci of constant hue. Errors in the spacing of the samples have been revealed by the scatter of the points relative to the smooth curve, and by the discrepancies between the visual estimates and the nominal specification as described above. The determination of the correct shape of any given iso-chroma contour was further assisted by comparison with neighbouring chroma contours of the same value and with the contours for the same chroma at different values.

The final report of the subcommittee* led to the definition of smooth contours of Munsell hue, value and chroma in terms of the C.I.E. co-ordinates. From these contours was derived the specification in Munsell co-ordinates which each pattern should possess to conform more closely to the regular contours of the ideal colour solid, and as new series of patterns were prepared, the colours have been adjusted to correspond more closely to the ideal smooth contours. In this re-notation, samples having the same Munsell value have approximately the same *Y* tristimulus value when measured under illuminant *C*. A new Munsell atlas of high-gloss samples based on the re-notation system has also been produced,† while a very impressive atlas, also based on the Munsell re-notation, has been prepared by the Japan Color Research Institute in a limited edition of 100 copies. This Institute is also using the Munsell Renotation System in sets of colour standards which it is issuing for medical applications. In a recent very elegant Skin Color Standard, several hundred chips, mounted both in book form and in a series of fans, are available for the medical identification of normal and abnormal skin colours. It is expected that the atlas will also have applications in the cosmetic and colour television fields.

The importance of this whole investigation lies not only in its application to the Munsell system, but in the evaluation of the re-

* S. M. Newhall, D. Nickerson and D. B. Judd: *J. Opt. Soc. Am.*, 1943, **33**, p. 385.

† H. R. Davidson, M. N. Godlove and H. Hemmendinger: *J. Opt. Soc. Am.*, 1957, **47**, p. 336.

lation between the visual attributes of a colour and its colorimetric specification. Three examples will serve to illustrate this wider interest. When the chromaticities of patterns having constant hue came to be plotted in the C.I.E. chromaticity chart, the loci were found to be curved with the exception of two of the hues, as illustrated in Fig. 7.5. Evidently only a very loose correlation exists between the colorimetric concept of dominant wavelength and the subjective attribute of hue, and if we are interested in the subjective appearance of a stimulus defined by its chromaticity co-

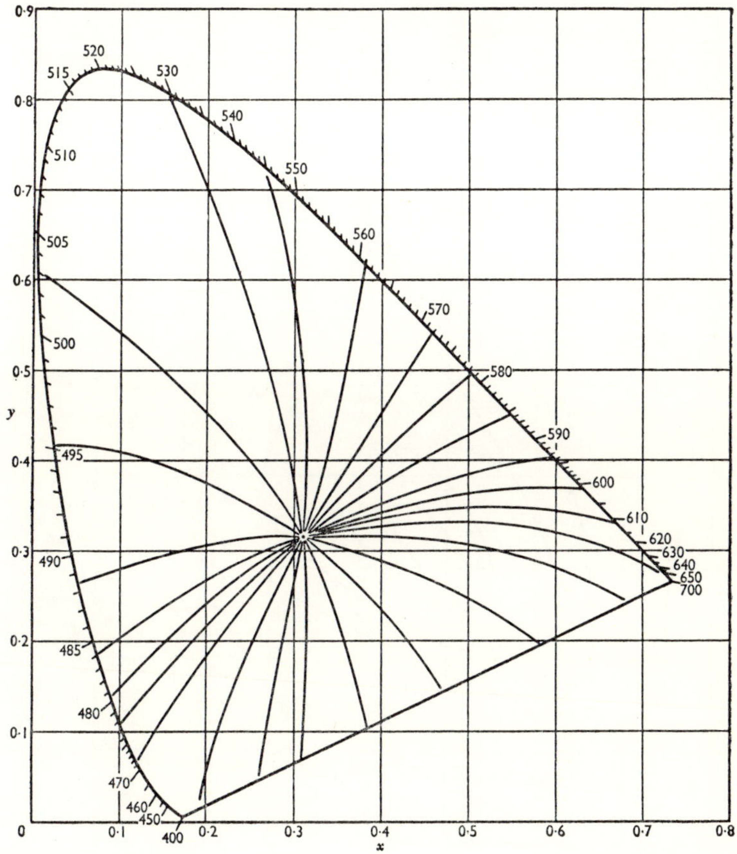

FIG. 7.5. A selection of the loci of surface colours of constant hue in the C.I.E. chromaticity chart, as determined by Newhall, Nickerson, and Judd

ordinates, account has clearly got to be taken of information such as that presented by Fig. 7.5.

Then again, the iso-chroma contours can be used to examine the uniformity of various projections of the C.I.E. chromaticity chart. Since loci of constant chroma are supposed to represent colours having a constant subjective difference from white, they should plot as circles in the ideal uniform chromaticity chart. The loci shown in Fig. 7.6 demonstrate once again the marked departure of the C.I.E. chart from uniformity and indicate the kind of oblique

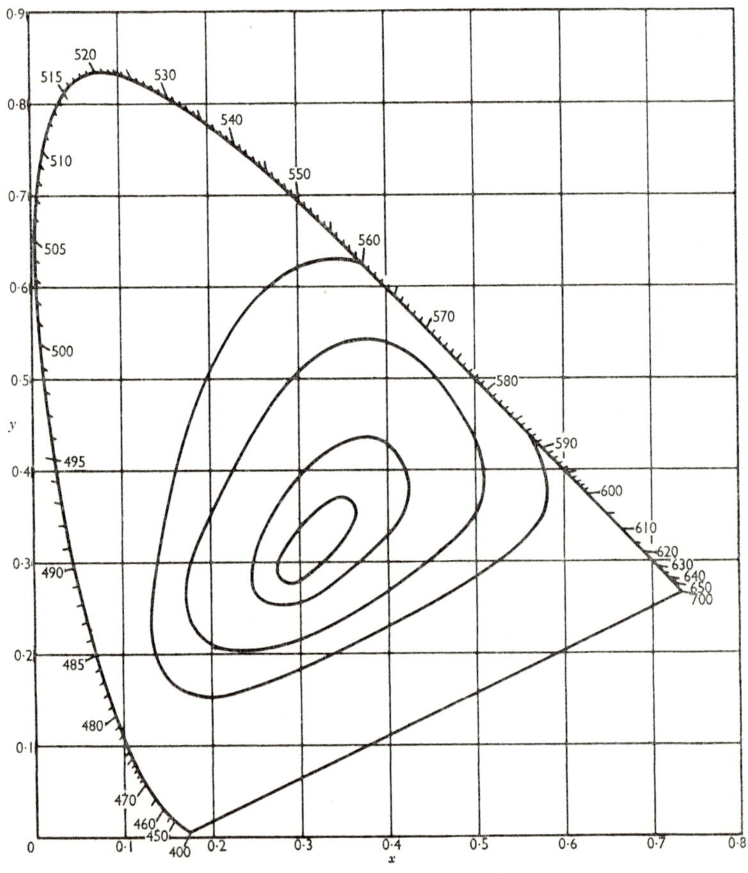

FIG. 7.6. A selection of the loci of surface colours of constant chroma in the C.I.E. chromaticity chart, as determined by Newhall, Nickerson, and Judd

projection necessary to produce a chart in which the loci would be more nearly circular. Incidentally, a uniform chromaticity chart based on discrimination data could, if desired, be used in reverse to study the spacing of the Munsell chips, instead of using the loci to confirm the uniformity of the chart.

A third point of general interest is the relation between the luminance factor (or the tristimulus value Y) and the corresponding Munsell values V for the neutral grey samples in the Munsell atlas as shown in Fig. 7.7. On the assumption that equal increments on the Munsell value scale correspond to equal intervals of lightness, this curve brings out very clearly the non-linear relation between lightness and luminance factor.

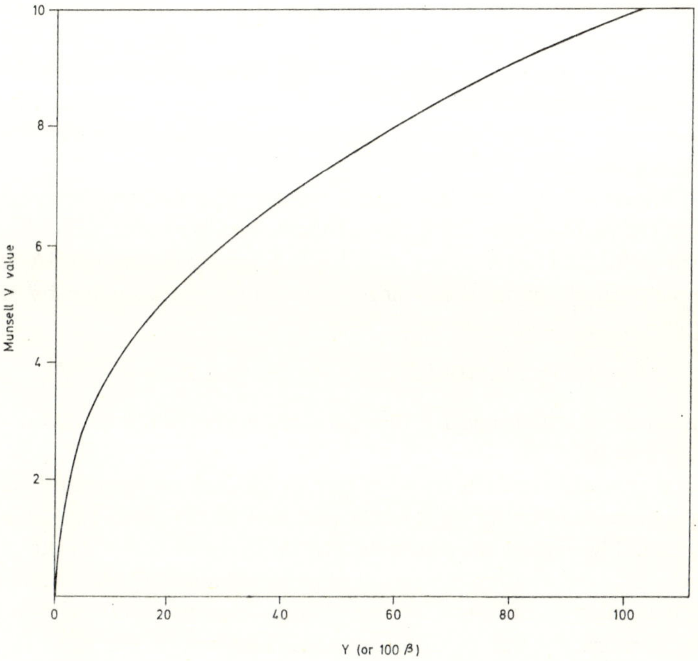

FIG. 7.7. Relation between the subjective assessment of the lightness of a surface and its luminance factor β, as shown by the relation between the Munsell value V of the surface and its tristimulus value Y ($=100\beta$) for the Munsell Re-notation System

Reference was made in the previous chapter to the chromatic-value diagram developed by Adams and based on the Munsell system.* He used the relation illustrated in Fig. 7.7 and extended it to the other tristimulus values X and Z. Thus if X_C, Y_C and Z_C are the tristimulus values of a sample under iluminant C, Adams derived three Munsell value functions V_X, V_Y and V_Z based on the relation between V and Y, with X_C, Y_C and Z_C used in turn in place of Y. The chromatic-value diagram is then obtained by plotting $(V_X - V_Y)$ as abscissa against $0{\cdot}4(V_Z - V_Y)$ as ordinate.

Since the relation between V and Y is non-linear, the chromatic-value diagram is a non-linear transformation of the C.I.E. co-ordinate system. It was originally based on Adams' theory of colour vision, which has not found wide acceptance, but there is no doubt that the diagram gives a surprisingly uniform spacing of surface colours. This is illustrated, for example, by the close approach to equality of spacing of the Munsell hue and chroma loci and the approximation of the chroma loci to circles, when samples of a given Munsell value are plotted in the chromatic-value diagram.

This transformation has found considerable use among colorimetrists especially in the United States, and extensive tables have been published for V_X, V_Y and V_Z.† A colour difference ΔE can be calculated using these chromatic value co-ordinates by the equation

$$\Delta E = \{(0{\cdot}23\,\Delta V_Y)^2 + [\Delta(V_X - V_Y)]^2 + [0{\cdot}4\Delta(V_Z - V_Y)]^2\}^{\frac{1}{2}}$$

One unit of difference by this formula corresponds to about 50 N.B.S. units.

An investigation which promises to lead to an important new type of atlas is being vigorously pursued at the Swedish Colour Centre. The system on which the atlas is being developed is known as the Natural Colour System of Hering-Johansson‡ and is somewhat similar in principle to the Colour Manual of Hesselgren to which reference was made on p. 180. The aim of the work is to produce an atlas that is based on an exclusively perceptive colour

* E. Q. Adams: *J. Opt. Soc. Am.*, 1942, 32, p. 168.

† D. B. Judd and G. Wyszecki: *Color in Business, Science and Industry* (Wiley: New York and London), 2nd Ed., 1963.

‡ A. Hard: *Internationale Farbtagung Luzern 1965*, vol. 1, pp. 357 and 367. (Musterschmidt-Verlag: Göttingen) 1966.

system. It is considered that two colours can differ in respect of five attributes, namely, hue, lightness, colour strength, cleanness and saturation. The reference points which define the structure of the colour solid are the six unitary colours white, black, yellow, red, blue and green. White is defined as a colour perception which is neither blackish nor chromatic; black as a colour perception which is neither whitish nor chromatic; yellow as a colour perception which is neither greenish nor reddish nor whitish nor blackish; and so on.

As with other colour solids, the neutral colours are located in the central vertical axis with black at the bottom and white at the top. Planes of constant hue then radiate round the vertical axis, with the four unique hues red, yellow, green and blue located 90° apart round the axis, with red and green, yellow and blue diametrically opposite each other. This means that opposing hues are not necessarily complementary to each other, neither will equal distances round the hue circle necessarily correspond to equal intervals of hue. Indeed, equality of spacing is not one of the criteria employed in the arrangement of the colours in the Natural Colour System. Instead, the scaling is based on the concept that it is possible to make a quantitative statement about the amounts of the unitary colours in any given colour. Thus the lightness scale is conceived as a linear scale between black and white so as to give the proportion (or percentage) of white at each point along the scale.

If this project is pressed to its conclusion and a new atlas is put on the market, it is likely to have a strong appeal to many users. It is probable that for certain applications the Natural Colour System will be particularly suitable and we must hope that its production will satisfy the other criteria to which reference was made earlier, such as the smoothness of the spacing, the accurate duplication of the patterns from one copy of the atlas to the next, and reasonable permanence of the samples.

CHAPTER 8

COLOUR MIXTURE DATA APPLIED TO THREE-COLOUR REPRODUCTION

THE PRINCIPLES OF THREE-COLOUR REPRODUCTION

A field in which colorimetry has been very vigorously applied in recent years is that of colour television. A massive literature has accumulated* in which the theory and practice of three-colour television has been elaborated, and colorimetric algebra exploited in a very remarkable way. Colour scientists can, indeed, take some pride in the fact that both the principles and the data were at hand when the television engineer found he needed them.

The elementary operation of a system of three-colour reproduction, either by colour photography or by colour television, may be studied by reference to the skeleton diagram in Fig. 8.1. Suppose *C*

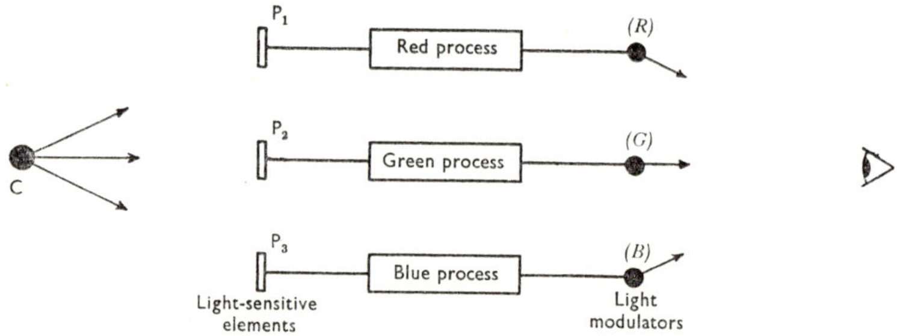

FIG. 8.1. Skeleton diagram of a three-colour system of colour reproduction

is some point in the scene which is being photographed or televised, and suppose P_1, P_2 and P_3 are three light-sensitive devices, either photographic emulsions or photoelectric cells, which are distinguished from one another by their spectral sensitivities. We will assume that P_1 has its maximum sensitivity in the red part of

* The basic series of papers appeared in two special colour television issues of the *Proc. I.R.E.*: 1951, Oct., **39**, and 1954, Jan., **42**. See P. S. CARNT and G. B. TOWNSEND: *Colour Television* (Iliffe: London), 1961 for a more recent account.

the spectrum, P_2 in the green and P_3 in the blue. In a first consideration of the system we may then consider that P_1 records the amount of red light radiated from C, P_2 the amount of green and P_3 the amount of blue. The activity in the P_1 system has then to be transferred to the red modulator (R) at the reproducing end of the process, P_2 has to be connected to the green modulator (G), and P_3 to the blue (B). Thus (R), (G) and (B) represent three light controls which radiate the appropriate amounts of red, green and blue light required to match and hence reproduce the colour C being transmitted. In photography, the light is recorded at P_1, P_2 and P_3 by the exposure of three emulsions, and the linkage with (R), (G) and (B) takes the form of the chemical processing of the emulsion and the printing of a positive. In the additive process, (R), (G) and (B) represent the densities of red, green and blue elements in the mosaic of the print, while in the subtractive process they represent the densities of the cyan, magenta and yellow layers in the positive. In television, the light is recorded at P_1, P_2 and P_3 by the liberation of electrons from the light-sensitive surface of the electron camera exposed through red, green and blue filters, the linkage with (R), (G) and (B) corresponds to the amplification, transmission and reception of the electrical signals, while (R), (G) and (B) themselves represent red, green and blue light-modulating devices whose output is determined by the incoming signals.

We shall assume for the moment that there is a strictly linear relation between the response in P_1, P_2 and P_3 and the output of the three light controls (R), (G) and (B), although this is an ideal condition which can never be satisfied in practice. The colour as seen by the observer is determined by the mixture of the light from (R), (G) and (B) which is focused on his retina, but as C is only one point in the complete scene, the analysis and synthesis represented by Fig. 8.1 have to be repeated for every picture element transmitted. If C is predominantly red, the P_1 process will be most active; if purple, P_1 and P_3 will be stimulated in excess of P_2; if white, all three processes will be more or less equally active, and so on.

The aim of a process of colour reproduction is, in a first analysis, to secure a mixture of (R), (G) and (B) which will give as exact a match of the original colour C as possible. In practice, considerable departures from this criterion may be desirable to take account of the viewing conditions under which the picture is seen, but for the moment we will ignore this aspect. The colorimetric problem is then to choose the particular red, green and blue colours for (*R*), (*G*) and (*B*), and to derive the spectral sensitivities of P_1, P_2 and P_3 so as to secure a faithful reproduction of *C*. It must, however, be emphasized that it is not the spectral composition of *C* that is reproduced, but the red, green, blue mixture that matches *C*. To that extent a colour reproduction system acts like a vast number of trichromatic colorimeters in which colour matches of all the elements of a scene have to be made.

The selection of (*R*), (*G*) and (*B*) will in part be a practical matter determined by the dyes, filters, phosphors or other light controls which are available. From the theoretical point of view, (*R*), (*G*) and (*B*) should be located on the chromaticity chart so that the triangle (*R*)(*G*)(*B*) (see Fig. 8.2) covers as large an area as possible. Since only positive mixtures of (*R*), (*G*) and (*B*) will exist in the reproduction, colours lying outside the triangle (*R*)(*G*)(*B*) cannot be correctly reproduced. It is therefore desirable that the line (*R*)(*G*) should lie as close to the spectrum locus as possible to ensure the correct reproduction of saturated reds, yellows and greens, but as saturated blue-greens are uncommon, a considerable separation between the line (*G*)(*B*) and the spectrum locus is of less consequence. A factor which may affect the choice of (*R*), (*G*) and (*B*), apart from their colour quality, is their intensity. If dyes or filters are chosen which give highly saturated colours, their light transmission will be low and the brightness of the picture will suffer.

Once (*R*), (*G*) and (*B*) have been selected, we can calculate the spectral sensitivity curves which P_1, P_2 and P_3 should have so as to reproduce correctly the chromaticities of all colours lying inside the triangle (*R*)(*G*)(*B*).* Suppose for the moment that P_1, P_2 and P_3 have sensitivity curves similar to the $\bar{x}_\lambda$, $\bar{y}_\lambda$ and $\bar{z}_\lambda$ distribution curves of the C.I.E. system and that (*R*), (*G*) and (*B*) are in fact the reference stimuli (*X*), (*Y*) and (*Z*). Then the reactions stimulated in P_1, P_2 and P_3 by any colour *C* could be calculated by multiplying the sensitivity curves of P_1, P_2 and P_3 by the energy curve of the light radiated from *C*, and integrating. But since this process is the same as multiplying the distribution curves by the energy

* A. C. Hardy and F. L. Wurzburg: *J. Opt. Soc. Am.*, 1937, **27**, p. 227.

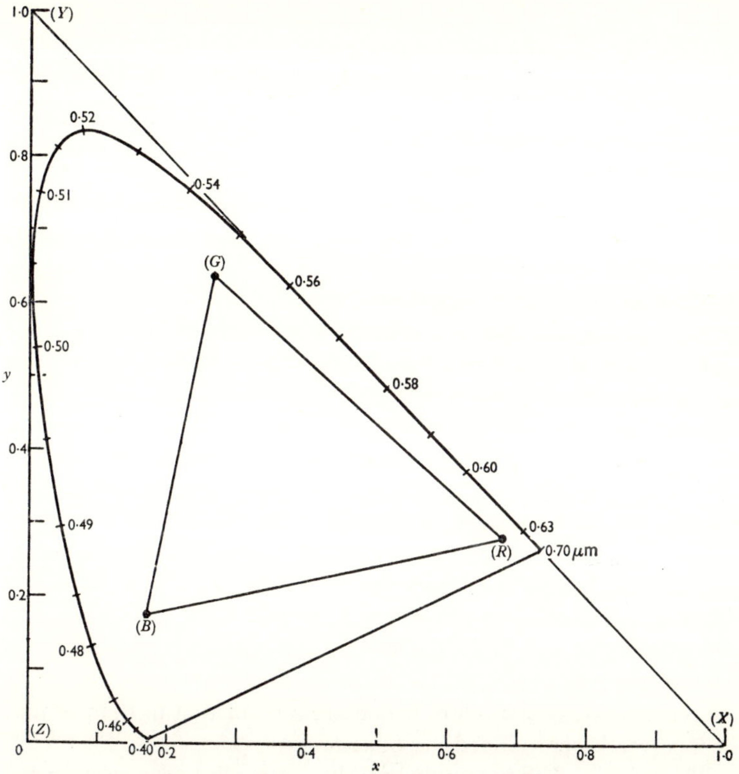

FIG. 8.2 Possible location in the chromaticity chart of the red, green, and blue stimuli of a colour reproduction system

curve, it is equivalent to the method discussed on p. 132 and illustrated in Fig. 4.14, by which the amounts of (*X*), (*Y*) and (*Z*) required to match or reproduce *C* can be calculated. Hence if the reactions in P_1, P_2 and P_3 cause a proportional radiation of (*X*), (*Y*) and (*Z*) in the reproduction, the colour *C* will be correctly reproduced for all chromaticities of *C*.

In practice, (*X*), (*Y*) and (*Z*) cannot be used because they are not real, physical, stimuli, but the same principle applies for other stimuli (*R*), (*G*) and (*B*). In that case, the sensitivity curves of P_1, P_2 and P_3 should correspond to the distribution curves of the standard observer, but transformed by the processes described in Chapter 3, so as to refer to (*R*), (*G*) and (*B*) instead of to (*X*), (*Y*) and (*Z*). It is thus evident that the laws and data of the trichromatic system enter into the heart of the theory of three-colour reproduction, as first shown by Hardy and Wurzburg, with many later studies.*

When, however, the C.I.E. distribution curves are transformed to refer to real stimuli such as (*R*), (*G*) and (*B*), we encounter one of the frequent departures from ideal colorimetric theory in an actual process of colour reproduction. For the distribution curves expressed in terms of real stimuli include areas which are negative in sign, as shown in Fig. 4.6, and this implies that the processes represented by P_1, P_2 and P_3 should respond in the reverse direction for certain wavelengths in the spectrum. If this requirement is ignored and all-positive sensitivity curves are used, the reproduction will tend to be desaturated, since (*R*), (*G*) and (*B*) will be less saturated than the theoretically correct stimuli associated with all-positive distribution curves. Distortions of hue of one kind or another will also occur if the sensitivity curves are of the wrong shape, or if their maxima are in the wrong part of the spectrum. Again, if the wrong sensitivity curves are used, pairs of colours of different spectral composition which match to the eye (metameric pairs) will not match to the colour cameras and no amount of subsequent tinkering with the system will correct their mismatch. Conversely, pairs of colours which are a mismatch to the eye may be incorrectly reproduced as a match.

Then again, there is unlikely to be a strictly linear relation between the light input at P_1, P_2 and P_3 and the light output of (*R*), (*G*) and (*B*). The characteristic curves of the three channels may not even be identical, but assuming that they are, if the slope or gamma of the characteristic curve obtained by plotting the

* A. C. Hardy and F. L. Wurzburg: *J. Opt. Soc. Am.*, 1937, **27**, p. 227.
D. L. MacAdam: *J. Opt. Soc. Am.*, 1938, **28**, p. 399, and 1938, **28**, p. 466.
J. A. C. Yule: *J. Opt. Soc. Am.*, 1938, **28**, p. 419, and 1938, **28**, p. 481. *Principles of Color Reproduction* (Wiley: New York and London), 1967.
H. D. Murray and D. A. Spencer: *Phot. J.*, 1938, **78**, p. 474.
W. D. Wright: *Phot. J.*, 1940, **80**, p. 25.
W. T. Wintringham: *Proc. I.R.E.*, 1951, **39**, p. 1135.
W. T. Hanson and W. L. Brewer: *J. Opt. Soc. Am.*, 1955, **45**, p. 476.
W. R. J. Brown: *J. Opt. Soc. Am.*, 1955, **45**, p. 539.
R. M. Evans, W. T. Hanson and W. L. Brewer: *Principles of Color Photography* (Wiley: New York. Chapman and Hall: London), 1953.
R. W. G. Hunt: *The Reproduction of Colour* (Fountain Press: London), 2nd Ed., 1967.

logarithm of the light output against the logarithm of the light input is greater than unity, the colours in the reproduction will be more saturated than in the original. For example, if a red colour were being reproduced, the signal initiated in P_1 would be greater than the signals in P_2 and P_3. By processing with a gamma greater than unity, the preponderance of the red signal would be accentuated and hence the redness, or saturation, would be increased. Conversely, a gamma less than unity will lead to desaturation. Minor changes of hue may also, of course, accompany these changes of saturation.

SUBTRACTIVE COLOUR PHOTOGRAPHY AND PRINTING

In subtractive processes of colour reproduction, the modulation of (*R*), (*G*) and (*B*) operates through absorption in cyan, magenta and yellow dyes or pigments. To understand this, we must first consider the general process of subtractive colour mixture as illustrated in Plate 9, p. 103, for a beam of light passing successively through yellow, magenta and cyan filters. An incident beam of white light, which we will assume initially possesses an equal-energy distribution through the spectrum, is represented by the white arrow on the left of Plate 9, while the energy distribution is shown in curve (*a*) on the right. The light is now assumed to pass through a yellow filter *Y*, which has the transmission curve (*b*); the beam will emerge coloured yellow and will now possess the energy distribution (*c*), as derived by multiplying (*a*) and (*b*) together. The shaded area above (*c*) represents the energy absorbed by the yellow filter and is seen to be predominantly from the blue end of the spectrum. The yellow beam as represented by the yellow arrow is now incident on a magenta filter M, whose transmission curve is shown by (*d*). The energy distribution of the light now transmitted is shown by (*e*), obtained as the product of (*c*) and (*d*), and it is evident that the light is now coloured red, as represented by the red arrow. The shaded area above (*e*) indicates the energy absorbed by the magenta filter, which in this example is largely in the green part of the spectrum. Finally, the light strikes the blue-green, or cyan, filter C, and since the function of the latter is to absorb the red radiations, the incident light loses its red quality to become approximately neutral in colour, as shown qualitatively by the grey arrow and quantitatively by the energy curve (*g*). The latter curve, as before, is the result of multiplying (*e*) by the transmission curve (*f*) of the cyan filter, while the shaded area above (*g*) represents the energy absorbed by the filter.

To control the colour of the emergent light we have to change the absorption in one or other of the three filters. In colour photography this means changing the densities of the dyes. If the density of any one of the dyes is increased or decreased, the shaded area in the corresponding energy curve is enlarged or diminished so that the contribution from that part of the spectrum to the emergent beam is correspondingly decreased or increased. Hence variation in the densities of the three dyes produces a variation in the amounts of the blue, green and red light in the transmitted beam. As we shall see below, if we wish to establish any simple relation between the densities of the dyes and the intensities of the red, green and blue components, we have to postulate highly artificial absorption characteristics for the three dyes. However, the merit of yellow, magenta and cyan as the colours for the dyes lies in the fact that with them the widest range of chromaticities can be produced. Thus red is obtained from the subtractive mixture of magenta and yellow, green from yellow and cyan, blue from cyan and magenta, black from all three in high density, and white when all three have very low densities. They are therefore known as subtractive primaries in pigment and dye mixture, in subtractive colour photography, in painting and in printing. The function of the yellow, magenta and cyan primaries and their relation to the three stimuli of additive trichromatic mixture is, however, most clearly brought out when they are referred to as the minus-blue, minus-green and minus-red dyes or pigments.

In an ideal process of subtractive colour reproduction, the yellow dye should control only the light in a block of blue wavelengths, the magenta a block of green wavelengths and the cyan a block of red wavelengths. However, as we have seen, the spectral absorption with real dyes and pigments is such that all three dyes affect to some extent all parts of the spectrum. The process does not, therefore, lend itself very readily to exact trichromatic analysis and we shall attempt here only a quite elementary treatment assuming simplified absorption curves. Yet such an analysis can provide a first understanding not only of the basic principles of the process but also of the corrections that have to be applied to allow for the unwanted and overlapping absorption that occurs in practice.

As shown by Hardy and Wurzburg* the ideal subtractive primaries should possess square-topped, interlocking spectral transmission or reflection curves such as those shown in Fig. 8.3. In the transmission bands, the dyes should be completely transparent, that is, $\tau' = 1{\cdot}0$, while in the absorption band the transmission τ will be given by

$$\tau = \alpha^{-cl}$$

where α is the absorption factor, c is the concentration and l the thickness of the dye, as discussed on p. 14.

Each dye will thus control the light flux F passing through its particular absorption band, F will be proportional to τ, and log F will be proportional to c. If the gamma of the process is unity, F will also be proportional to the exposure E in the red, green or blue process as the case may be. Variation of the concentrations of the three dyes will give completely independent control of the light flux $F_{(R)}$, $F_{(G)}$ and $F_{(B)}$ in the three absorption bands, the effective red stimulus (R) having the chromaticity of the light in the absorption band of the cyan dye, (G) the chromaticity of the green absorption band of the magenta dye, and (B) the chromaticity of the blue absorption band of the yellow dye. The system is then the complete equivalent of an additive process of colour mixture.

Suppose, now, that the absorption bands do not interlock exactly, but that the blue band is too broad and therefore overlaps the green as shown in Fig. 8.4. Evidently for weak concentrations of the magenta dye, the yellow dye will now control a band of wavelengths that is slightly greener than before, on account of its increased breadth towards the green wavelengths. On the other hand, for strong concentrations of the magenta, the yellow dye will control the same waveband as before. Similarly, for weak concentrations of the yellow, the magenta dye will control the same green band of wavelengths as before, whereas with strong yellow concentrations, the magenta will control a narrower, less blue, band of wavelengths. We can therefore conclude that any overlapping

* A. C. Hardy and F. L. Wurzburg: *J. Opt. Soc. Am.*, 1937, **27**, p. 227.

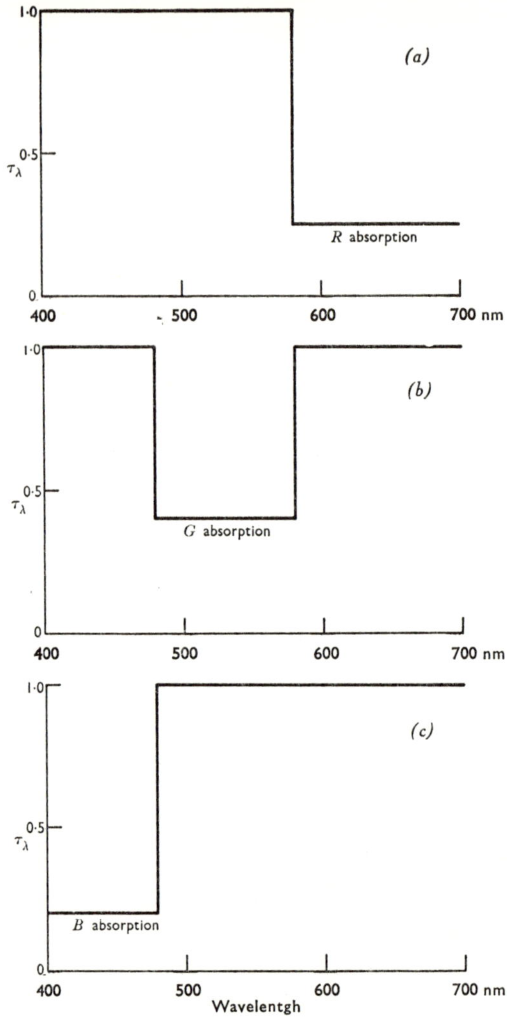

FIG. 8.3 Spectral transmission curves of ideal subtractive primaries. (*a*) Minus-red or cyan, (*b*) minus-green or magenta, (*c*) minus-blue or yellow

of the absorption bands will cause the chromaticities of (R), (G) and (B) to vary with the densities of the dyes, an effect which can be described by saying that the reproduction primaries are unstable.

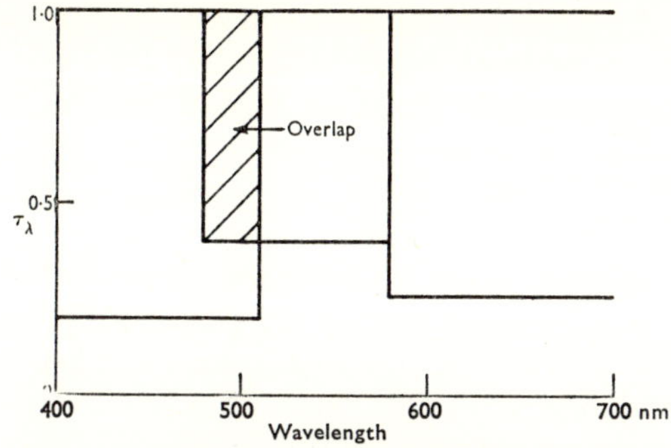

FIG. 8.4 Overlapping spectral transmission curves of hypothetical primaries

This will lead to some distortion of hue, while the additional absorption will also lead to some loss of light.

Next, suppose that the dyes have some unwanted absorption in the transmission bands, as illustrated in Fig. 8.5 for the case where

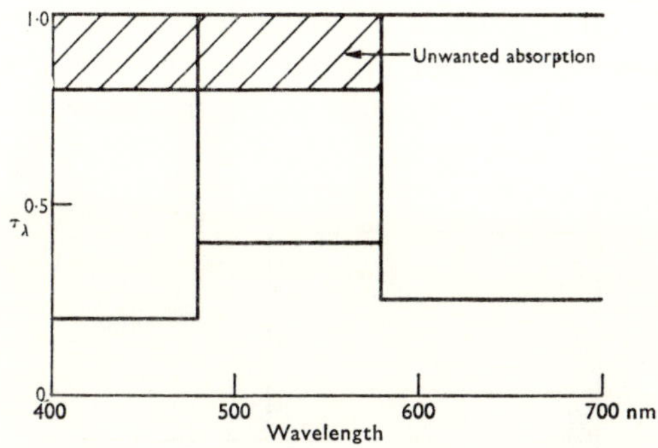

FIG. 8.5 Spectral transmission curves of hypothetical subtractive primaries with unwanted absorption in the cyan dye

the cyan dye partially absorbs in the green and blue bands. When the concentration of the cyan is very weak, the absorption will be negligible, so that all colours with a large red content, such as reds, yellows, purples, and white (which are the colours produced when the cyan is weak) will be correctly reproduced. But when the cyan concentration is high, i.e. when greens, blue-greens and blues have been photographed, there will be excess absorption in the green and blue. This will have the effect of making the saturated green, blue-green and blue colours duller than they should be, while the less saturated colours in these hues will suffer additional desaturation. In general, therefore, we can say that the effect of the unwanted absorption is to make the reproduction duller and less saturated, at least for some of the hues.

With real dyes, the absorption curves are neither square-topped nor straight-sided, the absorption and transmission bands do not interlock, and there is often a considerable amount of unwanted absorption. Reproduction can therefore never be perfect and one source of error has to be balanced against another to secure as good a compromise as possible. An exact analysis may be too complex to be worth attempting, but if a set of square-topped absorption curves giving as close a fit as possible to the actual curves are assumed, the types of error and their correction by masking or by control of the processing conditions can be studied. Successive approximations between theory and practice may then lead fairly rapidly to the best possible balancing of the various errors.

This is not the place for a detailed discussion of masking, but it may be helpful to include a simple account of the underlying theory. Suppose the cyan primary has the transmission curve shown in Fig. 8.6 for a particular concentration of the dye, with a

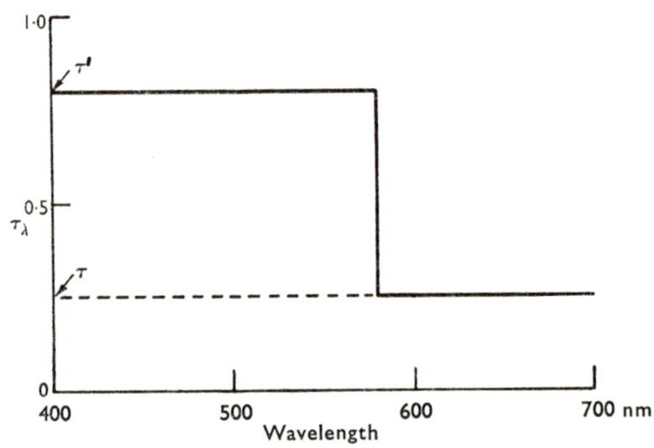

FIG. 8.6 Transmission curve of hypothetical cyan dye with unwanted absorption

transmission of τ in the main absorption band and τ', instead of the ideal 1·00, in the remainder of the spectrum. τ and τ' are related by Lambert's and Beer's laws as follows:

$$\tau = \alpha^{-cl}$$

and $$\tau' = \alpha'^{-cl}$$

where α and α' are the absorption factors for unit concentration and thickness for the two parts of the spectrum.

Since $$\log \tau = -cl \,.\, \log \alpha$$

and $$\log \tau' = -cl \,.\, \log \alpha'$$

$$\frac{\log \tau}{\log \tau'} = \frac{\log \alpha}{\log \alpha'}$$

giving $$\log \tau' = \frac{\log \alpha'}{\log \alpha} \,.\, \log \tau$$

Now $\log \alpha'/\log \alpha$ is a constant for a given medium, whatever units are chosen for c and l, and its value can be found by substituting known values of τ and τ' in $\log \tau'/\log \tau$.

To correct for the unwanted absorption, the transmission in the blue and green bands must be increased by $1/\tau'$. To do this we can make a positive transparency from the main 'red' negative and process it to have a gamma of $\log \alpha'/\log \alpha$. If the 'red' negative had a gamma of unity, the positive transparency or mask will have a density characteristic given by

$$D = \text{Constant} - \frac{\log \alpha'}{\log \alpha} \,.\, \log E$$

where E is the exposure of the original red negative and D is the density of the mask.

The mask is now applied to the main 'green' and 'blue' negatives, and positives are made from them under processing conditions giving a gamma of unity. The effect of the mask will then be to decrease the densities of both positives by D. D will be greater the smaller the value of $(\log \alpha'/\log \alpha) \log E$, that is the smaller the amount of red in the object being photographed. Since the density of the cyan dye will be high in such areas of the photograph, the mask will give the required type of correction for the unwanted absorption. Moreover, the transmission in the red band of the cyan dye is assumed to be directly proportional to the exposure of the red negative, i.e. τ is assumed proportional to E. [This is simply the equivalent of saying that the intensity of (R) in Fig. 8.1 is proportional to the activity in P_1.] Hence the absorption will be corrected by $(\log \alpha'/\log \alpha) \log \tau$, i.e. by $\log \tau'$. This means that the transmission will be increased by $1/\tau'$, which is exactly the correction necessary to compensate for the unwanted absorption. We must, however, note that this correction can operate only if there is some density in the magenta and yellow positives to be reduced. If, for example, we were photographing a saturated blue-green colour, there would be little or no density in these two layers and no correction would be possible.

This is a grossly simplified analysis, but it does illustrate the basic dependence of the gamma of the mask on the absorption characteristics of the offending dye, and it shows how the application of the mask will, in the ideal case, give perfect correction. In practice, of course, the absorption curve is not the simple kind of curve shown in Fig. 8.6, the other dyes will have unwanted absorptions, and the gammas of the various stages of processing will not have the theoretically correct values assigned to them. A whole sequence of compromises is therefore required, and theory must be combined with experiment to find the optimum solution.

Within the ambit of these compromises, the question of allowing for the negative areas of the ideal sensitivity curves of P_1, P_2 and P_3, already referred to in the previous section, has also to be considered. By a process somewhat analogous to the masking procedure just described, some degree of correction can be obtained, but it is unlikely to produce any very noticeable improvement in the colour rendition, relative to the other colour distortions present. Nevertheless, the direction of the error is towards desaturation and must be recognized as one of the factors contributing to the general degradation of the colours.

Because colour reproduction by photography depends basically on the laws of subtractive colour mixture, the C.I.E. system does not enter very directly into its specification or analysis. This is even more true in the case of colour printing and the graphic arts, where the spectral composition of the light entering the eye from each point of the print is controlled partly by subtractive mixture of the printing inks, partly by semi-additive effects of light scatter, and partly by the additive mixture of the fine mosaic of coloured dots of the printing process. The relative contributions of these factors will vary with the process; in gravure printing, for example, the subtractive effect is dominant, but in letterpress and lithography, additive mixture makes a very significant contribution to the colour mixing.

For the three-colour half-tone or letterpress process, Neugebauer* has shown that this can be treated as an eight-variable problem in additive colour mixture. The eight components consist of the coloured dots of the three printing inks alone, the areas of the dots where the inks overlap in pairs, the areas where all three inks overlap, and the white areas in the gaps between the dots. Eight simultaneous equations clearly do not lend themselves to easy solution and Neugebauer's analysis demonstrates that there can be no simple solution to what is inherently a many-variable problem. Certainly, as Yule emphasizes,† the C.I.E. system is not well adapted to the needs of colour printing theory, and more approximate and semi-empirical analyses in terms of ink densities are more meaningful to the colour printer. Yet the introduction of mass-production techniques for colour printing, in which photoelectric scanning equipment is used for continuous and automatic colour analysis and correction, implies that a rather detailed understanding of the colour mixture process is necessary. This includes a knowledge of the spectral composition of the light which is reflected from the print into the eye of the viewer, and since this light ultimately impinges on the retinal receptors as an additive mixture of the spectral elements of which it is composed, the final expression of the colour has to be in terms of the additive tristimulus system of the C.I.E. or something equivalent to it.

COLOUR TELEVISION

Several methods have been proposed for reconstituting the colour picture in a television receiver, but unlike photography they all involve the additive mixture of red, green and blue stimuli. In one method, a rotating disc made up of segments of red, green and blue filters is mounted in front of a white cathode-ray tube, and the picture on the tube is modulated by the incoming red, green and blue signals in synchronism with the passage of the corresponding colour segments in front of the cathode screen. In another, the virtual images of the pictures on three separate cathode-ray tubes

* H. E. I. NEUGEBAUER: *Z. wiss Photogr.*, 1937, **36**, p. 171
R. W. G. HUNT: *The Reproduction of Colour* (Fountain Press: London), 2nd Ed., 1967.

† J. A. C. YULE: *Principles of Color Reproduction* (Wiley: New York and London), 1967.

are seen superposed after reflection in dichroic mirrors. A third method uses three high-intensity cathode-ray tubes and wide-aperture lens or mirror systems to project three superimposed images on a screen. In yet another method, the cathode-ray screen consists of a mosaic of red, green and blue phosphor dots which are too fine to be resolved by the observer.

The relative merits of these different methods are of no concern to us here, nor the electrical means by which the scene is scanned, the signals transmitted and received, and the intensities of the three stimuli at the receiver modulated. For the purpose of our colorimetric analysis, we need only know that at the transmitter there are, in effect, three cameras—one red-sensitive, one green-sensitive and one blue-sensitive—and that the signals are used at the receiver to modulate red, green and blue lights which are additively combined to form the colour reproduction. This corresponds, of course, to the schematic arrangement represented in Fig. 8.1, but we must now consider the very interetisng way in which the signals in the three channels can be, and have been, interlinked at certain stages in the transmission between camera and receiver.

In the American N.T.S.C. system,* the principles of colorimetry have been applied in a most ingenious manner to the problem of colour television. Many television engineers have contributed to the system and Fig. 8.7 shows three stages in the

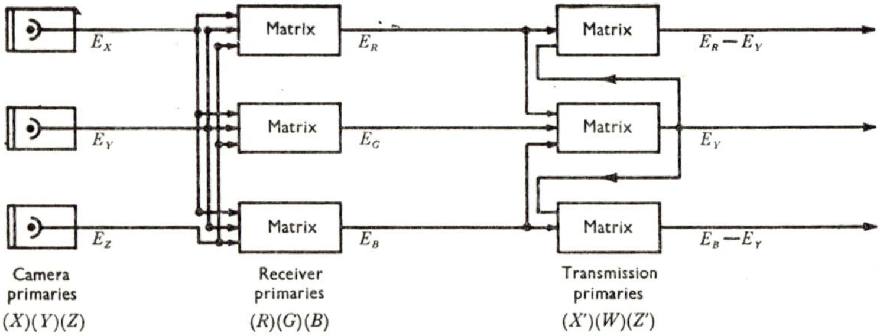

FIG. 8.7 Three stages in a colour-television linkage

* See Special Colour Television Issue, *Proc. I.R.E.*, 1954, **42**, No. 1.
D. G. FINK: *Color Television Standards: N.T.S.C.* (McGraw-Hill: London), 1955.
P. S. CARNT and G. B. TOWNSEND: *Colour Television: The N.T.S.C. System, Principles and Practice*, (Iliffe: London), 1961.

transmission as extracted from a diagram by Bingley.* The first stage represents the three cameras, which are here assumed to have spectral sensititivities corresponding to the $\bar{x}_\lambda$, $\bar{y}_\lambda$ and $\bar{z}_\lambda$ distribution curves. The signals which they initiate can therefore be given the symbols E_X, E_Y and E_Z, and the camera primaries, as we may call them, will correspond to (X), (Y) and (Z) of the C.I.E. system.

At the receiver the picture, as we have seen, is reproduced by the additive mixture of three red, green and blue lights, whose chromaticities will depend on the phosphors and filters used in the receiver. They can be symbolized by (R), (G) and (B). For exact chromaticity reproduction the camera sensitivities should correspond to the distribution curves $\bar{r}_\lambda$, $\bar{g}_\lambda$ and $\bar{b}_\lambda$, but since these curves are linear transformations of the $\bar{x}_\lambda$, $\bar{y}_\lambda$ and $\bar{z}_\lambda$ curves, correct reproduction can still be secured if the signals E_X, E_Y and E_Z can be transposed to refer to (R), (G) and (B). This is nothing more nor less than the transformation of a colour specification from the (XYZ) system to the (RGB) system, which by suitable matricing circuits, can be carried out electrically.

The transformation is represented by the second stage in Fig. 8.7, in which the signals E_X, E_Y and E_Z are shown feeding into three matrix boxes, with signals E_R, E_G and E_B emerging from them. The signals now refer to what we may call the receiver primaries (R), (G) and (B), and it will be realized that, by this procedure, the problem of negative camera sensitivities is neatly resolved.

In the third stage in Fig. 8.7 the colour signals are transformed into a luminance signal and two colour-difference signals. This has important advantages in the colour system itself and it also means that the luminance signal is available for use with monochrome receivers to give a black-and-white picture.

The luminance signal E_Y is compounded from E_R, E_G and E_B according to the equation

$$E_Y = V_{(R)} \cdot E_R + V_{(G)} \cdot E_G + V_{(B)} \cdot E_B$$

$V_{(R)}$, $V_{(G)}$ and $V_{(B)}$ are here the relative luminous efficiencies of (R), (G) and (B) for units based on some selected normalizing white such as S_C, and adjusted to give $V_{(R)} + V_{(G)} + V_{(B)} = 1{\cdot}0$. Thus E_Y is the appropriate signal for use in a monochrome receiver, while in

* F. J. Bingley: *Proc. I.R.E.*, 1954, **42**, p. 51.

the colour system it will represent the luminance of the picture. The colour information is given by two colour-difference or chrominance signals as they are called, $E_R - E_Y$ and $E_B - E_Y$, which are also extracted at the third stage of the transmission from the incoming signals E_R, E_G and E_B.

To understand the significance of these colour-difference signals, we must first consider the location in the chromaticity chart of what we may call the transmission primaries corresponding to these three signals, that is, the colours for which two of the three signals are zero. Fig. 8.8 shows the location of (R), (G) and (B) in the C.I.E. chromaticity chart, with (G)(R) extended to intersect the abscissa at (X') and (G)(B) extended to intersect it at (Z'). Now the transmission primary corresponding to the $E_R - E_Y$ signals can be found by putting $E_Y=0$ and $E_B - E_Y=0$, from which also $E_B=0$. Since E_Y is the luminance signal and this is zero, the primary must be located on the alychne (Z)(X). Since E_B is also zero, the primary must correspond to a mixture of (R) and (G) alone and must therefore be located somewhere along the line (G)(R). The only stimulus to satisfy these two conditions is (X'), which must therefore be one of the transmission primaries. By an exactly similar argument, the second transmission primary can be shown to be located at (Z').

The third primary will be given by putting both $E_B - E_Y = 0$ and $E_R - E_Y = 0$, giving $E_B = E_R = E_Y$. Since

$$E_Y = V_{(R)} \cdot E_R + V_{(G)} \cdot E_G + V_{(B)} \cdot E_B$$

we have

$$E_Y(1 - V_{(R)} - V_{(B)}) = V_{(G)} \cdot E_G$$

and since

$$V_{(R)} + V_{(G)} + V_{(B)} = 1$$
$$E_Y = E_G = E_B = E_R$$

Hence the third primary is the stimulus for which the red, green and blue signals are equal, that is the white stimulus (W) in Fig. 8.8, on which the units of (R), (G) and (B) have been normalized.

The transmission primaries (W) (X') (Z') therefore correspond to a white stimulus containing all the luminance, and to red and blue stimuli of zero luminance and therefore located on the alychne.

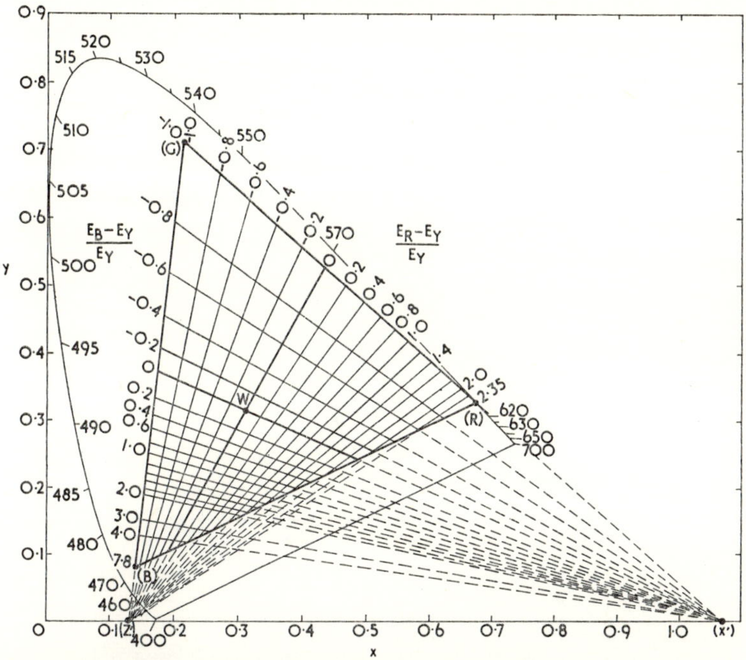

FIG. 8.8 Location on the C.I.E. chromaticity chart of the transmission primaries (X') (W) and (Z') of a colour television system. (R) (G) (B) are the receiver primaries, while the grid of lines within the triangle (R) (G) (B) shows the colour co-ordinates in terms of the normalized colour difference signals $(E_R - E_Y)/E_Y$, $(E_B - E_Y)/E_Y$

(After Carnt and Townsend, *Colour Television*, 1961)

Colours giving rise to signals for which $(E_B - E_Y)/E_Y$ is constant will lie on lines radiating from (X'), while colours for which $(E_R - E_Y)/E_Y$ is constant will lie on lines radiating from (Z'). It is therefore possible to construct a co-ordinate system in terms of $(E_B - E_Y)/E_Y$ and $(E_R - E_Y)/E_Y$ consisting of lines in the chromaticity chart radiating from (X') and (Z'). Such a grid is shown in Fig. 8.8 and it will be seen that the lines for which $(E_B - E_Y)/E_Y = 0$ and $(E_R - E_Y)/E_Y = 0$ pass through the point W. From a diagram of this kind the colour-difference signals can be read off for a colour of specified chromaticity and luminance.

The relation between the colour being transmitted and the

colour-difference signals can also be illustrated in a diagram of the type shown in Fig. 8.9 in which $E_B - E_Y$ and $E_R - E_Y$ are plotted on rectangular co-ordinates. The different hues are distributed around the origin and colours of constant dominant wavelength will lie on a line such as OC radiating from the origin O. The angle ϕ to the axis will thus define the dominant wavelength of the colour C. The distance of C from O will, however, increase both with its luminance and its purity. The point O itself corresponds to any neutral colour ranging from black to white.

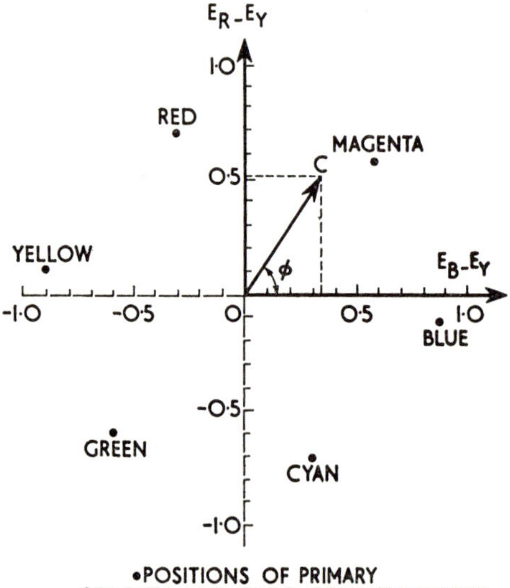

FIG. 8.9 Colour diagram using colour difference signals $E_B - E_Y$, $E_R - E_Y$ as axes
(After Carnt and Townsend, *Colour Television*, 1961)

One of the purposes of separating the luminance from the colour information is that it provides an opportunity to economize in the bandwidth required in a colour television system. It is found that little degradation either of colour quality or definition can be detected if the luminance signal alone carries the full range of frequencies. The bandwidth of the chrominance signals can therefore be reduced and in the American N.T.S.C. system, one of the signals is compressed more than the other. In fact, in this system a further change in the chrominance axes is introduced before the transmission of the signals and this is illustrated in Fig. 8.10. Here the colour-difference axes $E_B - E_Y$ and $E_R - E_Y$ have been weighted by the factors 1/2·03 and 1/1·14 respectively and new transmission axes I and Q at 33° to the vertical and horizontal axes respectively have been adopted. The location of the (I) and (Q) primaries associated with these new chrominance signals are shown on the chromaticity chart in Fig. 8.11.

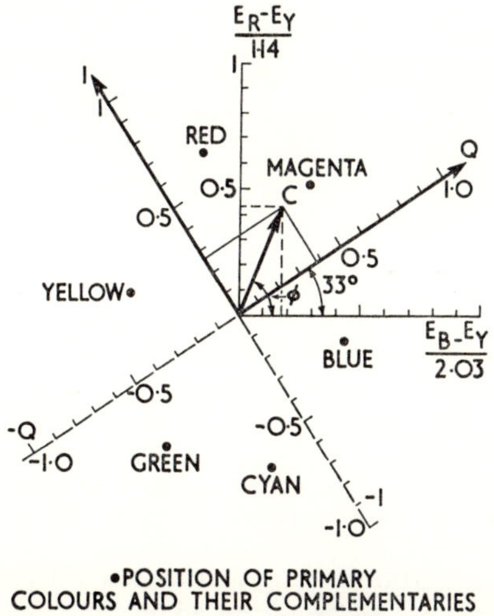

FIG. 8.10 Colour diagram using weighted colour difference signals $(E_B - E_Y)/2{\cdot}03$, $(E_R - E_Y)/1{\cdot}14$ as axes. The new transmission axes I and Q are also shown

(After Carnt and Townsend. *Colour Television*, 1961)

There is no occasion to discuss here the arguments leading to this particular choice of axes so far as they are based on signal-to-noise considerations and allied aspects of the system, but they do not seem to have been very well selected from the visual point of view. The loss of colour discrimination which occurs when small areas of colour are being viewed (p. 55) leads to confusions between such colours as blue and green, yellow and white, and pink and orange. This corresponds to the form of dichromatism known as tritanopia, for which the confusion zones in the chromaticity chart are as shown in Fig. 8.12.* Now it would be perfectly

* L. C. THOMSON and W. D. WRIGHT: *J. Opt. Soc. Am.*, 1953, **43**, p. 890.

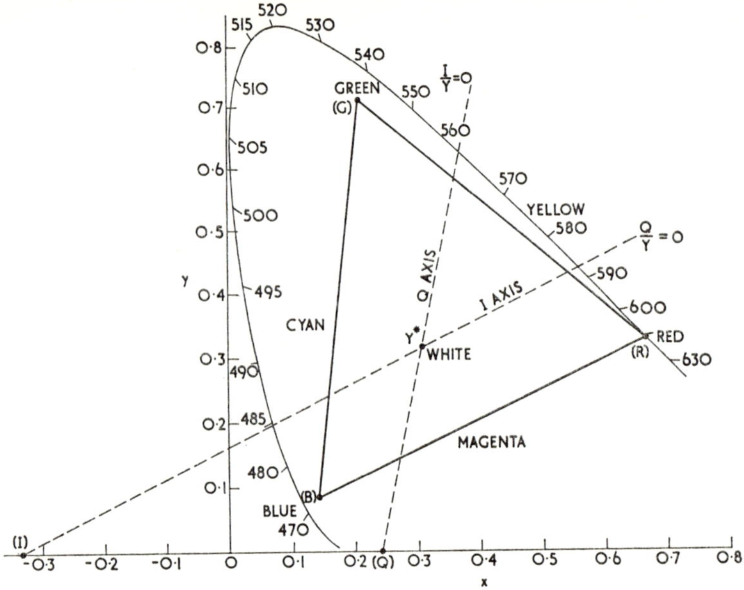

FIG. 8.11 The C.I.E. chromaticity chart showing the location of the *I* and *Q* transmission primaries and the *I* and *Q* axes (After Carnt and Townsend, *Colour Television*, 1961)

legitimate to argue from this that there was no point in transmitting colour-difference information along the direction of these zones for the fine detail of a colour television picture. Discrimination across the zones should, however, be maintained although not necessarily for the very finest detail in the picture, since colour discrimination in the red-green direction is likely to be less critical then luminance discrimination.

In the N.T.S.C. system, the *Q* axis has been chosen to carry the most restricted range of frequencies, while the *I* axis has the wider range. Although experiments have tended to support this choice of axes* they are certainly open to some criticism since the important direction for colour discrimination is transverse to the zones in Fig. 8.12, i.e. parallel to the red-green part of the spectrum locus. Perhaps too much should not be made of this, since the coordinate grid associated with the *I* and *Q* axes consists in practice of a series of curved lines as indicated in Fig. 8.13 rather than of

* K. HACKING; *Acta Electronica*, 1957, **2**, p. 18.

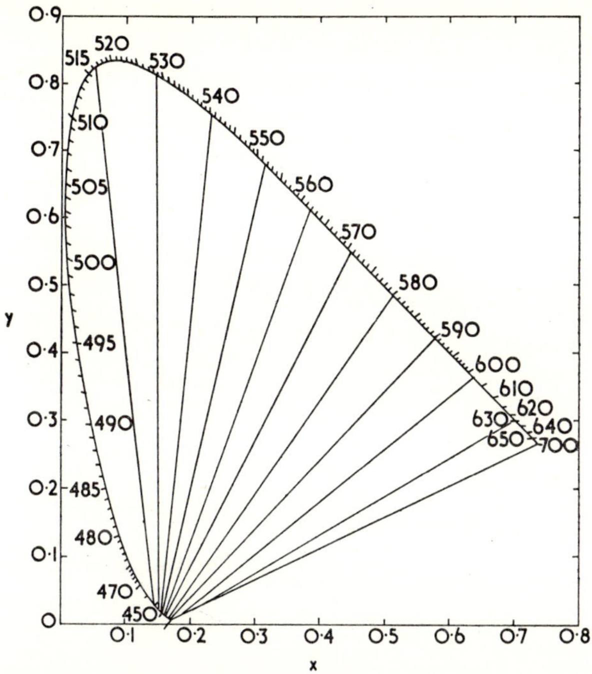

FIG. 8.12 The C.I.E. chromaticity chart showing the confusion loci associated with tritanopia

straight lines radiating from (*I*) and (*Q*), because of the need for gamma correction. Arguments for bandwidth saving based on experiments on small-field tritanopia are even less relevant under these conditions than if the grid were a series of radiating straight lines. In fact, in the PAL system the band-saving is applied equally to both chrominance signals.

A feature of the transmission that is of considerable colorimetric interest is that the *I* and *Q* components of the chrominance signal are transmitted by amplitude modulation of two sub-carrier frequencies which are 90° apart in phase. This means that the length and direction of the colour vector *OC* in Fig. 8.9 are determined respectively by the resolved components of the two amplitudes and by the phase angle between them. Hence to a first approximation, the sub-carrier phase determines the hue of the colour being reproduced, while its amplitude for a given lumin-

P

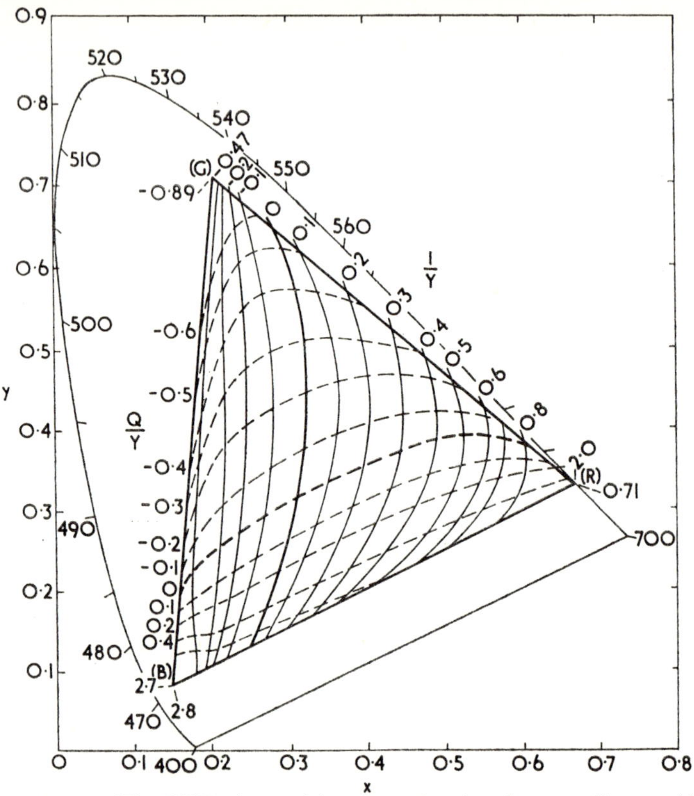

FIG. 8.13 The C.I.E. chromaticity chart showing the co-ordinate grid associated with the normalized I/Y and Q/Y axes in the N.T.S.C. system when a correction for a gamma of 2·2 has been made

(After Carnt and Townsend, *Colour Television*, 1961)

ance determines the saturation. The quality of a colour television picture is therefore determined to an important extent by the susceptibility of the phase and amplitude of the sub-carrier signal to interference and to distortions under difficult transmission conditions. The modified forms of the N.T.S.C. system developed in France (the sequential-memory system SECAM) and in Germany (the Phase Alternation Line System PAL) differ primarily from the N.T.S.C. system by the form in which the hue and saturation information is impressed on the sub-carrier. The adoption by the British authorities of the PAL system followed exhaustive tests of the relative merits of the three systems, particularly in regard to their capacity to retain the colour information correctly over long transmission distances and complex sequences of transmission links.

At the receiver the luminance and chrominance signals, in whatever form they have been transmitted, have to be decoded to modulate the $(R)(G)(B)$ receiver primaries. The system thus returns to a normal three-colour system, and it is of interest to note that although at no stage in the transmission are there more than three primaries, the system as a whole might be said to employ nine different primaries. This is a useful analogy to bear in mind when arguing about different theories of colour vision.

We have ignored in Fig. 8.7 important items such as the need for gamma correction of the signals to compensate for the high gamma modulation in the cathode-ray tube. Then in some systems a field sequential scanner (a rotating disk with red, green and blue filters) is used at the camera, and an additional stage introduced at which separate pictures are reproduced on red, green and blue cathode-ray tubes, and are viewed by storage pick-up tubes to convert the signals into dot-sequential form.

Some sources of colour distortion have already been mentioned, but departure from an overall gamma of unity, or the use of incorrect camera sensitivity curves, will introduce errors essentially similar to those found in colour photography. In addition, 'noise' in the system will affect the colour resolution, scattered light in the cathode-ray tube will tend to reduce the contrast in the picture, and transient signals may momentarily introduce very large colour distortions.

It is clear that the problems facing colour television are quite as tough as those facing colour photography, although their origin may be very different. It will prove a most interesting study in contrasting technologies to compare their respective progress towards perfection in colour reproduction.

THE ASSESSMENT OF A COLOUR PICTURE

There are certain straightforward photometric and colorimetric tests which can be and are applied to a colour reproduction. These include tests on the colour balance, that is, on the absence of hue in the greys and whites, tests on the chromaticity range, on the luminance range, on the gamma or contrast, and on the accuracy of chromaticity reproduction of certain key colours. The problem then arises of integrating the results of these piecemeal tests and assessing the general quality of the picture. This requires a knowledge of what makes a good picture.

In so far as this is regarded as an artistic question, we are unlikely to find an answer which everyone will accept. But just as the general characteristics of a monochrome picture can be specified in optical and photometric terms, so we should be able to find a basis on which to describe the characteristics of a colour picture.

The qualities of a colour reproduction embrace those of a monochrome picture, so that any comprehensive assessment must include size, definition, luminance, luminance range and luminance contrast. Of the additional items, perhaps colour balance is the most important, since the presence of any noticeable hue in the neutral greys and whites of a colour picture can be objectionable. The lack of neutrality is, however, a function of the reference white against which it is being judged. In the case of a colour print, for example, this will generally be the white border and the judgement will be quite critical. In a colour television picture, on the other hand, the whites will be judged by comparison with the ambient lighting around the set, and this may range from the blueness of daylight to the yellowness of tungsten light; hence, under one condition the balance of the picture might be thought to be too yellow and under another condition the same picture might be regarded as too blue.

Chromaticity range is an attribute that obviously has to be considered, although it is perhaps less important than might be expected. Pictures are more likely to be criticized for the garish use of colours that are too vivid than for the restrained use of desaturated colours. Nevertheless, unless the chromaticity range is adequate, the brilliance of some scenes cannot be faithfully recorded. Consistency of colour quality is also important. If some hues are reproduced in their full vividness, while others are degraded, the result will be unsatisfactory,* and a consistent mediocrity in all hues is to be preferred. Major hue distortions will be objectionable, but small errors of hue are likely to pass unnoticed

* R. M. Evans: *J. Opt. Soc. Am.*, 1943, **33**, p. 579.

except with familiar objects, of which flesh tints are the outstanding example.

Surprisingly little attention appears to have been paid in the literature to the specification of the colour gradations in a picture. Recognition of the importance of the luminance contrast or gamma has been carried over from monochrome reproduction, but as we saw in our discussion of colour atlases in the previous chapter, gradations of hue, gradations of chroma, gradations of white content, and so on, are also visually significant. Certainly saturation gradients occur with great frequency in nature, owing to the desaturation associated with diminishing concentration of a pigment, to the diminishing thickness of a colorant layer, to the increasing effect of mist with distance in a landscape, and to the change of surface reflection with the angle of illumination and the angle of view, and with the bloom on a surface.

The author has suggested* that the ability of a system to reproduce these gradients could be specified by a purity characteristic curve and perhaps by a purity gamma. Thus, if a series of patterns of increasing saturation were selected from a colour atlas and used as a test chart, their purities and those of the reproduced colours could be measured and the one plotted against the other to give a curve of the type shown in Fig. 8.14. Experience might show that some other co-ordinates were to be preferred, but the desirability of establishing some such relationship seems unquestionable.

Allied to the reproduction of colour gradients is the colour resolution or colour discrimination of a system. This, too, seems to have received insufficient attention, but just as the eye has a limited ability to discriminate one colour from another, so a colour system can reproduce a given colour only to a limited precision and consistency, and can resolve colour differences only down to a certain minimum value. The reasons for this limited discrimination are not unlike those in vision—fluctuation in the sensitivity of the colour processes, non-uniformity of photographic emulsion, photoelectric mosaic or colour-phosphor layer, noise in the transmission system, flattening of the characteristic between input and output, etc. For given values of these factors, the colour discrimination of a system is likely to improve the closer (R), (G) and

* W. D. Wright: *Brit. Kinemat.*, 1948, **13**, p. 1, and *Acta Electronica*, 1957, **2**, p. 26.

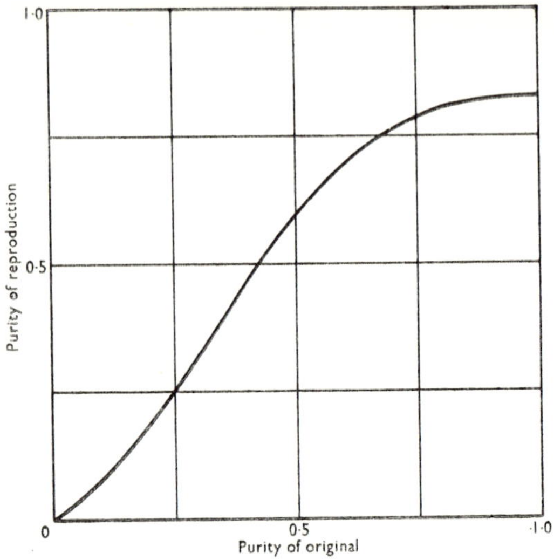

FIG. 8.14 Possible specification of the saturation characteristics of a colour reproduction system by means of a purity gamma curve

(*B*) approach to each other in the chromaticity chart, although the colour gamut of the system will, of course, contract. This would be analogous to using a colorimeter with desaturated matching stimuli to make more accurate measurements of near-white colours, and it follows that to some extent the requirements for a wide colour range in a system are likely to conflict with the requirements for good colour discrimination.

The colour resolution of a system is one of the most important factors determining its ability to reproduce surface texture successfully, and surface texture in turn is one of the most vital items in an aesthetically satisfying picture. The texture of skin, for example, is revealed by very minute gradations of hue and saturation and lightness, in addition to its fine detailed structure, and these gradations are easily lost in a colour reproduction on account of the limited colour resolution of the process. Fig. 8.15 shows the very small range of chromaticities represented in the various parts of a typical face, while over the area of a given part such as the cheek, the gradations of colour are very much smaller than this.

Discrimination tests can be very readily applied to a colour system, and the problem is in some ways analogous to making a

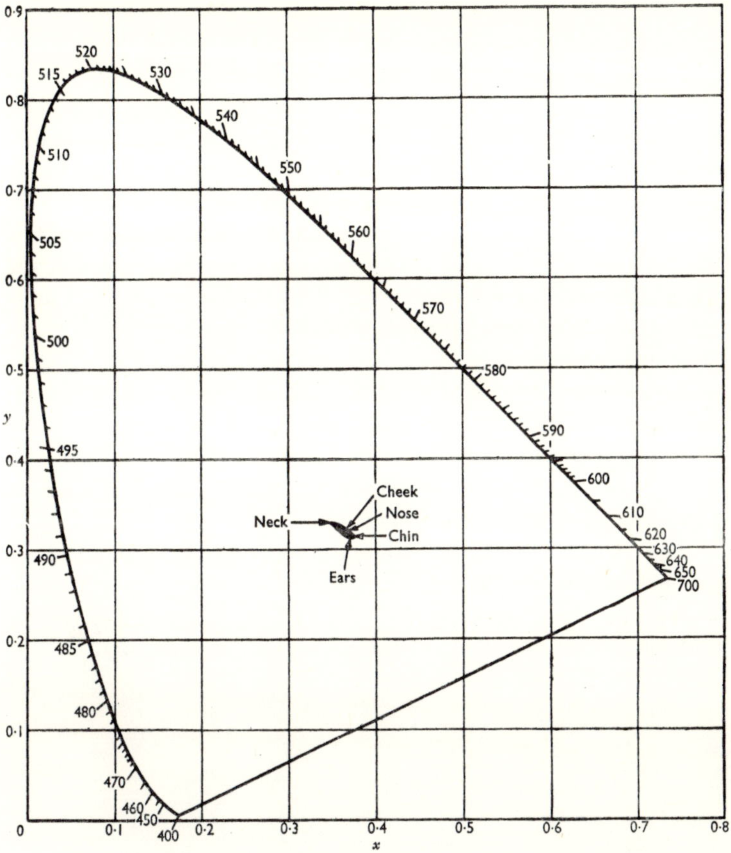

FIG. 8.15 The chromaticities of a typical face under illuminant *Sc*

test for defective colour vision. Indeed, some of the tests employed in colour-vision testing might well prove of use in colour photography and television. If they were combined with tests on the resolution of fine detail, which is often one of the first casualties in colour television although not of colour photographic transparencies, then it might be possible to derive a useful figure of merit for texture rendition.

On the other hand, the colour distortions in a system are less easy to assess, since the chromaticity differences between the original and the reproduction, as revealed by colorimetric tests, may not give a fair idea of the subjective distortion of the colours. The subjective appearance is very materially affected by the viewing conditions under which the picture is seen. To take two examples, if the picture is bright and is seen against a dark surround, the saturation of the colours will be enhanced; or if the luminance contrasts which exist in a three-dimensional object are faithfully represented in a two-dimensional reproduction of the object, the subjective contrast will be too great. Exact objective reproduction will rarely provide the most realistic or successful presentation, and perhaps the next big advance in this field will be to establish more clearly the relationship between objective measurement and subjective appearance.

Without objective measurements and data, progress in colour reproduction would be seriously hampered, and we can be satisfied that colorimetry is necessary to the development of improved systems. Yet we must agree with Evans, Hanson and Brewer* that no simple set of colorimetric relationships between the original scene and the reproduction is likely to be found, and that, until adequate relationships are established, an empirical approach is inevitable.

* R. M. Evans, W. T. Hanson and W. L. Brewer: *Principles of Color Photography* (Wiley: New York. Chapman and Hall: London), 1953.

CHAPTER 9

COLORIMETERS AND SPECTROPHOTOMETERS

PRACTICAL METHODS OF COLOUR MEASUREMENT

The measurement of colour may be undertaken for a variety of reasons. For some purposes a specification which has a national or international status may be required, as in the standardization of signal glasses; in some commercial transactions the description of a sample in unambiguous terms is necessary to ensure that the products of an industry satisfy a customer's requirements; again, colour may be an important guide to the quality of a raw material, and of the grading of the product on some system which is accepted by both producer and consumer is necessary; the measurement of colour can be used to control industrial processes; in colour photography and television the assessment of the performance of a system requires the measurement of the colours of the original and of the reproduction; in the lighting industry, colorimetry and spectrophotometry have figured prominently in the development of the fluorescent lamp and the study of its colour-rendering properties; the results of subtractive colour mixture must be predicted largely on the basis of spectral transmission and reflection data; and in scientific research, colour measurements can provide effective means of recording and analysing natural phenomena.

Needs such as these can sometimes be met adequately by measurements or descriptions of a colour which have a purely local significance. If some colour-measuring device is employed in a factory to determine when a particular stage in the manufacture of a product has been reached, the only condition necessary for the successful operation of the device is that the readings given by the instrument can be calibrated relative to the changes occurring in the manufacturing process, so that the one can be followed in terms of the other. Where, however, results have to be compared between different factories or with other organizations or groups of workers, the measurements must be on some more standard basis. They might be obtained with identical instruments under exactly the same conditions, so that a specification in terms of the scales used in each instrument had the same meaning for all the individuals carrying out the particular set of measurements, but the confusion that could result from the extension of this principle to several different types of instrument can readily be imagined. It was to avoid such confusion that the C.I.E. system was created.

The fundamental instrument for colour measurement is the spectrophotometer, since the information which it gives about spectral composition is not only essential for the solution of many colour problems but also permits the standard C.I.E. specification to be derived. In consequence, we find that the spectrophotometer, especially the recording instrument, is becoming more and more important as an industrial tool. Yet not every laboratory needs an instrument of this complexity or can afford the initial cost and the need for skilled operation and maintenance. Hence colorimeters of various types are still widely used.

No attempt has been made in this chapter to describe all the colour measuring instruments on the market. Instead, the aim has been to bring out the basic principles of representative types of equipment, with special emphasis on the optical and photometric features of the design. However clever the electronics, the measurements will be in error if proper attention has not been paid to the optics. For this reason descriptions of some visual instruments have been included even though they are unlikely in these days to occupy a very prominent position on the laboratory bench. It is no bad thing to go back to first principles in the search for new ideas to meet the demand for increasingly sophisticated equipment.

ADDITIVE COLORIMETERS

We shall first consider instruments in which a colour is matched and measured visually by the additive mixture of light stimuli. Three such instruments, those of Guild and the author, and the new equipment developed by Stiles, have already been described in Chapter 3. The additive colorimeter which has been most widely used in industry, however, is that developed by Donaldson* and illustrated in principle in Fig. 9.1. As in the Guild instrument, colour filters are employed to select the red, green and blue radiations, while the amount of light in the three beams is controlled by varying the area of each filter through which the light has to pass. The method of mixing, however, is quite different, since the light from the projector lamp *S* after passing through the colour filters is

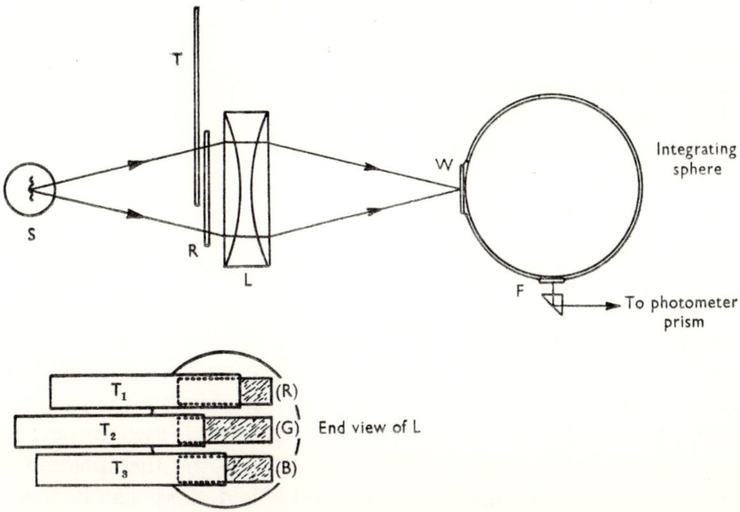

FIG. 9.1 The colour-mixing system of the Donaldson colorimeter

focused by the double condenser lens L on the window W of an integrating sphere. The filament image at W thus consists of three differently coloured, superimposed images, but within the sphere the three beams suffer repeated diffuse reflections and hence produce uniform colour mixing on the surface of the sphere. A second window F allows light from the opposite wall of the sphere to emerge and be reflected into the photometer prism for observation in the usual way.

The amounts of the three radiations entering the photometer prism are approximately proportional to the areas of the filters (R), (G) and (B) exposed by the shutters T_1, T_2 and T_3, but allowance has to be made for departures from perfect uniformity

*R. DONALDSON; *Proc. Phys. Soc.*, 1935, **47**, p. 1068 and 1947, **59**, p. 554.

of illumination over the face of the lens L and for reflection and absorption losses in the condenser lenses. The filters themselves have to be chosen for uniformity and permanence, and for general colorimetry they will be saturated red, green and blue in colour.

The three-stimulus instrument, however, suffers from the limitations of all trichromatic colorimeters, namely, limited precision of setting and large variations from observer to observer. Ives* suggested that for special purposes such as the matching of near-whites, the ordinary red, green and blue matching stimuli could be replaced by desaturated stimuli whose positions in the chromaticity chart are very near together. This would overcome these difficulties, but only at the sacrifice of the chromaticity range of the instrument. As an alternative solution, Donaldson introduced a six-stimulus instrument acting on the same principle but incorporating six filters—red, orange, yellow-green, green, blue-green and blue—in front of the condenser lens, with six shutters to control the areas of the filters which are exposed. Through this means, an approximate match between the spectral composition of the instrument stimuli and the test colour can be secured by preliminary setting of the shutters, the final colour match being made, as with the three-stimulus instrument, by operation of the red, green and blue controls. Two important advantages follow from this arrangement. In the first place, the macular spot is no longer obtrusive and a field of 15° can be used, giving greatly increased precision of setting. This was further enhanced by the inclusion of a contrast field of the Lummer-Brodhun type. Secondly, the approximate energy match very largely eliminates the observer trouble, since the more similar the energy compositions, the more nearly will all observers agree on the colour match, however pigmented their eyes.

The extent to which the spectral composition can be simulated is illustrated in Fig. 9.2, which shows the transmission curves of the six filters and the spectral composition of the six stimuli combined to give an approximate energy match with illuminant S_B. The other extreme has been shown in Fig. 5.1 for a trichromatic instrument using spectral matching stimuli. Here the effect of observer differences will be at their worst, although for the study of individual variations of colour vision, this is an advantage.

* H. E. Ives: *J. Opt. Soc. Am.*, 1923, **7**, p. 243.

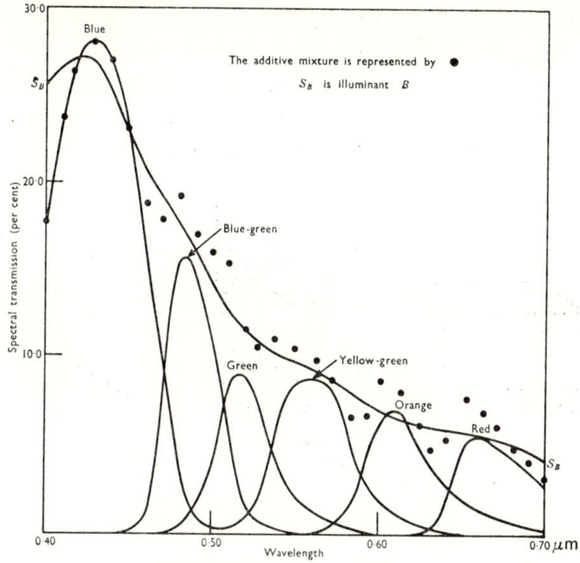

FIG. 9.2 The spectral composition of the two halves of the field in the Donaldson six-stimulus colorimeter when light from the illuminant S_B is matched by a mixture of the six instrument stimuli
(*Proc. Phys. Soc.*, 1947, 59)

The colour specification itself has to be derived from the conversion equations supplied by the makers of the apparatus; these specify the tristimulus values of the instrument stimuli in terms of (X), (Y) and (Z). The calibration of the instrument is by no means easy, but Donaldson's paper indicates that a precision of about 0·002 in x and y and an accuracy of the same order is possible in the measurement of typical samples. The significance of these values varies with the chromaticity of the sample, and a lower accuracy is naturally to be expected in the colorimetry of light sources such as gas-discharge lamps or even fluorescent lamps, where spectral lines form part of the energy composition, since these cannot be closely imitated in the colorimeter. Nevertheless, this instrument must be regarded as the most successful visual colorimeter yet designed for accurate colour measurements and one capable of giving results having an accuracy that would satisfy many of the demands likely to be met in industry.

Instruments of a lower precision and accuracy are sometimes

called for, and a design described by Burnham* has been applied by the author to the semi-portable instrument illustrated in Plate 10(*a*) (p. 230), which may be taken as typical of this class of instrument. The colour-mixing principle as described by Burnham is shown in Fig. 9.3(*a*). A colour-filter assembly is mounted so that it can be

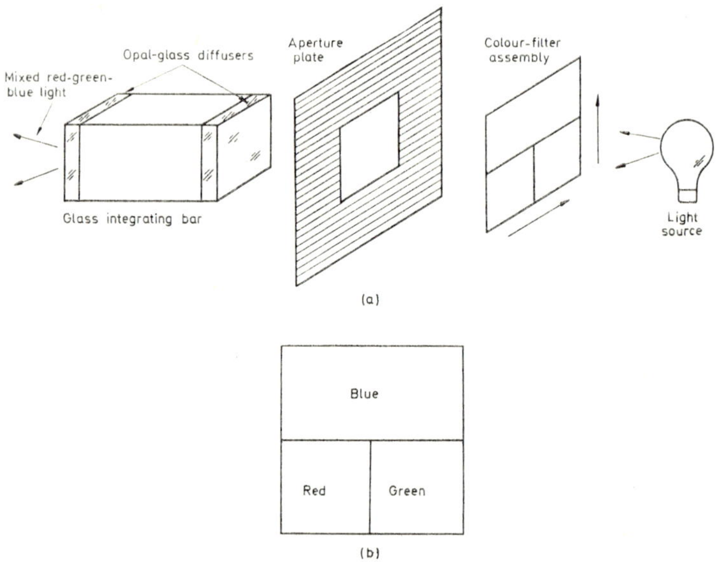

FIG. 9.3 (*a*) The separate elements of the colour-mixing system described by Burnham and used in the colorimeter shown in Plate 10(*a*). (*b*)The filter assembly

moved in two directions at right angles across a square aperture in an aperture plate. In this way, varying areas of the filter may be exposed, giving a composite light which is mixed by diffusion in the two opal glass plates and by internal reflection within the rectangular glass integrating bar. The loss of light in this device is considerable, but the mixing is very effective.

One possible filter arrangement is shown in Fig. 9.3(*b*). Movement in one direction varies the proportion of red to green in the mixture, and in the other direction the proportion of blue to the red-green mixture; the two colour controls on the instrument will then enable the chromaticity chart to be traversed in the directions

* R. W. BURNHAM: *Am. J. Psych.*, 1952, **65**, p. 603.

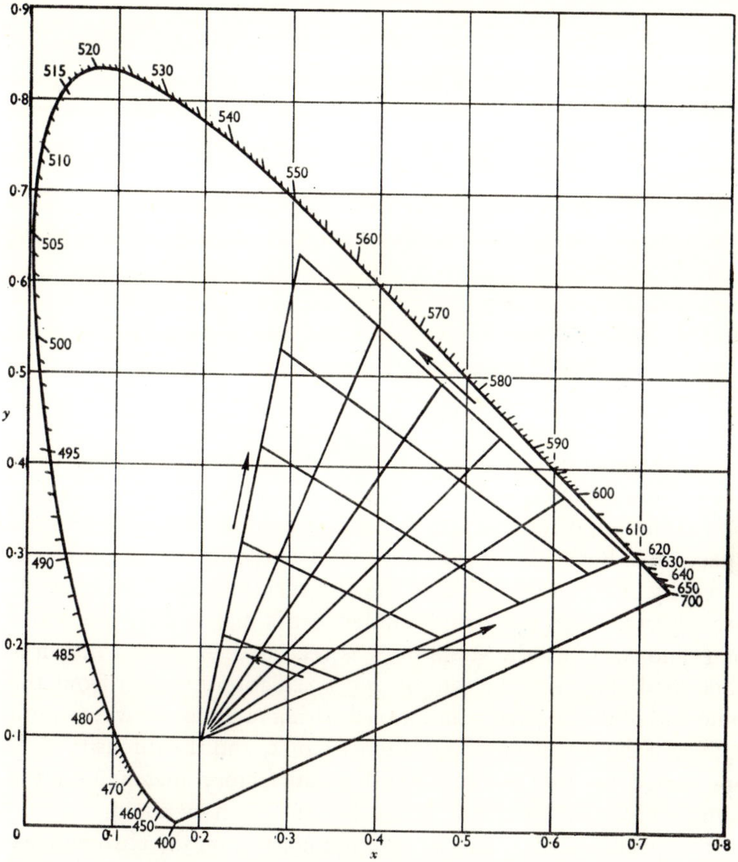

FIG 9.4. Grid showing the traverse of the chromaticity chart produced with the two colour controls of the colorimeter of Plate 10(*a*)

shown by the grid in Fig. 9.4. An independent intensity control has to be provided, and in the author's instrument the distance of the light source from the filter assembly can be varied, with neutral filters to extend the luminance range. The size of field can also be varied, up to a maximum angle of 10°, by means of an iris diaphragm.

One convenient feature of the instrument in Plate 10(*a*) is the inclination of the viewing tube at a comfortable angle for the observer looking down it, and another is the provision of a mirror below the tube which can be set at any angle about a vertical or horizontal axis. In this way the light from an object located at almost any position in space can be directed into the instrument and its colour measured. This particular instrument was designed initially to measure the colour of the aurora, but it was first used during the total eclipse of the sun on the 30th June, 1954, as described in Chapter 10, and has since proved a convenient general-purpose laboratory colorimeter and colour comparator.

Reference should also be made in this section to an elegant method of vector colorimetry described by Guild,* in which no measurements of the intensities of the matching stimuli are required. Although the method does not appear to have been employed extensively, it deserves mention partly because there may be special investigations for which it might prove especially useful and partly because of the interest in the underlying principle.

In the apparatus, provision has to be made for a radiation from the red end of the spectrum, P in Fig. 9.5, to be mixed with a second radiation, of wavelength λ_1, taken from any other point in the spectrum. Both the wavelength and intensity of λ_1 and the intensity of P can be varied by the observer, and the three controls are adjusted until the sample under test is matched by a mixture of P and λ_1. Only the value of λ_1 is recorded. A second match is then made between the sample and a radiation Q taken from the blue end of the spectrum and mixed with a radiation of wavelength λ_2; again the wavelength and intensity of λ_2 and the intensity of Q are varied by the observer until a satisfactory match with the sample has been obtained. The value of λ_2 is now noted.

The positions of λ_1 and λ_2 are then marked on the chromaticity chart. From the geometry of the chart, any colour which is matched by a mixture of P and λ_1 must lie on the straight line joining P and λ_1; similarly, a colour matched by mixing Q and λ_2 must lie on the line joining Q and λ_2. The location of the sample C on the chart is therefore found simply as the intersection of the lines $P\lambda_1$ and $Q\lambda_2$. The co-ordinates of C can then be read off the diagram.

Experimentally, the method has the great advantage that no accurate quantitative intensity controls are required. Algebraically, the system has the advantage of locating the colour directly on the C.I.E. chart without requiring any transformation calculations.

* J. Guild: *Trans. Opt. Soc.*, 1925, **27**, p. 139.

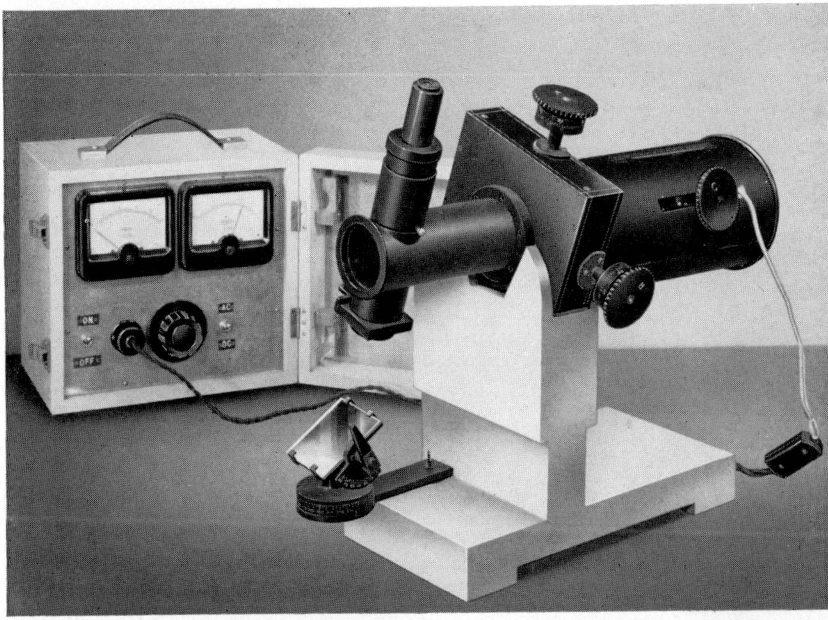

(*a*) Semi-portable filter colorimeter designed by the author on a principle described by Burnham (see page 228)

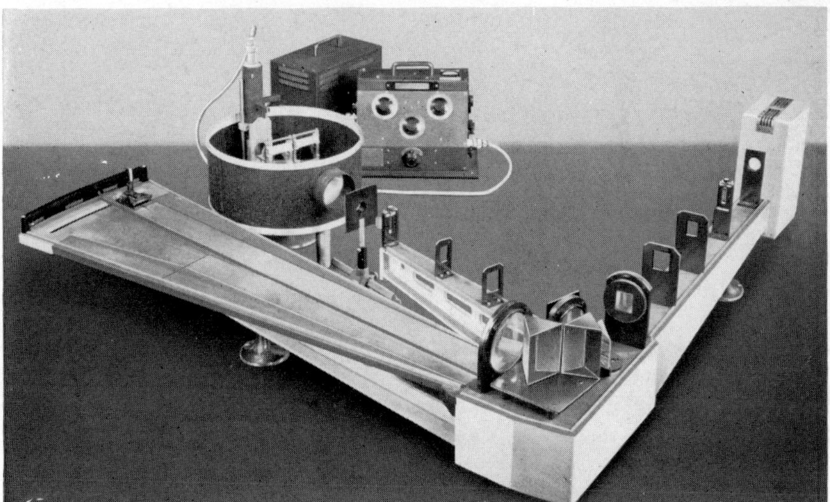

(*b*) General view of the Wright spectrophotometer (see page 248)

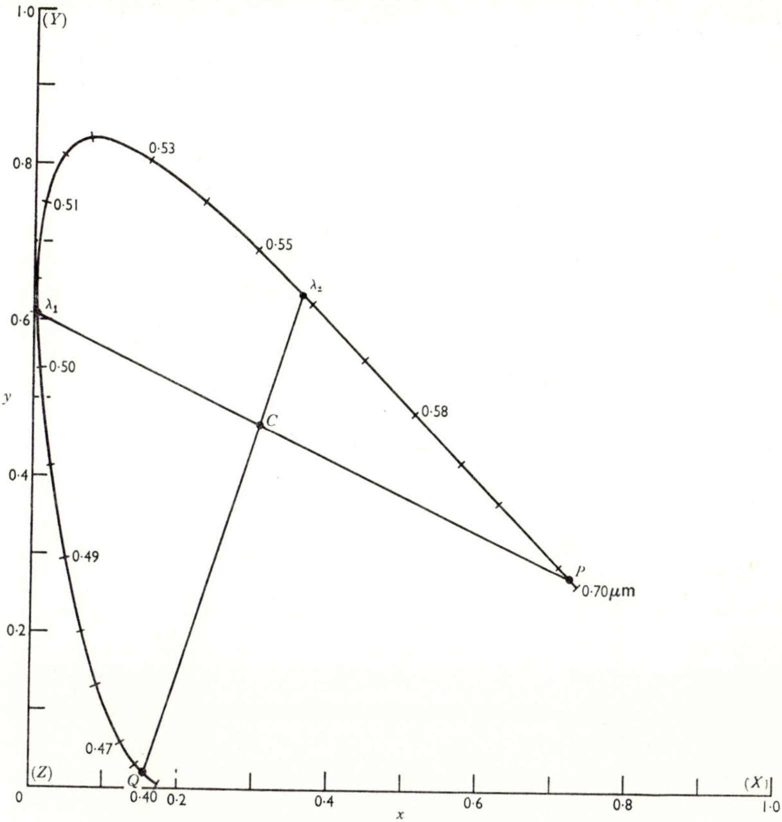

FIG. 9.5 The location of a colour in the chromaticity chart using Guild's vector principle

The accuracy is limited by the accuracy associated with graphical methods and visual observations, but for some applications this may not be an important limitation. A vector colorimeter working on these lines has been described by Guild in the reference quoted above, while the Wright colorimeter (Fig. 3.2) can be adapted to operate in the same way. It is, however, also possible to visualize the same principle employed, say, in measuring white surfaces, in which *P* and *Q* are no longer monochromatic radiations, but are near-white surfaces of slightly different chromaticity whose C.I.E. co-ordinates are accurately known. An arrangement of this kind might well form the basis of a differential colorimeter having promising possibilities.

Apart from its merits as a method of measurement, the vector principle can be used in connection with the transformation of a colour specification from one trichromatic system to another, provided the spectrum locus in each system is known. Thus two monochromatic radiations, P and Q, can be chosen as the origins of the two vectors and their positions marked on the chromaticity charts of the two systems. The specification for the colour C is given in one system and can therefore be plotted in the corresponding chart. A line from P through C can be drawn to determine λ_1 (Fig. 9.5) and a second line from Q through C to give λ_1. Then, λ_1 and λ_2 can be marked on the spectrum locus on the other chart, and the two intersecting lines $P\lambda_1$ and $Q\lambda_2$ can be drawn to find C. A geometrical transformation can never be carried through to the same accuracy as the corresponding algebraical transformation, but where speed is more important than accuracy it has the advantage. It can also serve as a check against gross numerical errors in the algebraical method.

A visual differential reflectometer has been developed by Ishak and the author* which can also be used as a differential colorimeter. The basic optical system is illustrated in Fig. 9.6 in which the two samples C_1 and C_2 to be compared are illuminated by the direct light from the source S but with supplementary illumination from the virtual images S_1 and S_2 of the source formed in the two mirrors M_1 and M_2. The relative illumination on C_1 and C_2 is varied by increasing M_1C_1 and decreasing M_2C_2 or vice versa. The light from the two samples illuminates the two halves of a photometer prism, and the observation consists of setting the positions of M_1 and M_2 with the instrument control until the differential illumination of the two samples just compensates for their difference in luminance factor, as judged by the match between the two halves of the photometric field. The samples are then interchanged in position and the observation repeated. The difference in position of M_1 and M_2 in the two observations then enables the difference in luminance factor to be calculated from the geometry of the system and the differential application of the inverse square law. The observation is capable of high precision and the principle of the

* I. G. H. Ishak and W. D. Wright: *Optica Acta*, 1966, **13**, p. 139.

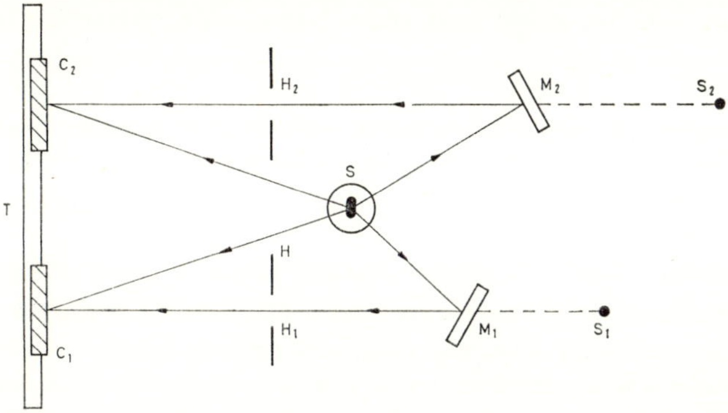

FIG. 9.6. Optical system of differential reflectometer and colorimeter (Ishak and Wright). (The lateral separation between the two samples C_1 and C_2 relative to their distance from the source S has been exaggerated in this diagram for the convenience of illustration.)

system leads to a truly differential measurement of high accuracy. If the observer's V_λ curve is reasonably normal, an observation made with no colour filter interposed gives the difference ΔY between the Y tristimulus values of the sample. With the appropriate orange and blue filters interposed to convert the observer's spectral sensitivity to the $\bar{x}_\lambda$ and $\bar{z}_\lambda$ curves, the differences ΔX and ΔZ can also be measured. The same optical principle can be employed with a photoelectric matching system in place of the eye, although no instrument of this type has yet been constructed.

SUBTRACTIVE COLORIMETERS

Colorimeters acting on the subtractive principle operate by the absorption or subtraction of light from an initially white beam. The outstanding example of such an instrument is the Lovibond Tintometer, in which sets of red, yellow and blue glass slides provide the subtractive primaries, and it is of interest to note that the matching is again basically a three-variable process as with the additive trichromatic colorimeter.* There is in fact a very close parallel between additive and subtractive colorimeters and additive and subtractive methods of colour reproduction. The latter have already been described in Chapter 8 and illustrated in Plate 9.

* Strictly the red slides should be described as magenta (or minus green) and the blue as cyan (or minus red) to correspond to the true subtractive primaries.

The success of the Tintometer* lies in the production of the graduated series of standardized glass slides ranging from very slightly tinted specimens to deeply coloured slides. Each slide is numbered according to its depth of colour on a scale which has two properties. First, each series is additive, so that a slide of a given value in a series is equivalent to two or more slides in the same series having the same total value. Second, the three series are related together in such a way that, when red, yellow and blue slides of equal value are superposed and viewed against a white background, they look approximately grey or neutral in colour.

In the instrument itself, provision is made for illuminating the sample to be measured and viewing it in one half of a 4° field of view, while the other half is provided by an illuminated white surface, in front of which the appropriate coloured glasses are inserted until a match is obtained. Matching is facilitated by the filters being mounted in racks so that they can be readily traversed in turn across the field of view. For the many occasions when definite, though arbitrary, colour measurements are required, a specification in terms of the Lovibond scales may be perfectly satisfactory, and products are often standardized for colour on this basis.

In the Lovibond-Schofield system,† the instrument has been modified to give the Tintometer reading in a form which can be readily converted to the equivalent C.I.E. specification. The same series of glass slides are employed as in the ordinary Tintometer, but not more than two of the series are used at any one time when making a colour measurement. Thus, in measuring green samples, the yellow and blue series are used together; for orange samples, the red and yellow, etc. The use of the two series alone allows the sample to be matched in respect of hue and saturation, while the lightness is matched by means of a vane which controls the relative illumination of the two halves of the field.

A conversion graph has been prepared from which the x and z chromaticity co-ordinates can be read off directly from the readings on the two Lovibond scales. Tables are also available to give

* G. S. Fawcett: *Proc. Phys. Soc.*, 1944, **56**, p. 8.
D. B. Judd, G. W. Haupt and G. J. Chamberlin: *J. Opt. Soc. Am.*, 1962, **52**, p. 813.

† R. K. Schofield: *J. Sci. Instr.*, 1939, **16**, p. 74.

the luminance factor of the sample, and hence, if required, its complete tristimulus specification.

The accuracy of measurement with the Tintometer is limited, of course, by the fact that the colour of the matching field can be varied only by the discrete steps which separate one glass slide from the next in its series. Nevertheless, the number of different colour combinations obtainable with the slides is several million. The colour vision of the observer will also affect the accuracy in much the same way as in additive colorimetry, although the subtractive colorimeter has something of an advantage, especially for desaturated colours, since the spectral composition of, say, a near-white test surface will differ little from that of the near-white obtained with slightly tinted Tintometer glasses.

One of the most practical applications of the Tintometer system is to the grading of coloured products where the grading of colour is likely to be confined to a relatively small colour range. This can be covered by quite a small number of coloured glasses, for which a very simple form of colour comparator has been developed. Extensive use of this equipment is made in the testing of oils, food, agricultural products and factory-produced goods, in pharmaceutical and medical work, and the like.

PHOTOELECTRIC COLORIMETERS

The principle of the photoelectric colorimeter is very simple. Three photocells (X), (Y) and (Z) are required (Fig. 9.7) with spectral sensitivity curves respectively similar to the $\bar{x}_\lambda$, $\bar{y}_\lambda$ and $\bar{z}_\lambda$ equal-energy distribution curves. The photocurrents have to be amplified and measured to give readings X, Y and Z which are linearly related to the light input in the three cells. With the three units so balanced, or their readings so adjusted that equal values would be obtained if the light to be measured had an equal-energy distribution, the readings X, Y and Z obtained on a sample will be directly proportional to the tristimulus values of the sample.

The feature of this elementary system that has so far proved very difficult to realize in practice is the development of a photocell with exactly the required spectral sensitivity. This applies both to the photo-emissive type of cell and to the photovoltaic cell, and it would be quite unrealistic to expect that, in the future, enough variables in the preparation of the light-sensitive surface will be discovered so that cells of the correct spectral sensitivity can be designed at will. If, however, a colour filter is placed in front of each of the photocells in Fig. 9.7, the spectral sensitivity of each cell and filter considered as a unit will be given by the product of

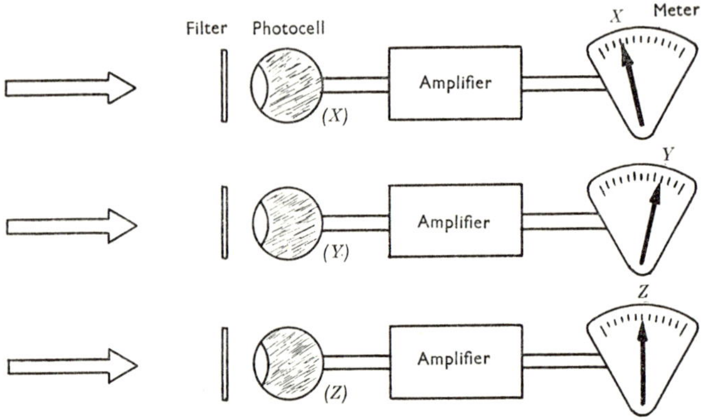

FIG. 9.7 The principle of the photoelectric tricolorimeter

the sensitivity curve of the cell and the transmission curve of the filter. The inclusion of a filter greatly increases the chances of obtaining any desired sensitivity curve, but since the absorption of a filter depends, as indicated in Chapter 1, on complex resonance systems within the molecules of the absorbing medium, the chemist is also faced with a difficult problem to produce a filter having some specified absorption curve. Hence we must anticipate that even the cell and filter combination will give only an approximation to the $\bar{x}_\lambda$, $\bar{y}_\lambda$ and $\bar{z}_\lambda$ sensitivity curves.

Actually the system need not be restricted to units capable of giving readings directly in terms of (X), (Y) and (Z), since the sensitivity curves could be linear transformations of the C.I.E. distribution curves and still give standard data. Something of the simplicity and directness of the measurement would, however, be lost, although the extra freedom implicit in the possibility of using a transformed system should not be forgotten; indeed, we have seen that advantage of this freedom has in fact been taken in colour television systems.

If the sensitivity curves have no relation at all to the C.I.E.

distribution curves, the readings can never be manipulated to give a true C.I.E. specification. However, in an instrument intended, say, for the control of some technical process, there may not be any point in striving for C.I.E. data; it may be sufficient to use three photocell-filter combinations which respond maximally in the red, green and blue regions of the spectrum. But we are here primarily concerned with instruments which are intended for colour measurement rather than just colour control.

Four of the most widely used photocell-filter colorimeters are the Elrepho Colorimeter made by Zeiss, the Colormaster differential colorimeter* made by Manufacturers Engineering and Equipment Corporation, the Color Eye made by Instrument Development Laboratories and the Colorcord made by Joyce, Loebl. The Elrepho instrument is one of the few instruments that uses diffuse illumination from the walls of an integrating sphere; the sample is placed at the base of the sphere and the measurement is made in a direction normal to the surface of the sample by a photocell mounted at the top of the sphere. The tristimulus values are derived directly from readings of the photocurrent. In the Colormaster instrument light from a single source is divided into two beams which illuminate the test sample and a white reference sample respectively. Matched photocells associated with each of the samples collect the reflected light and the two photo-currents are equated in a highly sensitive balancing circuit giving a critical null point. The actual control to give the balance and to provide the measurement is a variable aperture which adjusts the light flux falling on the reference sample. The measurement has, of course, to be repeated for the three red, green and blue filters to give the C.I.E. tristimulus values. A repetition precision of about 0·1 N.B.S. unit is claimed for the instrument. In the Color-Eye the sample and the reference standard are illuminated by diffuse illumination from the two halves of an integrating sphere and a photomultiplier flicker photometer system measures the light reflection from the sample relative to that from the standard for each of the tristimulus filters. The Colorcord instrument uses near-normal illumination and diffuse collection of the reflected light in

* L. G. Glasser and D. J. Troy: *J. Opt. Soc. Am.*, 1952, 42, p. 652.
L. G. Glasser: *Congrès F.A.T.I.P.E.C.*, III, 1955, p. 47.

an integrating sphere. The measurement is made with a photomultiplier detector system. An unusual feature of this instrument is the design of the red, green and blue filters, since each filter is built up from a number of sections of different colours in the form of a mosaic. In this way, a much closer approach can be achieved to the required $\bar{x}_\lambda$, $\bar{y}_\lambda$, $\bar{z}_\lambda$ sensitivity curves of the detector unit.

Hunter has developed a specialized photoelectric Color-Difference Meter,* which is proving very well adapted to the grading of a wide variety of products. The instrument employs photovoltaic cells and incorporates circuits by means of which the signals can be combined to record colour differences in terms of the NBS unit. A noteworthy adaptation of this instrument is the automatic colorimeter for raw cotton † (see p. 274) with which the lightness and yellowness of the cotton can be tested; the appropriate readings are derived from the photo-currents and projected on a two-dimensional grid on a panel at the top of the instrument.

In an alternative design of colorimeter the normal type of colour filter has been replaced with a spectroscopic system incorporating a templet in the plane of the spectrum. The principle is illustrated in Fig. 9.8. Light entering the slit S is collimated

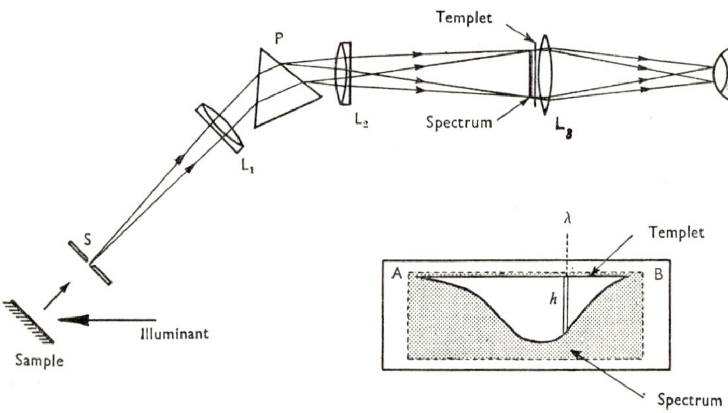

FIG. 9.8 The principle of the spectrum and templet colorimeter

* R. S. HUNTER. *J. Opt. Soc. Am.*, 1958, **48**, p. 985.
† D. NICKERSON, R. S. HUNTER and M. G. POWELL: *J. Opt. Soc. Am.*, 1950, **40**, p. 446.

by the lens L_1, dispersed by the prism P and focused by L_2 to form a spectrum. A mask or templet, shown also in the lower diagram, is placed in the plane of the spectrum and the light passing through the aperture is reunited by the lens L_3 and condensed on the photocell C.

The complete optical system performs exactly the same function as a colour filter, and the spectral transmission of the system can similarly be defined by means of a spectral transmission curve. Since the shape of the curve will be governed mainly by the height h of the aperture in the templet at each wavelength λ, it becomes a purely mechanical problem of cutting the aperture in the templet to the right shape to secure any desired transmission curve.

The complete colorimeter includes the light source, the specimen to be measured and the photocell, in addition to the spectrum and mask system. The specimen may be mounted either in front of the entrance slit of the prism system, when it will be illuminated by the white light from the standard source, or in front of the photocell, when it will be illuminated by the integrated light transmitted through the templet. The first arrangement has the advantage that any fluorescence in the specimen will be correctly recorded, but it is likely to be less efficient from the point of view of light collection.

The major difficulty in the construction of the instrument is the cutting of the templet. Aperture plates with adjustable leaves have been used, but these are not now considered very satisfactory. The height h required at each wavelength can be calculated from the photocell response when a slit is allowed to traverse the spectrum, or, if the height of the slit is adjustable, h can be found by experiment from the setting required to give the response corresponding to each distribution curve. When the templet has been cut, it can be checked by again traversing a slit through the spectrum, noting the response and correcting any residual errors. It has also been suggested that colour filters might be introduced to give approximately the correct output; the function of the templet would then be merely to correct for any differences that remained between the actual and the required spectral sensitivities.

A templet colorimeter and photometer which has been described by Winch and Machin,* had some success as a commercial instrument. In spite of the rather small spectrum—about one inch long—chromaticity measurements accurate to about 0·002 in x and y were possible. The instrument employed the very satisfactory photocell system developed by Winch and Machin† and marketed as the G.E.C. Precision Photoelectric Photometer Unit in which the photocell itself consists of a potassium-on-silver, photo-emissive cell, with the grid leak and an electrometer triode housed with the photocell in an evacuated glass outer jacket.

The spectrophotometer developed by the author‡ and described in the next chapter has also been used as a templet colorimeter. A spectrum some 8 inches long is reunited by means of a curved mirror which reflects the light back through the dispersing system. A G.E.C. photometer unit of the type just described was used to record the light output, and templets shaped to give the $\bar{x}_\lambda$, $\bar{y}_\lambda$ and $\bar{z}_\lambda$ distribution curves can be placed in front of the spectrum mirror. This instrument has proved a very effective tool, both for teaching and academic research, although the need to check the calibration at fairly frequent intervals can be tedious. This is of less concern when the instrument is used as a differential colorimeter.

However, it must be admitted that while the spectrum-templet system is attractive in principle, its exploitation has been disappointing in practice. The arrangement provides all the flexibility necessary to give the correct $\bar{x}_\lambda$, $\bar{y}_\lambda$, $\bar{z}_\lambda$ sensitivity curves, but the elaboration of the optical train and the need for frequent calibration checks have mitigated against the development of a commercially viable instrument. Also, the system is less efficient in light collection than a photocell-filter instrument and must, therefore, have a lower light sensitivity. Perhaps some new thinking may lead to a more successful design, and one near-equivalent of the same principle would be the use of a multilayer interference filter which varies in path retardation along its length to give an interference spectrum. A templet placed in contact with such a spectrum filter gives a fair measure of control over the spectral

* G. T. Winch and C. F. Machin: *Trans. Ill. Eng. Soc.* (Lond.), 1940, **5**, p. 93.
G. T. Winch: *Trans. Ill. Eng. Soc.* (Lond.), 1956, **21**, p. 91.

† G. T. Winch and C. F. Machin: *G.E.C. Journal*, 1935, **6**, p. 205.

‡ W. D. Wright: *Optica Acta*, 1954, **1**, p. 102.

transmission curve of the filter and templet considered as a single unit.

VISUAL SPECTROPHOTOMETERS

Nowadays visual spectrophotometry plays only a negligible part in colorimetry, and the reasons for its decline are not hard to seek. In the first place, the observations themselves are lengthy and fatiguing, and in general neither their precision nor their accuracy now compares favourably with photoelectric instruments. It is true that the spectral transmission of a filter can probably be measured to the same order of accuracy with a visual as with a photoelectric instrument, but with reflecting surfaces there is insufficient light for accurate visual measurements. Again, if the photoelectric instrument is a recording instrument, there is an immense saving of time. The potential advantages of photoelectric spectrophotometers are therefore speed, sensitivity, comfort, precision and accuracy.

There are, nevertheless, a number of precautions that have to be observed if all these advantages are to be realized in practice, and there is perhaps a greater danger that they will be overlooked in photoelectric, than in visual, equipment. In the author's laboratory, for example, the colorimeter described on p. 75 can be used as a visual spectrophotometer for transmission measurements, and if for some reason it were necessary to make a rapid check on the transmission of a filter at one or two wavelengths, this instrument could be used as quickly and effectively as the photoelectric instrument described on p. 247 with much less likelihood of any incorrect adjustment or other source of error passing undetected.

Hence, as an introduction to spectrophotometry, it will be as well to describe at least one visual instrument, if for no other reason than that it embodies the experience accumulated by designers of optical equipment in the early days of spectrophotometry. For this purpose we will choose the Guild instrument,* the chief components of which are illustrated in Fig. 9.9.

The apparatus consists essentially of a Hilger constant-deviation spectrometer to which a second collimator has been attached. Two

* J. Guild: *Trans. Opt. Soc.*, 1924, **26**, p. 74.

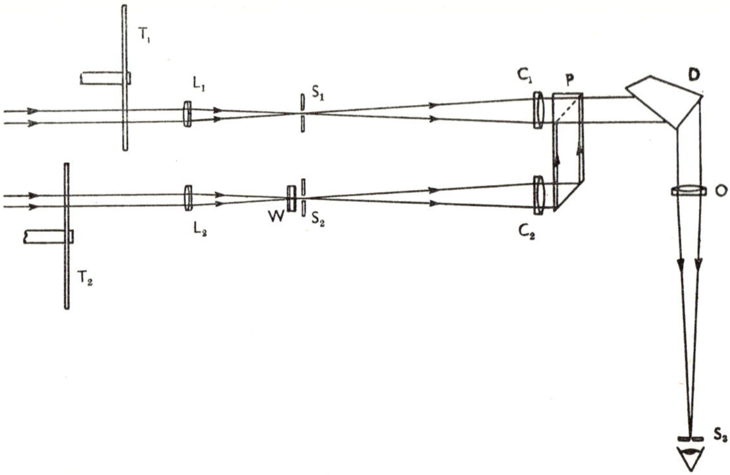

FIG. 9.9 Optical system of the Guild spectrophotometer

parallel beams of light are derived from a single source and focused by lenses L_1 and L_2 to form images of the source on the two collimator slits S_1 and S_2. The light is then collimated again by C_1 and C_2, and the two beams are brought into juxtaposed positions by passage through the photometer prism P; after refraction in the dispersing prism D, the beams are focused by the objective O to form two superimposed spectra in the plane of the slit S_3. When the observer places his eye behind S_3, he sees the photometer field illuminated by the two beams in monochromatic light whose wavelength is determined by the position of S_3 relative to the spectra formed in its plane. This wavelength can be varied by rotation of the prism D using the calibrated wavelength drum attached to the instrument. The field itself is of the contrast pattern, with which greater sensitivity can be obtained by setting for equality of contrast than by the simple setting for equality of luminance. The design of the prism P to give this type of field is fully described in the original paper. Intensity control is secured through rotating sectors which can be introduced into either beam at T_1 or T_2. A series of fixed aperture sectors is provided, the final setting being made by adjustment of a double photometer wedge W mounted in front of S_2.

In using the apparatus, Guild describes a number of points of technique which should be followed to obtain the highest accuracy. The specimen whose transmission is to be measured is placed in the beam that illuminates the slit S_1; when large enough, it is placed to the left of L_1, but when small and thin, it may be mounted immediately in front of the slit. A match is made with the specimen in position by adjusting the photometer wedge. The specimen is then removed and a sector introduced in the same beam and a new match made, again by adjustment of the wedge. If the proper sector is chosen, the densities of the two wedge settings need not differ by more than about 0·02. From the aperture of the sector and the wedge readings corrected for wavelength, the transmission of the filter for the wavelength concerned can be calculated, and similar observations can be repeated for other wavelengths through the spectrum to give the complete spectral transmission curve. When the transmission of the filter is fairly low, it is convenient to use sectors in both beams; where the transmission is varying rapidly with wavelength, the final setting can be made with the wavelength drum instead of with the wedge, so that in effect the wavelength for a given transmission is determined, rather than the transmission for a given wavelength.

The widths of the slits can be adjusted separately, and in general S_1 and S_3 are made as wide as considerations of purity will permit. When the absorption is changing rapidly with wavelength the slits must be narrow, but they can usually be wider for blue wavelengths than for red because of the increased dispersion at shorter wavelengths. The slit S_2 is adjusted merely to bring the illumination to the correct level; purity is unimportant in this beam, since its only function is to provide a comparison field of colour identical to that of the test field. Serious errors may arise in the transmission curve of a filter if its surfaces are not reasonably parallel since this may cause a translation of the image of the source across the slit when the specimen is inserted. This can lead to a marked change in the flux of light through the slit, unless the source is uniform over its surface; a ribbon-filament lamp is probably the most satisfactory source in this respect at the present time. Errors due to stray light are most serious at either end of the spectrum, since the scattered light from the brighter parts of the spectrum will be relatively large. The stray light can be effectively removed by the use of a colour filter placed in front of the observer's eye, if chosen to transmit maximally in the part of the spectrum being measured and to absorb in other regions.

The measuring equipment itself includes the wavelength scale, the sectors and the wedge. The wavelength scale can readily be checked by illuminating the slits with sources emitting line spectra, such as the mercury lamp. The larger sectors need no calibration if they are made carefully, since the errors in their manufacture are considerably smaller than anything that can be observed photometrically, while the sectors with openings of only a few degrees can, if desired, be checked photometrically; if this is done, it rules out any question of the validity of Talbot's law for very small openings. The wedge can be calibrated with the sectors, and a correction curve for wavelength can be found by recording the density on the wedge which is equivalent to a particular sector at various wavelengths through the spectrum.

We should note that Guild used the Maxwellian view with the observer's eye more or less in the plane of the spectrum. This permits a larger and more uniform field of view than if a strip of the spectrum is focused with an eyepiece, although in general the latter arrangement gives the higher level of illumination.* A less satisfactory feature of the design is the use of a single instead of a double monochromator. A second monochromator is now the recognized method of eliminating errors due to stray light and is to be preferred to the use of colour filters.

It is also worth noting that Guild used sectors and wedges in preference to polarizing prisms, which, although mechanically easy to operate, absorb light and may be inaccurate on account of partial polarization in other parts of the optical system. On the other hand, the König-Martens spectrophotometer, in which polarizing prisms are employed for the photometric control, was used for many years at the National Bureau of Standards for accurate transmission measurements.†

Most of the sources of error noted by Guild exist in photoelectric spectrophotometers as well as visual instruments, although

* See J. Guild: *Proc. Opt. Convention*, 1926, **1**, p. 81.
F. Twyman and G. F. Lothian: *Proc. Phys. Soc.*, 1933, **45**, p. 643.

† K. S. Gibson: Chap. 5 on 'Spectrophotometers' in *Analytical Absorption Spectroscopy*, ed. M. G. Mellon (Wiley: New York. Chapman and Hall: London), 1950.

the techniques necessary to eliminate them may differ depending on the type of instrument. One particular source of error, namely stray light, is likely to make a different impact if the spectral sensitivity of the photocell extends outside the visible spectrum. Thus the error due to stray light from the violet and ultra-violet parts of the spectrum will be negligible in a visual observation because the eye is very insensitive to these short wavelengths. For some photocells, on the other hand, this may be a wavelength region of high sensitivity and special precautions may be required to minimize the error.

PHOTOELECTRIC SPECTROPHOTOMETERS

Photoelectric spectrophotometers are of two types, non-recording and recording instruments. In a non-recording instrument, measurements have to be taken wavelength by wavelength through the spectrum, and either the photocell is used to test for equality between two beams when their relative intensities are varied with some calibrated optical device, or the output from the cell is recorded and used as a measure of the amount of light entering the cell. A number of instruments of this type have been developed and two examples will be described here to illustrate their general features and potentialities. In the recording instrument, the results are plotted automatically through a suitable optical, mechanical and electrical linkage to give a continuous record of the variation of reflection or transmission with wavelength. This is a remarkable achievement in view of the small amount of light available, and the development in the 1930s of the Hardy or General Electric spectrophotometer, which we shall describe in some detail, marked a new stage in the application of spectrophotometry to colorimetric problems.

The Beckman spectrophotometer,* which is typical of the single-monochromator non-recording class of instrument, has been chosen for description here in view of its early appearance in the field. Other designs, such as the Unicam and the Uvispek, have also proved eminently sound instruments both for industry and research.

The monochromator consists of a slit, a concave mirror, and a

* H. H. Cary and A. D. Beckman: *J. Opt. Soc. Am.*, 1941, **31**, p. 682.

30° quartz prism, from which the dispersed light is reflected back at the second face along nearly the same path as the incident light. After emerging from the exit slit the light passes to the sample and thence to the photocell. Two photo-emissive cells are provided, one a red-sensitive caesium-oxide cell for use at wavelengths longer than 620 nm, and the other a blue-sensitive caesium-antimony cell for shorter wavelengths. The photocurrent is balanced, and thus measured, against a potentiometer, any lack of balance being amplified electronically and indicated by a milli-ammeter to give greater sensitivity of setting at the null point.

The instrument has good light sensitivity, and for transmission measurements, a slit width of only 0·5–1·5 nm need be used. For reflection measurements the special attachment shown in Fig. 9.10

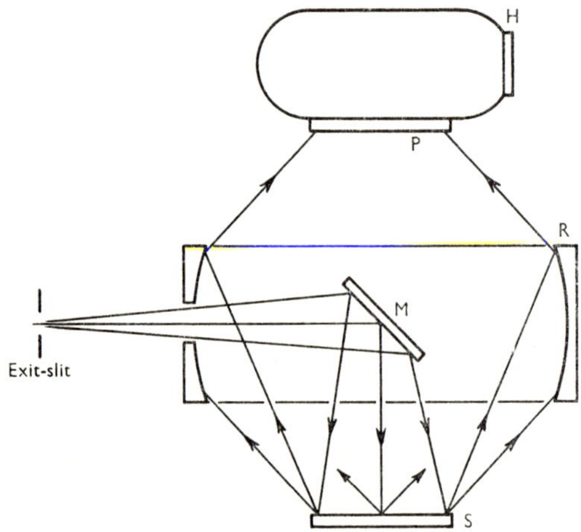

FIG. 9.10 Attachment for reflection measurements with the Beckman spectrophotometer

is provided. The monochromatic light is reflected by the mirror M down on to the specimen S. An annular cone of the light diffusely reflected by S is then picked up in the ellipsoidal metal mirror R and reflected on to the diffusing plate P, behind which the photocell H is mounted. Although the incident illumination is normal and the viewing angle is at approximately 45° to the normal, there is a 20° angular spread from 35° to 55° in the light collected by R.

Since less light is available for reflection measurements, the slit width has to be increased to some 5–10 nm depending on the wavelength region. Very dark samples are difficult to measure accurately and in spite of the extreme simplicity of the optics of the monochromator, a single dispersion instrument is bound to include some stray light. Nevertheless, equipment of this type has proved a very effective tool for colour research and colour control,

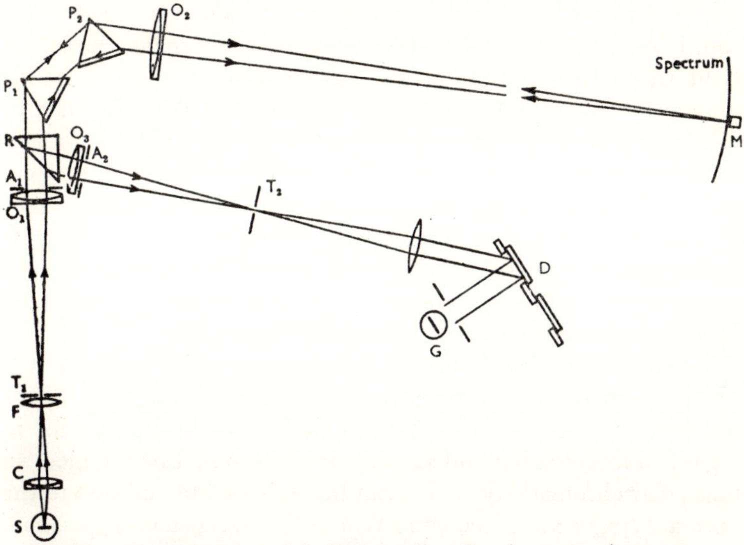

FIG. 9.11 Optical system of the Wright photoelectric spectrophotometer

and has the advantage of being compact in size and moderate in price.

An instrument on an entirely different scale has been developed by the author,* and is intended primarily as a research and teaching instrument. The optical system, as shown in Fig. 9.11, calls for no comment except that in place of a slit in the 8-inch long spectrum, a mirror strip is used to select and reflect the required waveband back through a lower level of the prism to be focused at the exit slit T_2. The instrument is thus a double monochromator with the consequent freedom from stray light which that implies, and since it is a large instrument, there is a corresponding increase in the amount of light available.

* W. D. WRIGHT: *Optica Acta*, 1954, **1**, p. 102.

The sample holder is located beyond the exit slit and as it can be rotated about a vertical axis, any angle of incident illumination can be provided. A G.E.C. Precision Photoelectric Photometer Unit, already described on p. 240, was used for a number of years very successfully but this has now been replaced with a photomultiplier and digital voltmeter. As the photocell unit can also be rotated about a vertical axis, the light reflected from the sample can be measured at any angle to the surface so that the instrument can, if required, be used as a goniospectrophotometer. Sufficient light from the monochromator and sufficient sensitivity in the photocell are available for the measurement of densities of 5 or 6 in a filter, or luminance factors as low or lower than 0·005 in a reflecting surface. Mirrors of various widths from 2 mm to 5 cm can be placed in the spectrum, and where extremely dark samples or small areas of a sample are being studied, it may be legitimate and desirable to sacrifice purity and increase the amount of light by using one of the wider mirrors. The bandwidth normally used for reflection instruments is of the order of 5–10 nm.

Reference has already been made on p. 240 to the conversion of the instrument into a templet colorimeter with a long mirror mounted in the spectrum. The flexibility of the instrument in regard to illumination and viewing angles can still be retained, so that polar chromaticity curves can be measured in addition to the normal type of measurements. Where fluorescence of a specimen is present, provision has to be made for the sample to be illuminated and mounted in front of the exit slit, although the light efficiency of the system is thereby reduced. A general view of the instrument is given in Plate 10(*b*), p. 230.

The recording spectrophotometer developed by Hardy* and manufactured by the General Electric Co. in America, was redesigned a few years ago † and a diagram of the new system is shown in Fig. 9.12. Two prism systems in series constitute a double monochromator, from which monochromatic light free from stray light emerges through the exit-slit 3. The monochromatic light is polarized by passage through a Rochon prism and then falls on a Wollaston prism, which serves to split the beam into two

* A. C. Hardy: *J. Opt. Soc. Am.*, 1935, **25**, p. 305.
J. L. Michaelson: *J. Opt. Soc. Am.*, 1938, **28**, p. 365.

† B. S. Pritchard and W. A. Holmwood: *J. Opt. Soc. Am.*, 1955, **45**, p. 690.

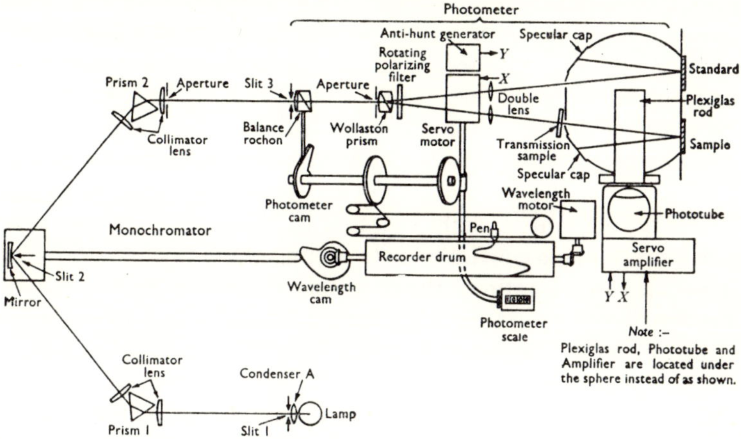

FIG. 9.12 Schematic diagram of the General Electric Hardy recording spectrophotometer (Pritchard and Holmwood)

plane polarized components with their vibration directions at right angles to each other. The relative intensities of the two components will depend on the orientation of the Rochon prism with respect to the Wollaston prism. When the vibration of the beam incident on the Wollaston prism is symmetrical in relation to the vibration directions of the two beams produced by passage through the Wollaston prism, the two beams will be of equal intensity. But if the Rochon prism were rotated 45° on either side of this position of symmetry, one beam would increase to a maximum intensity while the other would be reduced to zero intensity, and vice versa. In practice, the Rochon prism is permitted a rotation of 45° on only one side of the position of symmetry, so that the light flux in one of the beams can be made to vary from an average value, say F, up to a maximum flux approximately equal to $2F$, while the other beam simultaneously decreases from the average value F down to zero light flux.

After leaving the Wollaston prism, the two beams pass through a rotating polarizing filter whose presence we will for the moment ignore, and are then focused into two openings in an integrating sphere; they thus illuminate respectively the sample and the standard magnesium-oxide surface, which are mounted at the back of the sphere opposite the apertures. Suppose, at some wavelength λ, that the reflection factor of the sample is ρ_λ compared with a factor of unity for the oxide surface; then if the amounts of light reflected by the sample and by the standard are to be equal, the light flux in the beam falling on the sample must be greater than the flux in the beam falling on the standard in the ratio of $1 : \rho_\lambda$. This ratio can be varied from unity up to infinity by rotation of the Rochon prism, and so at some position of this prism the amounts of light reflected by the sample and the standard can be made equal. Since the setting of the Rochon prism can be related to the ratio of the light flux in the two beams, the reflection factor ρ_λ can be measured, provided that some test for equality in the two reflections can be incorporated in the instrument. To achieve this the two beams are arranged to enter the integrating sphere in rapid alternation by means of the rotating polarizer, which is driven by a synchronous motor so as to modulate the two beams at a rate of 60 c/s and in opposite phases. After reflection alternately from the sample and the standard the light is collected in the integrating sphere and piped down through a Plexiglas rod to a photocell mounted outside the sphere.

In general, there will be fluctuation of light owing to the inequality of the reflections from the sample and the standard, and this light ripple will be converted into an electrical ripple in the photocell. By means of a servo system, this ripple is made to control and set the Rochon prism to the position giving minimum ripple. When the instrument is put into operation the wavelength is changed continuously from one end of the visible spectrum to the other, and the setting of the Rochon prism will vary continuously, corresponding to the varying reflection factor ρ_λ of the sample. A suitably calibrated cam is linked to the Rochon prism so that the complete spectral reflection curve can be drawn on a recording drum.

Accurate results on an instrument of this complexity can be obtained only by careful calibration and by the application of suitable checks during a series of measurements.* These include the checking of the wavelength scale by recording the absorption curve of a sample of didymium glass, which possesses several sharp maxima and minima at known wavelengths in the spectrum,

*K. S. Gibson and H. J. Keegan: *J. Opt. Soc., Am.*, 1938, **28**, p. 372.
H. Hemmendinger and H. R. Davidson: *Proc. Symp. Colour Measurement in Industry* (Colour Group (G.B.): London), 1967, p. 1.

measurement of the 100-per-cent-reflection curve by using magnesium oxide as both sample and standard, and measurement of the zero-reflection curve by completely obstructing the sample beam. The photometric scale may be checked by measuring neutral filters or grey opaque samples which have been accurately calibrated for transmission or reflection by other methods.

The time required to measure a single curve is governed by the amount of light available. This in turn depends on the lightness of the sample and the width of the slit, which is normally set at 10 nm. If the specimen is dark or the slit is narrow, a lower speed of recording must be used to compensate for the small flux of light in the photocell. Similarly, where a specimen possesses sharp absorption bands, a high speed or a wide slit will tend to lower the resolution of the instrument and may also cause a lateral displacement of the curve. Even in the most difficult circumstances, however, a satisfactory record can usually be obtained with a run of two or three minutes. In comparing this time with that required by other methods, allowance should also be made for the time required to make any additional runs that may be considered necessary to check the calibration of the instrument.

In a number of important industrial problems the interest is centred on the measurement of small differences in colour rather than on the absolute measurement of a colour. Although the Hardy instrument was not designed specifically for differential measurements, a number of samples can be recorded rapidly and compared with one another without elaborate checks on the calibration.

There are two problems which cannot be satisfactorily handled with the instrument as described. The first of these is the testing of samples which are subject to some degree of fluorescence. When a sample is illuminated by one of the standard sources, any fluorescence within the visible spectrum will be indistinguishable from the ordinary reflected light, and a measurement with, say, a trichromatic colorimeter will include both the fluorescent and reflected components and give a correct record of the colour as actually observed. When, however, the sample is illuminated by each part of the spectrum in turn, then if wavelength λ gives rise to fluorescence at some other wavelength λ', the λ' radiation will be added to the reflected portion of the λ radiation and will be recorded by the photocell system as if all the light were of wavelength λ. This can lead to important errors in the spectral reflection curve, although the difficulty is one shared with any spectrophotometer in which the sample is illuminated at the exit, rather than the entrance, slit of the instrument.

The other difficulty with which the Hardy instrument is faced is the separation of the specular from the diffuse component of the light reflected from a sample. An integrating sphere has no directional properties, and as a consequence the instrument tends to measure the reflection factor of the sample rather than its luminance factor. With perfectly diffusing surfaces this will be immaterial, but with glossy and semi-glossy specimens the difference in the two factors can be most marked. A partial solution to the problem has been to provide black caps for insertion on the two sides of the sphere at the points where the specular reflection, if any, would strike the wall of the sphere. This device will be most effective with highly glossy specimens, but with semi-glossy surfaces, the light that is measured will still consist of a rather indeterminate mixture of the specular and diffuse components. We must therefore anticipate that with such samples the results may differ significantly from measurements made with other types of instrument.* This has in fact been borne out by recent interlaboratory comparisons of colour measurements made using a number of G.E. and other types of spectrophotometer.† It is to be hoped that if the conditions for illumination and viewing geometry recommended by the C.I.E. Colorimetry Committee in 1967 (see p. 28 and Appendix I, Part 2) are adopted by instrument manufacturers, this will help to minimize the discrepancies in the measurements made with different instruments.

The G.E. spectrophotometer normally records a linear plot of the reflection values against wavelength. Other photometric cams can, however, be employed to plot values on a non-linear or on an expanded scale. A density cam is also available and leads to more accurate readings where the reflection values are low.

One very important addition can now be purchased, namely an Automatic Tristimulus Integrator,‡ by means of which the

* W. D. WRIGHT: *Proc. Conf. Optical Instruments and Techniques*, 1961, p. 3 (Chapman and Hall: London), 1962.

† A. R. ROBERTSON and W. D. WRIGHT: *J. Opt. Soc. Am.*, 1965, **55**, p. 694.
F. W. BILLMEYER: *J. Opt. Soc. Am.*, 1965, **55**, p. 707.

‡ H. R. DAVIDSON and L. W. IMM: *J. Opt. Soc. Am.*, 1949, **39**, p. 942.

products $P_\lambda . \beta_\lambda . \bar{x}_\lambda$, $P_\lambda . \beta_\lambda . \bar{y}_\lambda$ and $P_\lambda . \beta_\lambda . \bar{z}_\lambda$ are summated continuously as the curve is being recorded. The great saving of time and effort which this represents does not need to be stressed. The addition means, of course, that there is yet another link in an already complex instrument, and the need for perfect adjustment is still further increased. The more complicated an instrument becomes and the more interrelated its various parts, the less easy is it for a correction to be made for any inaccuracy of setting or maladjustment that may develop in the mechanism. However, the use being made of equipment of this type in colorant formulation (see p. 264) indicates that when the instrumental complexities are matched by experimental skill, the industrialist has to hand a tool of great commercial potential.

There is a second type of recording spectrophotometer in which the photometric measurement is obtained from the value of the photo-current itself instead of from an optical light attenuator. In this type of instrument the light from the monochromator is again sub-divided into two components by means of an optical beam-splitter, with one beam illuminating the reference white and the other the sample. Some optical device such as an oscillating mirror or a light chopper is introduced so that the light reflected by the reference standard and the sample is fed in alternation into a photo-cell. By means of a phasing device the alternating photo-currents are separated and amplified so that the signals can be compared in a ratio-recording potentiometer. The setting of the potentiometer at the balance point is then recorded continuously while the wave-length is varied through the spectrum. The Beckman DK–2 spectrophotometer is of this type.

One spectrophotometer may differ from another in a variety of ways, for example, by incorporating either a single or double monochromator and the use of either a prism system or diffraction grating; by the method of illumination and light collection from the sample; by the facility for expansion either of the wavelength scale or the photometric scale, and the possibility of measuring small differences on an expanded scale; by the range of densities that can be measured; by the attachment of an automatic tri-stimulus integrator to the recording device, and so on. The number of recording spectrophotometers now on the market has increased considerably in the last few years, but it would be invidious here to attempt any assessment of their relative merits. They include, for example, the Zeiss PMQ II spectrophotometer, the Bausch & Lomb Spectronic, the Cary spectrophotometer and the Pretema Spectromat. In the Spectromat the normal type of monochromator is replaced by 25 interference filters which isolate in turn 25 wavebands through the spectrum. Another special feature is the display on a screen of the spectral transmission or reflection curve as given by the height of 25 vertical lines at the appropriate wavelengths.

Reference must also be made to a new Japanese spectrophotometer and digital tristimulus integrator* on account of its very novel features. A long photometric comb (Fig. 9.13) is inserted in the monochromatic beam which illuminates the white reference standard surface, and the setting of the comb is controlled by a

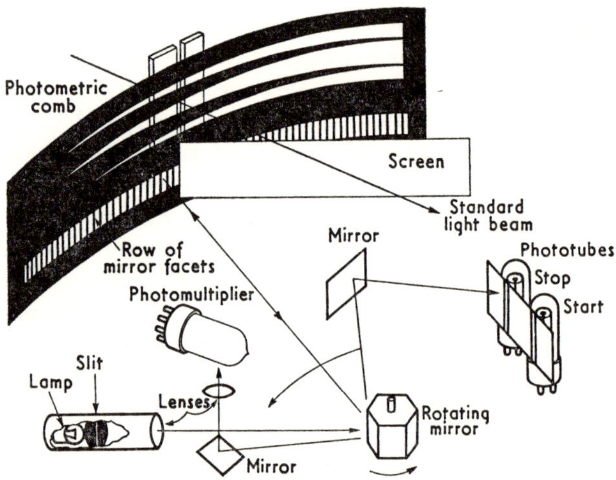

FIG. 9.13 Schematic arrangement of the photometric control in a new Japanese spectrophotometer and the auxiliary optical system for registering the setting of the comb by electronic counting.
(Mori, Nikura and Azuma, *C.I.E. Compte Rendu*, 1963)

servomotor until balance is secured between the light reflected from the standard and from the sample. On the same cylindrical member in which the comb is cut, there are 250 mirror facets mounted, and the position of the comb at the balance point is

* L. Mori, I. Nikura and T. Azuma: *C.I.E. Compte Rendu*, 1963.

determined by electronic counting of the number of facets exposed. To do this a rotating hexagonal mirror sweeps a light beam across the mirrors and back into a photo-multiplier. At the same time another beam is reflected into two photomultipliers to generate stop-and-start signals defining the beginning and end of each photometric pulse group. This method of recording is well adapted to the selected-ordinate method of tristimulus integration which is employed. A selected-ordinate signal generator linked to the wavelength motor ensures that the pulse counts are made at the correct wavelengths so that their sum gives the X, Y and Z values. Different selections of ordinates are available corresponding to different illuminants. This is an ambitious and expensive instrument, but is symbolic of the modern trend in optical instrumentation.

We can look forward with confidence to continuing advances in the design of spectrophotometers to meet the changing demands of industry. In particular, we can expect that there will be more frequent provision for the reversal of the optics for the measurement of samples which fluoresce. Similarly, arrangements will be more common for the digital output of the data and facilities for the on-line computation of the tristimulus values X, Y, Z. Accurately calibrated standards (see p. 258) to check the performance of individual instruments will also come into wider use.

CHAPTER 10
PRACTICAL APPLICATIONS OF COLORIMETRY

THE CLASSIFICATION OF COLOUR PROBLEMS

Several references have been made in the preceding chapters to the uses to which measurements of colour can be put, but when we attempt any logical classification, we are foiled by their great variety. We might attempt a broad division into measurements made (*a*) when we are interested in the visual quality of a colour for its own sake or (*b*) when we wish to use the colour of an object as an indicator of other properties which it may possess. Decorative colours, such as the colours of fabrics, paints, cosmetics, wood, tiles, leather, laundry, etc., obviously fall in the first class. So, too, do the colours occurring in nature, for example, those of flowers, foliage, birds or insects, although there are many examples where the recording of such colours is of interest not only from the aesthetic point of view, but also because the colour is related to the natural growth and development of the plant or creature concerned. On the more technical side, the colours of signal glasses, light sources and camouflage materials, and the quality of reproduction in colour photography and colour television, are examples which would also be included in the first group, since our interest in each lies essentially in the visual appearance of the product. On the other hand, when we use the colour of fruit as a guide to its sweetness, we have an application which falls in the second class. In the second group also would come the testing of the end-point in a titration, the estimation of the strength of a solution, and tests on the quality of oils or the refining of sugar. Similarly, in photo-elasticity, the colour of the fringes is used as a measure of the strains within the specimen, while in the heat treatment of steel the colour may be watched to judge the temper of the metal.

With such diverse applications—and the list is obviously not exhaustive—the broad division in this way may be interesting but is hardly of much value as a guide to the type of measurement or of apparatus which may be required. For some purposes a specification on an arbitrary scale is adequate, while in others a standard specification of the highest possible accuracy may be demanded. Again small differences between colours are important in some measurements; in others, it is the absolute colour quality that matters. For certain applications the spectral composition of the light has to be determined, but for others, a knowledge of the integrated colour alone is sufficient. Each problem must therefore be treated on its merits, and in the subsequent sections of this chapter typical problems have been selected from various industries and sciences to illustrate the different ways in which colour measurements can be applied.

COLOUR STANDARDS IN INDUSTRY

The general practice in industry is to define the colour of a product by a colour standard, while colour matching and inspection is normally made by eye. Nevertheless, the maintenance of the colour standard itself may call for some absolute method of colour measurement, while the colouring process and the colouring materials may also be controlled by colour measuring instruments.

If we take the automobile industry as an example, we can define six stages in the colouring of a car body as illustrated in Fig. 10.1.* Each of these stages involves some colour observation, colour control or colour definition by sample or instrument, but it is at stage 3 where colour control is most critical. It has been estimated

1	2	3	4	5	6
Stylist's colour scheme →	Vendors' paint samples →	Master colour panels →	Paint supplies →	Paint application →	Finished product
Defined by colour chips and panels	Laboratory tests including stability of colour with weathering, stoving, etc.	Maintenance against colour drift	Check against master colour panels	Control of paint viscosity, spraying time, conveyor speed, baking schedule, etc.	Inspection for colour defects, etc. Spot repairs to give accurate colour match

FIG. 10.1 Six stages in the colouring of a car body

that the chromaticity of the master panels needs to be held constant at least to within ± 0·0005 in x or y for near-neutral colours and to within ± 0·005 in luminance factor. The paint laboratory which is

* W. D. WRIGHT: *J. Opt. Soc. Am.*, 1959, **49**, p. 384.

supplying the paint may have to maintain a master 'master panel' to even closer limits if it is to insure itself against drift in its supplies.

At the moment it is hardly possible to make absolute spectrophotometric or colorimetric measurements to this order of accuracy, but it sets a goal for the designer of such instruments. If effectively permanent samples are available as reference standards, then working standards can be checked to the required precision by using a tristimulus colorimeter as a differential instrument to measure the small difference between the working standard and the reference standard. Differential measurements would also be of value in defining and checking the colour tolerances that could be permitted at stages 2 and 4.

Colour standards are also in demand as a means of checking and calibrating colorimeters. The inter-laboratory comparisons referred to on p. 252 have shown a depressingly large spread in colour measurements made with instruments of the same design, let alone with different types of instrument. One way to reduce this spread is to make available colour standards that have been calibrated to a high accuracy at a standardizing laboratory; these can then be used to check and, if necessary, correct the instrument scales of a colorimeter. For some purposes, indeed, it may be sufficient to provide sets of identical standards that have not been given their C.I.E. specifications, if the main aim is to lock a group of instruments together so that all give identical, but not necessarily C.I.E., readings.

One project undertaken at Imperial College* had the dual aim of comparing the accuracy of spectrophotometric measurements made in different laboratories and of providing an extensive series of calibrated grey tiles that could be used for differential measurements with a spectrophotometer. The spectral reflection curves of the tiles are shown in Fig. 10.2. Another very important development on these lines has been the production of sets of twelve ceramic colour standards specially selected for their value in diagnosing and checking the performance of colorimeters and spectrophotometers. They have been accurately calibrated at the National Physical Laboratory and are available from the British Ceramic Tile Council.†

* S. Johnson, D. H. Phillips, A. R. Robertson and W. D. Wright: *J. Soc. Dyers and Colourists*, 1963, **79**, p. 731.
A. R. Robertson and W. D. Wright: *J. Opt. Soc. Am.*, 1965, **55**, p. 694.
† F. J. J. Clarke and P. R. Samways: N.P.L. Report No. MC2, August, 1968.

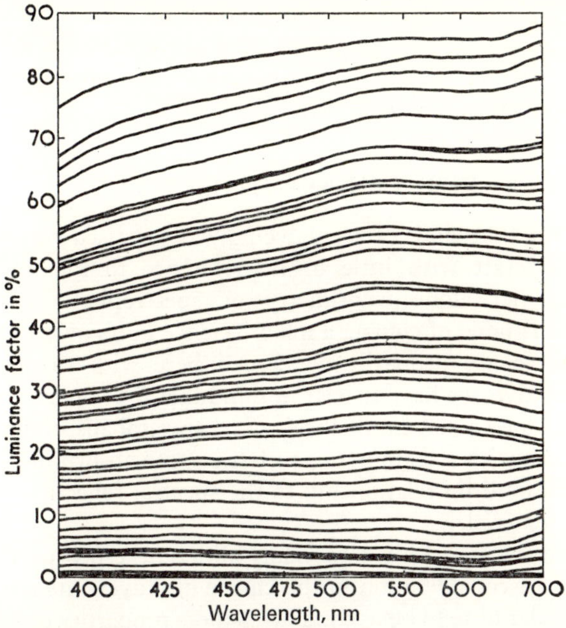

FIG. 10.2 Curves of spectral luminance factor against wavelength for a set of 46 tiles proposed as reference standards for use in differential spectrophotometry
(Johnson, Phillips, Robertson and Wright, *S.D.C. Symposium*, 1963)

The ceramic industry itself is an example of the need for close colour control, since a tiled wall provides the ideal viewing conditions for the detection of any colour differences that may exist among the component tiles. Malkin and Dinsdale* have described an interesting study to determine customer tolerances for colour variation in white, grey, pink, blue and yellow tiles. The tiles themselves were measured with a Colormaster colorimeter and some 700 observers who inspected panels of tiles were required to indicate which panels were unacceptable because of their colour variation. The results were then analysed by relating the colour differences ΔE expressed in N.B.S. units to the percentage rejection vote. From several points of view this investigation is typical of the way colorimetry can and should be applied to the solution of important commercial problems. It also, incidentally, high-lighted

* F. MALKIN and A. DINSDALE: *Proc. Symposium on Colour Measurement in Industry* (Colour Group (G.B.): London), 1967, p. 221.

some of the uncertainties still surrounding the subject, particularly the discrepancies that emerged in the interpretation of the results depending on the choice of uniform colour scale and colour difference formula from which ΔE was calculated.

THE DYEING INDUSTRY

For some thousands of years dyeing was practised very successfully as a craft with little or no reference to the underlying chemistry, let alone colorimetry or spectrophotometry. Dye chemistry is now, of course, a major science in its own right* and although we are primarily concerned here with the colouring power of a dye, we have to bear in mind that the choice of a dye is not determined solely on the basis of its spectral absorption curve. In any case, with certain dyes and fabrics, the absorption curve of the dye bears little relationship to the spectral reflection curve of the dyed material.†

Since dyes are applied to many different types of fibre, an obvious first consideration in the choice of a dye is its affinity for the particular fibre to be dyed. The fastness properties of a dye will be another very important factor in its selection.‡ The dye may have to be fast to light or fast to washing, or it may have to resist the action of organic acids in the form of perspiration. The selection of the dye may be governed by the texture of the material to which it is to be applied, since a tough compact material is difficult to penetrate and if a dye were used which fixed rapidly, the dye would be mainly on the outside of the material, leaving the inside of the threads almost white. With pale colours, too, the dye must not fix too rapidly, for otherwise it will be difficult to produce an even level dyeing. Allowance has to be made for the effect which the texture of a surface has on its colour quality, since the manner in which the light is reflected from, say, the woven fibres in a piece of silk will be radically different from the reflection given by the rougher structure of a tweed jacket or by the pile of a carpet. In

* See, for example, T. VICKERSTAFF: *The Physical Chemistry of Dyeing* (Oliver and Boyd: London), 2nd Ed., 1954.

† H. W. ELLIS: *Proc. Phys. Soc.*, 1941, **53**, p. 288.
J. G. GRUNDY: *Proc. Phys. Soc.*, 1942, **54**, p. 1.

‡ C. H. GILES and R. B. McKAY: *Textile Research Journal*, 1963, **33**, p. 528.
C. H. GILES: *J. App. Chemistry*, 1965, **15**, p. 541.

many instances the colour of a dyestuff in solution is very different from the colour of the finished dyeing, since the final shade may be the result of chemical reactions occurring in the dye-bath itself.

One of the more specifically colorimetric problems of interest to the dyeing industry has been the scope which exists for the production of brighter dyes. These are of commercial importance in their own right and also as components in a dye mixture. There is a theoretical limit to the luminance factor which a coloured surface can have for a given chromaticity (assuming that it does not fluoresce) and in some hues, particularly the yellows, existing dyes approach more closely to this limit than others. The dye manufacturer may therefore wish to concentrate his research on those hues where there is the greatest scope for improvement.* To locate these, a comparison has to be made between the chromaticities and luminance factors of actual dyed surfaces with the chromaticity loci as derived by MacAdam† for the so-called optimal colours. These loci are shown in Fig. 10.3 for surfaces with luminance factors ranging from 0·10 to 0·95 when illuminated by the tungsten source S_A.

Another colorimetric problem is the degree of colour constancy of a dyed fabric with change in the colour of the illuminant. Some reference to the phenomenon of colour constancy has already been made in Chapter 2 and its relation to metamerism was discussed in Chapter 5. The fabric represented by curve A in Fig. 5.6 is an extreme example of poor colour constancy since its colour is green in daylight and brown in tungsten light. Such dyes must have mystified the old craftsmen who had no knowledge of the spectral energy distributions of their light sources or the spectral reflection characteristics of their fabrics. Even today, surprise is often expressed at patterns which change their colour appearance in such a dramatic fashion. For a dyehouse equipped with a spectrophotometer there is no great problem in avoiding the use of dyes which are likely to show serious lack of colour constancy, although the situation has been complicated somewhat by the need to include the fluorescent lamp as one of the illuminants for which colour constancy has to be ensured.

* T. Vickerstaff: *Proc. Phys. Soc.*, 1945, **57**, p. 15.
E. Atherton and R. H. Peters: *Congrès F.A.T.P.E.C.*, III, 1955, p. 147.

† D. L. MacAdam: *J. Opt. Soc. Am.*, 1935, **25**, pp. 249 and 361.

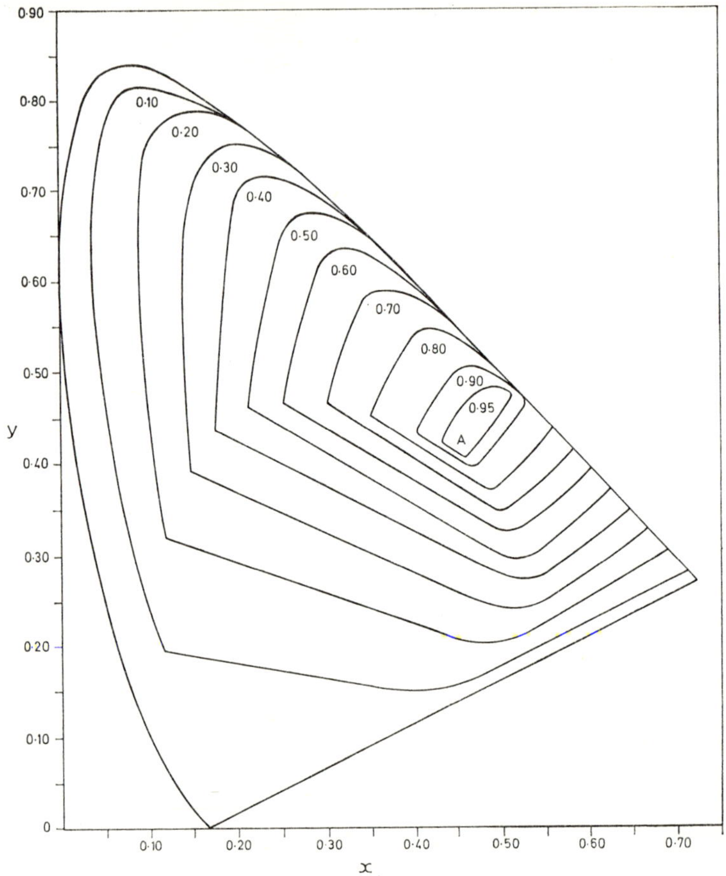

FIG. 10.3. Loci of optimal colours in the 1931 C.I.E. chromaticity chart for surfaces under illuminant S_A, as determined by MacAdam. These loci are boundaries which define areas on the chromaticity chart within which surface colours of a given luminance factor must be located. The figures against each locus give the luminance factor for that locus

However, the real challenge in the dyeing industry is the problem of calculating a dye recipe that will enable a fabric to be dyed to match a sample submitted by a customer. This has in the past been the key task of the trained colourist who, from his store of experience and by reference to his file of previous dye recipes, has often been able to choose a dye formula that would give a close approach to the colour required. Further, he has usually possessed a remarkable ability to decide the direction and magnitude of the changes necessary in the concentrations of the component dyes to improve the match until it was effectively perfect. Nevertheless, there is inevitably an element of trial and error in such a procedure and this can prove time-consuming and expensive. Moreover, it can mean that a dyehouse is dependent almost entirely on the skill of a master craftsman who may not always be available and whose eyesight and judgement may in due time begin to fail. There has therefore been intense activity in the dyeing industry in recent years to develop more scientific methods of colorant formulation based on the principles of colorimetry and spectrophotometry. In the previous edition of this book it was stated that the field in which colorimetry had been most vigorously applied was that of colour television. Pride of place has now almost certainly passed to the dyeing industry.

There are a number of different approaches to the practical problem of colorant formulation, but they all seem to be based on the fundamental work of Kubelka and Munk.* They established a relation, for monochromatic light, between the reflection factor $R\infty$ of an infinitely thick colorant layer and the absorption and scattering coefficients, K and S, of the colorant. As described on p. 285 in an account based on a paper by Duncan, these coefficients are a function of the physical properties of the colorant particles and the medium in which they are immersed. In the Kubelka-Munk theory K and S are additive, which means that the K and S of a mixture can be calculated from the K and S of individual components in the mixture and their volume concentrations. Other simple deductions from the theory are that in order for a surface colour to have a high reflection, the ratio of K/S should be low. Further, for a colorant to have a large tinting strength or colouring power, K/S should be high at certain wavelengths. The ratio K/S of a colorant at a given wavelength is therefore its most significant property and this is the basis of the single-constant theory used most widely in computing procedures for colorant formulation.

* P. Kubelka and F. Munk: *Z. Tech. Physik*, 1931, **12**, p. 593.
P. Kubelka: *J. Opt. Soc. Am.*, 1948, **38**, p. 448.
For an excellent summary and discussion of the literature and the problem see D. B. Judd and G. Wyszecki: *Color in Business, Science, and Industry* (Wiley, New York and London), 2nd Ed., 1963.

However, as Davidson and Hemmendinger have shown,* a two-constant theory involving the determination of separate K and S values leads to more accurate colour predictions, especially in the case of pigment mixtures.

The physical theory underlying light scatter in a colorant layer is complex and does not lend itself to exact solution. Moreover, from the colorimetric point of view, it is a many-variable problem since the final colour is governed by the proportions in which all the wavelengths in the visible spectrum are present in the light finally reflected from the dyed layer. Hence the commercial exploitation of the Kubelka-Munk theory had to await the advent of the electronic computer which alone could handle the massive calculations and the iterative procedures involved in deriving a dye formula and which, because of its speed, could do so in a period of time that would be short enough to be of any economic interest to industry. If the calculations had lasted hours or even days, then the trained craftsman could easily have competed with the computer scientist.

Duncan, and Park and Stearns† were among the first to appreciate the technical and commercial significance of the Kubelka-Munk analysis, although at that time the resources of the modern computer were not available. Subsequently Davidson and Hemmendinger‡ introduced the first practical computer capable of handling the mathematics involved. This is now a fully operative service in which the central equipment is an analogue computer known as the COMIC. The spectral reflection curve of the sample to be matched is first measured on a spectrophotometer, usually the G.E. instrument or the IDL Color Eye used as an abridged spectrophotometer, and values of the reflectance at 16 wavelengths through the spectrum are read off. The 16 dials of the computer are then set to the values of the Kubelka-Munk absorption function corresponding to these reflectances, as read off from a table. Plug-

* H. R. Davidson and H. Hemmendinger: *J. Opt. Soc. Am.*, 1966, **56**, p. 1102.

† D. R. Duncan: *Proc. Phys. Soc. London*, 1940, **52**, p. 380.
J. Oil and Colour Chemists' Assoc., 1949, **32**, p. 296.
R. H. Park and E. I. Stearns: *J. Opt. Soc. Am.*, 1944, **34**, p. 112.

‡ H. R. Davidson and H. Hemmendinger: *J. Opt. Soc. Am.*, 1948, **48**, p. 218A.
H. R Davidson, H. Hemmendinger and J. L. R. Landry: *J. Soc. Dyers and Colourists*, 1963, **79**, p. 577.

in units representing the colorants to be used in the formulation are then inserted in the computer and their concentration dials are adjusted to give the same 16 reflectances as those of the sample. This is judged by observing the sixteen dots on an oscilloscope screen which have to be located on the zero line. If this adjustment can be achieved it means not only that will the formula lead to a colour match but that the metamerism will be effectively zero. If the dots cannot all be brought simultaneously on to the zero line, further subsidiary calculations have to be made on a Tristimulus Difference Computer. This is used to reduce the difference in colour between the sample and the dye mixture to zero, while the metamerism is checked by repeating the calculation for a second illuminant.

In the Instrumental Match Prediction (IMP) service introduced by Imperial Chemical Industries in 1963, a digital computer technique is used, again based on Kubelka-Munk formulae as modified by Atherton and further elaborated to suit various dyeing systems.* The basic procedure is, first, to measure the tristimulus values X, Y, Z of the pattern to be matched using a Colormaster colorimeter. These values are fed into the computer together with the spectral absorption values α_λ of the three dyes to be used in the formulation and the spectral reflectance values $R_{\lambda s}$ of the undyed substrate. The computer is then programmed so that, starting from arbitrary concentrations c_1, c_2 and c_3, of the three dyes, the tristimulus values of the dye mixture are evaluated, the differences ΔX, ΔY and ΔZ from the pattern are calculated, concentration changes Δc_1, Δc_2 and Δc_3 are worked out to minimize ΔX, ΔY and ΔZ and the whole calculation repeated. This iterative process is repeated through a series of successive approximations until ΔX, ΔY and ΔZ are reduced to zero and the computer prints out the required recipe. A simplified computer flow diagram for this match-prediction programme is shown in Fig. 10.4.

The programme has proved highly successful in predicting

* E. Atherton: *J. Soc. Dyers and Colourists*, 1955, **71**, p. 389.

J. V. Alderson, E. Atherton and A. N. Derbyshire: *J. Soc. Dyers and Colourists*, 1961, **77**, p. 657.

J. V. Alderson, E. Atherton, C. Preston and D. Tough: *J. Soc. Dyers and Colourists*, 1963, **79**, p. 723.

T. L. Dawson, A. N. Derbyshire and D. R. Lemin: *J. Soc. Dyers and Colourists*, 1965, **81**, p. 306.

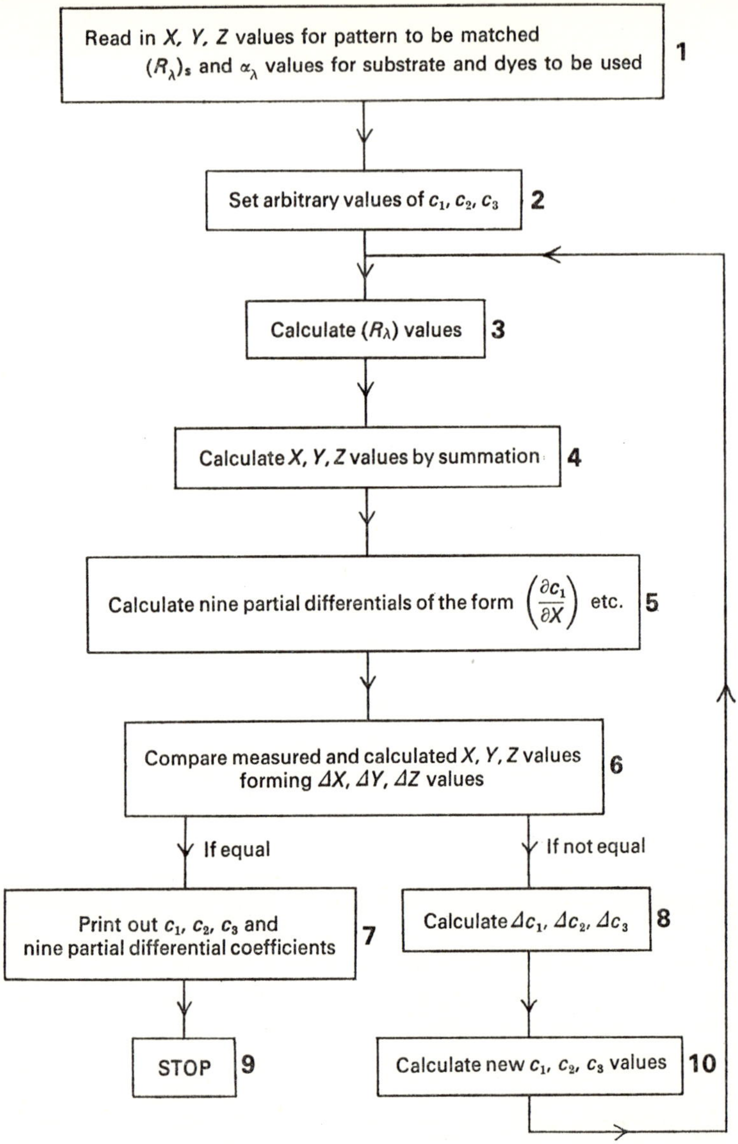

FIG. 10.4. A simplified computer flow diagram as used in the Instrument Match Prediction process of colorant formulation operated by Imperial Chemical Industries (Alderson, Atherton and Derbyshire)

recipes that in most cases give dyeings very close to the pattern to be matched. The residual difference can usually be removed by a single correction to the first predicted recipe by applying the differential correction coefficients calculated by the computer. If additional information is available of the tristimulus values of the pattern under illuminant A and illuminant C, the computer can also be programmed to select the dye combination that will give minimum metamerism with the pattern.*

Prediction systems of this complexity are not only a challenge to the theoretical colorimetrist. They also present a very real challenge to the instrument manufacturer since a system such as IMP, which offers a world-wide service through the Telex transmission of tristimulus values, requires identical colour measurements to be made in various parts of the world. It is likely, incidentally, that the 10° colour-matching functions of the 1964 C.I.E. system will find their most important application in this general field of dye and pigment mixture calculation. It is here that large areas of colour are involved and as the accuracy of the computing techniques increases, so does the need to employ colour matching functions as accurately representative of the viewing conditions as possible.

Several other methods of colorant formulation have been developed† of which the most novel uses a system of complementary tristimulus colorimetry.‡ Instead of the usual summation for tristimulus values

$$X=\sum(\tau_\lambda \,.\, P_\lambda \,.\, \bar{x}_\lambda), \text{ etc.}$$

a new value Q_x is calculated by the summation

$$Q_x=\sum[\log(1/\tau_\lambda) \,.\, P_\lambda \,.\, \bar{x}_\lambda]$$

with similar values for Q_y and Q_z. This in effect turns the chromaticity chart inside out, with the spectrum locus collapsing on to the black point in the centre of the chart and the various hues distribu-

* E. ATHERTON and F. NORTH: *Internationale Farbtagung Luzern* 1965, 2, p. 823. (*Die Farbe*, 1966, **15**.)

† See, for example, Section I—Calculation of Dye Recipes, in *Internationale Farbtagung Luzern* 1965, 2, pp. 811–79. (*Die Farbe*, 1966, **15**.)

‡ C. N. REILLY, H. FLASCHKA, S. LAURENT and B. LAURENT: *Anal. Chem.*, 1960, **32**, p. 1218. I. F. TROTTER: *J. Soc. Dyers and Colourists*, 1962, **78**, p. 76. E. GANZ: *Internationale Farbtagung Luzern* 2965, 2, p. 831. (*Die Farbe*, 1966, **15**.)

ted in directions complementary to those in the normal chart. For a dye, log $(1/\tau_\lambda)$ is proportional to $\epsilon.c$, the product of the extinction coefficient and the concentration, as shown on p. 15. The most important property of the system is its additivity, so that the Q-values of a dye mixture can be calculated from the Q-values of the component dyes on the same centre of gravity principle as in the normal chromaticity chart.

There is, of course, an inherent error of principle in the system, since two dyes which are metameric and which therefore have the same tristimulus values X, Y and Z, will not normally have the same Q-values. The simplicity of the calculation of dye mixtures on this system therefore has to be offset against the approximate nature of the result. Nevertheless, since no system of colorant formulation leads to an exact analytical solution, it could well be that at least under certain conditions the degree of approximation is sufficiently close to provide a commercially viable service.

LIGHTING

In Chapter 1 we considered the spectral energy distribution of various light sources and showed in Fig. 1.1 how the distribution changed in the case of Planckian radiators as the temperature of the source increased. The colorimetric counterpart of Fig. 1.1 is the black-body locus in the chromaticity chart as shown in Fig. 10.5, in which the colour of the source is seen to move across the chart in the general yellow-to-blue direction as the temperature is increased. This locus is useful for determining the correlated colour temperature of a source whose chromaticity lies near but not on the locus. This correlated temperature, as defined on p. 6, is the temperature of the black-body nearest in colour to the source in question. In determining the relevant point on the locus account has to be taken of the non-uniform discrimination in the x, y chromaticity chart. This means that it would not necessarily be correct in Fig. 10.5 to regard the perpendicular to the locus as defining the nearest colour, even though it gives the shortest line; instead isotemperature lines as calculated by Judd and revised first by Kelly and then by Robertson should be used.*

* D. B. Judd: *J. Opt. Soc. Am.*, 1936, **26**, p. 421.
K. L. Kelly: *J. Opt. Soc. Am.*, 1963, **53**, p. 999.
A. R. Robertson: *J. Opt. Soc. Am.*, 1968, **58**, p. 158.

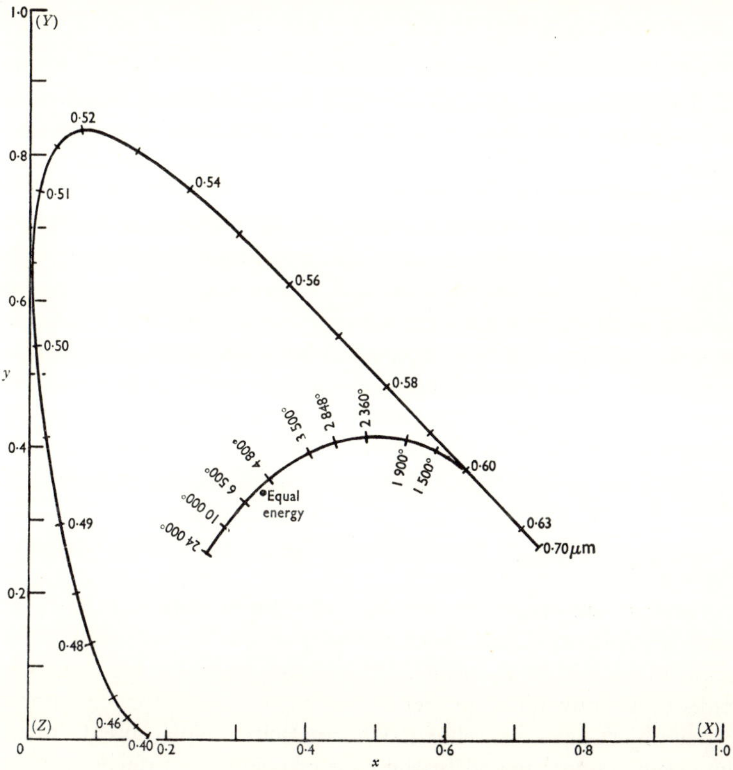

FIG. 10.5 The locus of Planckian (black-body) radiators in the chromaticity chart. Absolute temperatures are shown against the locus

In the development of the fluorescent lamp, the chromaticity chart played an important part both in classifying the various colours of source that were produced, some of which were at a considerable distance from the black-body locus, and in the actual choice of phosphors used in the lamps. Indeed, the mixing of fluorescent powders under an ultra-violet lamp is an unusual and fascinating experiment in additive light mixture.

Much bigger issues than the colour of the lamp itself have, however, been raised by the introduction of the fluorescent lamp. As indicated in Chapter 1, small departures from colour constancy of illuminated objects are quite common with these lamps and they have frequently been criticized on account of the colour distortions which they cause, even though their own colour may be perfectably acceptable. The unsatisfactory colour rendering arises from the abnormal spectral distribution which the fluorescent lamp possesses, but there has been great difficulty in establishing a generally acceptable procedure for specifying their colour-rendering performance.

The problem can be appreciated by considering the example of a fluorescent lamp designed to have the same colour as daylight—for convenience we will assume that its chromaticity is the same as that of the standard illuminant S_C—and that S_C is regarded as having satisfactory colour-rendering properties against which the fluorescent lamp can be compared. Adaptation under the two illuminations will be determined primarily by the colour of the source itself, but partly also by the colour of the integrated light from the illuminated objects in the scene. The latter may not be quite identical under the two illuminants, but if we assume that the differences are averaged out by the variety of reflections that occur in a typical scene, then general adaptation will be eliminated as a variable in the problem.

Next, white objects under the two illuminations will have the same chromaticity, and since the adaptations are the same, their appearance will be the same. Whites and neutral greys thus provide very convenient reference points to which the colours of other objects can be related. Chromaticity displacements must, however, be anticipated in the case of coloured surfaces because of the difference in spectral distribution of the light from the fluorescent lamp and that from S_C. Hence in order to assess the colour-rendering properties of the fluorescent lamp, we have to assess the significance of these displacements.

This means that tolerances have to be established for permissible colour distortions, and we may note in passing that the problem is somewhat similar to that of assessing the quality of a picture in a colour-reproduction system, although in the fluorescent lamp the colour distortions tend to be of a particular type since they originate mainly in a deficiency of energy at the red end of the spectrum, and in the presence of the mercury lines. We can expect that the tolerance limits will vary greatly according to the objects used in the test, and since there is no doubt that the colour of foods of various kinds, and of the human complexion, give rise to the most frequent comment, these should be used in any tolerance tests. Even with this limitation, a chromaticity displacement in one direction may be acceptable with one foodstuff and objectionable with another, since the significance of the change may be different. Much will depend on whether the colour change is in a direction to make the food look more appetizing or less appetizing, or fresh or stale or even bad, and while additional redness may be desirable for meat or beetroot, this may not necessarily be true for greenstuffs or butter or fish.

Evidently these questions can be put to experimenal tests, and several investigations on these lines have been reported. For example, Sanders* studied the colour tolerances for a variety of foods and for the human face, keeping the adaptation constant; in some experiments the colour changes were produced by the introduction of dyes into the food itself, but in most of them by variation of the quality of the light in a specially constructed booth. He then evolved a procedure for weighting the colour distortions produced by a fluorescent lamp and assessing its general colour-rendering property.

After prolonged study and many different investigations had been carried out, the C.I.E. Committee on Colour Rendering in 1963 recommended the adoption of a colour-shift method for the specification of the colour-rendering properties of a light source.† A selection of eight Munsell chips is used in this rating and their chromaticities are determined for the test source and for a reference illuminant. A General Colour Rendering Index is then derived from a formula incorporating the mean chromaticity difference ΔE for the eight samples. Certain restrictions have to be imposed in this derivation, notably in the choice of reference illuminant and the size of the colour difference between the test and reference sources for which the method will give a valid colour rendering assessment. The oustanding difficulty is how to take account of the different states of adaptation when comparing two sources which are markedly different in colour. This problem has still not been completely solved.

The subtle and intractable nature of the problem can be judged

* C. L. Sanders: *Ill. Engineering*, 1959, **54**, p. 452 and p. 640.

† *C.I.E. Compte Rendu* 1963.
D. Nickerson and C. W. Jerome: *Illum. Eng.*, 1965, **60**, p. 262.

by the fact that a period of 15 years had elapsed since a previous provisional recommendation was adopted by the C.I.E. that colour

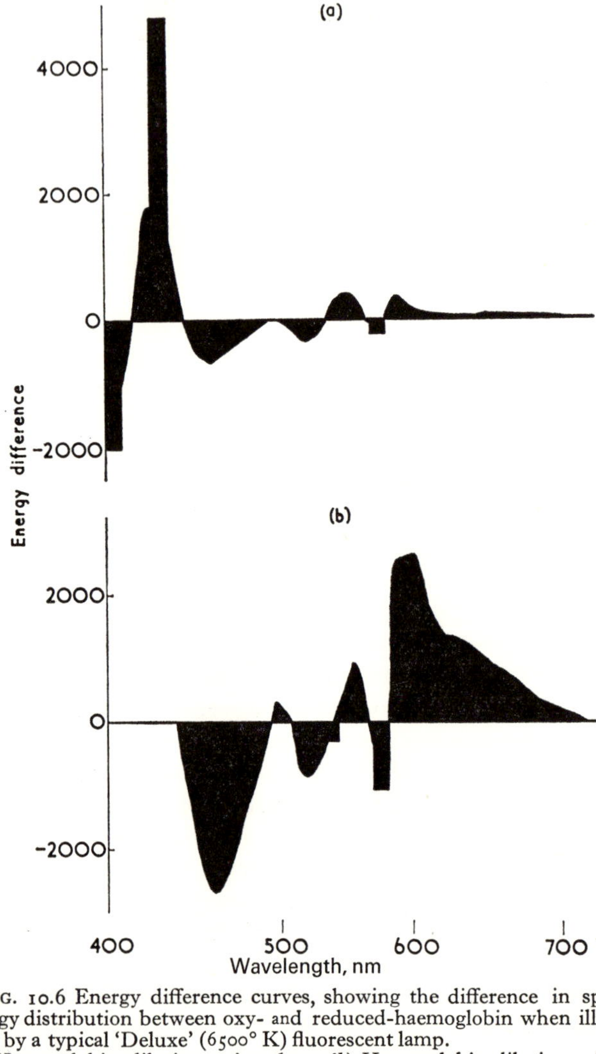

FIG. 10.6 Energy difference curves, showing the difference in spectral energy distribution between oxy- and reduced-haemoglobin when illuminated by a typical 'Deluxe' (6500° K) fluorescent lamp.
(*a*) Haemoglobin dilution 1 in 1600; (*b*) Haemoglobin dilution 1 in 100 (Wright, 1963).

rendering should be assessed on the basis of an approximate spectral energy distribution as given in 8 defined spectral bands.

The difficulty is, of course, that the colour distortions are a function of very many variables, embracing both the spectral energy distribution of the source at all wavelengths through the spectrum and the spectral reflection characteristics of the illuminated objects, also throughout the spectrum. For special problems, neither the use of 8 Munsell chips nor 8 spectral bands will give sufficiently detailed information to determine the acceptability or otherwise of a particular source for a particular job. For example, in hospital lighting the colour appearance of a patient's skin may be an important indication of his condition, and in the case of cyanosis, the blue colour of the lips or the ears is a result of the difference in absorption between oxyhaemoglobin and reduced haemoglobin. The interplay of the rather sharp irregularities of the spectral absorption curve of haemoglobin with the fluorescent lamp energy distribution which includes the mercury lines, makes any prediction of the colour change based on abridged spectrophotometry highly uncertain.* Thus when a sample of haemoglobin is illuminated by a typical fluorescent lamp and viewed by transmission, the change in energy distribution of the transmitted light when the haemoglobin changes from the oxy to the reduced form is given by the energy difference curves in Fig. 10.6 for two concentrations of the haemoglobin. Clearly any interpretation of what this means in terms of a change of colour calls either for accurate spectrophotometry and colorimetry or, more probably, for assessment by the medical profession under actual working conditions.

AGRICULTURE AND FOOD

The extent to which colour is involved in assessing the quality of agricultural products is extremely wide. Everyone automatically judges the ripeness of fruit by its colour, while the quality of cheese, butter or bacon and other similar products is to some extent estimated in the same way.

The value of colorimetry in agriculture depends very largely on the scale on which the industry is organized. Small farming communities cannot be expected to pay much attention to scientific descriptions of colour, but where the products form the raw material of a major industry, as with tobacco, wool or cotton, it is as reason-

* W. D. Wright: *Nature*, 1963, **198**, p. 1239.

able to apply accurate checks to the raw material as to the later stages in the manufacturing process. Again, where individual farming units sell their products through Government-controlled stations, or where the products are officially tested and graded before being sold to the consumer, colour may be one of the characteristics which have to be recorded. Cotton is an interesting example of this, since its colour, which is an important factor in the grading of the raw material, varies according to the previous history of the plant.* Thus in the early part of the season the cotton is generally of high grade and is bright in colour, but cotton from bolls that open after frost may become stained a yellow colour. Again, open cotton which has been left in the field and exposed to the weather becomes darker and of a lower grade. If cottons of different grades are mixed, the resultant mixture is likely to appear spotted or mottled. Sometimes the colour may provide evidence of attacks on the bolls by insects, fungi, etc. The grading of the cotton is therefore an essential service to the cotton industry, not only because it separates the high from the low grades, but also because it enables bales of cotton of similar quality to be accumulated and made available to the cotton manufacturer, who is thus ensured of a supply of material of reasonably uniform quality.

In applications of this kind, colour tests will be acceptable to the growers and manufacturers only if the results are described in terms which they already employ or can easily understand. Where colour standards are required, they are best provided by specially selected samples of the product or by suitably chosen standards from a colour atlas or from the Tintometer slides. As mentioned on p. 238 the Hunter Color-Difference Meter† is a very useful instrument in this field, especially since it can be modified to suit the special criteria of particular products, as has been done with the automatic cotton colorimeter.‡

For some of the less critical applications it may be feasible to specify a colour merely by describing it in words. The obvious difficulty is that different names are likely to be given to the same colour and the same name used by various people for different

* *The Classification of Cotton,* U.S. Dept. Agriculture, Misc. Pub., No. 310, 1938.

† R. S. Hunter: *J. Opt. Soc. Am.*, 1958, **48**, p. 985.

‡ D. Nickerson, R. S. Hunter and M. G. Powell: *J. Opt. Soc. Am.*, 1950, **40**, p. 446.

colours. In course of time, as the study of colour becomes more systematic, a greater uniformity of terminology may develop and in this connection the ISCC–NBS method of designating colours should be noted.* The classification originated in connection with the description of drugs and chemicals and has been adopted for use in the U.S. Pharmacopœia and National Formulary. In agriculture the system has been applied to the naming of soils, and a group of some 50 or 60 names has been selected to cover the range of soil colours.† Charts of standard patterns to correspond to the selected descriptions have been prepared in pocket size for convenience in field work, and provided that the charts can be kept clean, any soil can be compared with the chart and described by names which have a generally accepted significance. This is a logical development of colour specification where the range of colours in a particular type of object is limited and where, as is obviously the case in soils, very exact specification is not demanded. Whether a single system of terms for all applications can ever come into general use is perhaps debatable, and the kind of difficulty that may arise has been indicated by Nickerson in connection with soils.‡ Thus, soils which are reddish compared with other soils are referred to by soil men as 'red', although the correct description of their colour would be 'reddish-brown'. There exists an inevitable tendency to expand the colour scale when dealing with materials which vary in colour over a comparatively limited range, a tendency which is also exhibited in the paper trade, where papers only slightly tinged with colour may be referred to as 'blue' or 'yellow'. It is unlikely that a contraction of the colour scale in such instances would be either welcome or desirable.

The colour of foods has become a major field for study§ and has justified the publication of a book on the subject written especially for the food technologist.|| The appearance of food bears an

* *Nat. Bur. Stds. Circ.*, No. 553, 1955.

† T. D. Rice, D. Nickerson, A. M. O'Neal and J. Thorp: *U.S. Dept. Agriculture, Misc. Pub.*, No. 425, 1941.

‡ D. Nickerson: *Soil Science Soc. Am.*, 1941, **6**, p. 392.

§ *Color in Foods* (Nat. Acad. of Sciences—Nat. Res. Council: Washington), 1954.

See also W. A. Gould, R. B. Davis and J. O. Mavis: 'Color Grading Agricultural Products', *Am. Soc. Testing Materials Bull.*, No. 202, Dec. 1954. p. 19.

|| G. MacKinney and A. C. Little: *Color of Foods* (Avi Publishing Co: Westport, Conn.), 1962.

obvious and direct relation to its palatability, while the artificial colouring of food to assist in its marketing has become widely practised. With many foods the colour is an important element in its official grading, especially with processed foods. This may well necessitate colour checks at various stages in the production process, from the picking and grading of fruit, tomatoes, vegetables or tobacco, to the manufacture of jam, butter, cheese or tomato sauce, and the dehydration of fruit and vegetables.

As an example of the kind of study that is being made we may quote the relationship between the maturity of tomatoes and their colour, that has been investigated by Hutchings, Wood and Young at the Unilever Research Laboratory.* A maturity scale from 1 to 10 was established covering the 42–47-day maturity span. At 5 on the scale the tomato was just beginning to change in colour from the full green, while at 10 on the scale the tomato was overripe. Some 50 or 60 greenhouse tomatoes were measured on a Colorcord tristimulus colorimeter and Fig. 10.7 shows the mean chromaticities of the tomatoes at successive points on the maturity scale from 2 to 10. The maturity was assessed subjectively by the operator and one of the aims of the investigation was to determine the validity of the maturity scale and the accuracy to which the maturity could be judged by eye. As shown in the original paper, there was a considerable spread in the chromaticities for each maturity number and there was also, of course, some non-uniformity of colour over a given tomato. It is interesting to note that the authors of the paper comment on the difficulty which an operator who was colour defective would have in tomato grading by colour. This was a point that was noted some years ago in a report on Defective Colour Vision in Industry by a Committee of the Physical Society Colour Group.†

Another characteristic of tomatoes of great optical interest if not directly concerned with their colour is the surface structure of a tomato when examined under the interference microscope. The surface is seen to consist of a vast number of very small indentations, each of which is only a few wave lengths deep as judged by the number of Newton-type interference rings in each indentation.

* J. B. Hutchings, F. W. Wood and R. Young: In the press.

† *A Report on Defective Colour Vision in Industry* (Physical Society: London), 1946.

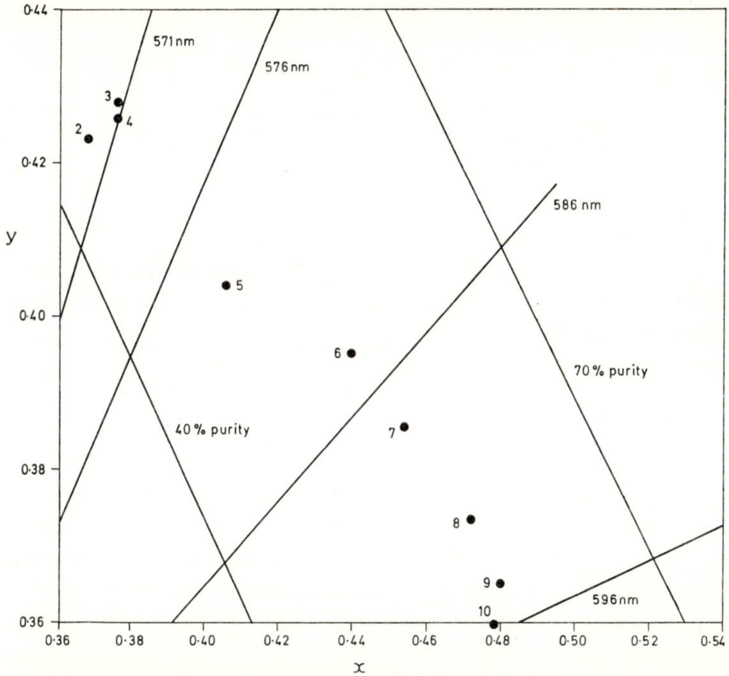

FIG. 10.7. Mean chromaticities of tomatoes at various stages in the ripening process, indicated by points 2 to 10 on a maturity scale, as measured at the Unilever Research Laboratory (Hutchings, Wood and Young). Lines of constant dominant wavelength and purity are shown in the diagram

This effect was first shown to the author by Professor S. Tolansky and illustrates the optical fascination in the whole subject of surface texture.

Chamberlin* has described some entertaining experiments to show the vital part played by colour on the appetising appearance of our meals. This, of course, is determined both by the spectral reflection characteristics of the food and the colour rendering properties of the lighting, and for this reason particular attention has to be paid to the lighting of restaurants. Accurate colour reproduction is also of key importance in food advertisements.

Consumer preferences have been studied by comparing the scores awarded by a panel of observers with the colour of the product as measured by colorimetery and spectrophotometry. Here

* G. J. Chamberlin: *Die Farbe*, 1957, **6**, p. 133.

considerable difficulties are encountered because of the nature of the samples to be measured. These will not normally be in the form of flat, uniform surfaces which can be placed against the port-hole of a colorimeter or spectrophotometer. Peas or potato chips, for example, do not provide easy material for the normal spectro-photometer to handle. Further, inconsistency in the measurements can occur between one instrument and another because of the varying influence of such factors as the turbidity, translucence and light scatter on the measurements on account of the different illumination and viewing conditions employed in different instruments.

CHEMICAL TESTS: INDICATORS

Many chemical reactions can be followed by the colour changes which accompany them, and the measurement of the colour may be used to provide quantitative information about the products of the reaction. Colour tests of this type are now being employed for very varied purposes, including *p*H determinations, testing the quality of milk, the estimation of sulphanilamide in biological fluids, the estimation of toxic gases in the air, and so on.

Regarded as colorimetric problems, the tests all have very similar principles. They may be illustrated by reference to a test which is carried out in the refining of coal-tar products. The test, which can be applied to hydrocarbons boiling below 200° C and which are substantially free from phenols and pyridine bases, consists of shaking together equal volumes of the sample and sulphuric acid of specified strength for a given time and then measuring the colour of the acid layer after a stated interval. If the colour is specified in standard terms, it can be used as a measure of the degree to which the refining has been carried. When the range of colours involved in tests of this kind is limited, it is convenient to define the standard colours in terms of samples rather than in terms of the C.I.E. system, and for this purpose the permanent Tintometer slides have proved very useful. Thus in the tests on coal-tar products just referred to, the colours may, as described in British Standards 135, 458, 479, and 875 (1939), be recorded in terms of the Lovibond scale. A simple form of comparator containing the necessary range of slides can then be employed to make the colour measurements.

A very interesting application of colorimetry and colour discrimination has been reported by King* in connection with chemical indicators. Indicators are widely used to record changes in hydrogen-ion concentration, and the efficiency of an indicator for this purpose will depend partly on the rate at which the chromaticity of the solution changes with change of *p*H, especially at the point of chemical neutrality, and partly on the ease with which the colour changes can be memorized.

From the spectral transmission curves of various indicators such as phenolphthalein, Tashiro's indicator and methyl red at various *p*H values, King calculated the chromaticities, and then expressed the changes in terms of the just perceptible colour differences discussed in Chapter 6. In this way it was possible not only to rate the sensitivity of different indicators, but also to describe the spectral characteristics which an ideal indicator should possess for maximum sensitivity. The possibility of improving the efficiency of existing indicators by the addition of chemically inactive dyes was also considered.

THE PULP AND PAPER INDUSTRY

Several references have been made in earlier chapters to the special importance of the measurement of white surfaces, a type of measurement which occurs more frequently in the paper industry than in any other. The desire to maintain the exact lightness and hue of any given type and grade of paper is self-evident; in times of acute paper shortage the uniformity of whiteness from page to page of a book may have to be sacrificed, but under normal conditions it is the natural aim of a publisher to produce a book, or possibly a series of books published at quite long intervals, in which the pages are all of the same quality. The same is generally true of paper used for other purposes, such as drawing paper, notepaper, photographic paper, paper for wrapping and packing, wallpaper, and so on. The paper manufacturer therefore has so to control the various processes in his trade from the pulping to the dyeing and finishing as to ensure that each grade of paper which he manufactures is up to the standard which he has set for himself or which his customer desires. Selected samples might be kept as

* J. King: *The Analyst*, 1952, **77**, p. 742.

standards, except that they are liable to become discoloured in course of time, and thus lose their value as permanent records. Specification of the colour and lightness of a paper in terms of the C.I.E. system is therefore much to be preferred from the point of view of certainty and permanence, although the use of samples as working standards is probably a more convenient method in the paper mill itself.

The problem is complicated by the very fine limits to which the paper manufacturer is expected to work. These narrow limits arise from the great sensitivity of the eye to small differences in the whiteness and lightness of near-white surfaces and to the critical conditions under which the surfaces are compared. According to Harrison,* the luminance factor needs to be measured to within 0·002–0·003 and the chromaticity co-ordinates to the fourth decimal place. Visual measurements to this order of accuracy may just be possible, by using apparatus in which both halves of a large matching field have closely similar energy distributions. Alternatively, photo-electric instruments of the type described on p. 237 may be used, and the Zeiss Elrepho instrument is finding increasing use in the paper industry.

Whiteness has been the subject of extensive study in the paper industry and a variety of formulae have been suggested by which it might be defined. These have generally been in terms of the Y tristimulus value of the surface and some measure of the reflection in the blue region of the spectrum relative to that in the yellow or red. A more specifically colorimetric formula has been suggested by Croes.† Friele,‡ on the other hand, has proposed a formula based on the dimensions of a discrimination ellipsoid centred on the standard white. The use of fluorescing dyes as optical bleaches raises a number of new problems, in respect of both the specification of whiteness and its instrumental measurement. As Grum§ has pointed out, this is a case where it is essential for the sample to be measured with a spectrophotometer with the optics reversed, that is, with the illuminated sample located at the entrance slit,

* V. G. W. Harrison: *Proc. Phys. Soc.*, 1942, **54**, p. 86.
† A. W. Croes: *J. Opt. Soc. Am.*, 1959, **49**, p. 830.
‡ L. F. C. Friele: *Die Farbe*, 1959, **8**, p. 171.
§ F. Grum: *Proc. Symposium on Colour Measurement in Industry* (Colour Group (G.B.): London), 1967, p. 194 and p. 109.

instead of the exit slit, of the monochromator. He illustrates a Beckman DK-2 spectrophotometer with this arrangement in which the light travels in the reverse direction to that in the normal use of the instrument. It is also essential to use a light source with an adequate ultra-violet content. This should preferably correspond as closely as possible to the new standard illuminant D_{6500}, and the measurement of the whiteness of surfaces containing optical bleaches is one of the most pressing examples of the need for the realization of a laboratory source to represent the new standard illuminants defined by the C.I.E.

THE PAINT INDUSTRY

Many of the colour problems which arise in the paint industry are similar to those which occur in dyeing. Thus, the identification of pigments by their spectral reflection curves, the problems of pigment mixture and of metamerism, the measurement and matching of pigments, have been surveyed recently in a discussion between paint technologists and colorimetrists.*

Because the colour of a paint before application is closely similar to its colour after it has been applied to a surface, its description in terms of the C.I.E. system is a more useful and more widely used method of specification than in the case of dyes. Trichromatic colorimetry is also used in the grading of oils and varnishes and in fading tests.† When the fastness of a pigment or, for that matter, of a dye has to be investigated and the relative merits of different pigments or dyes have to be compared, the conditions to which the pigment is to be exposed must be accurately controlled. If fastness to light is being tested, the level and quality of the illumination must be specified and maintained at a constant level. Yet unless the degree of fading can be described quantitatively in terms which have an exact and accepted significance, the value of the tests is largely nullified.

Another instance in which colour changes can be usefully recorded on the trichromatic system has been described by Tilleard.‡ This is in connection with phenomena observed when bronzing

* Fundamentals and Problems of Color, *J. Paint Technology*, 1967, **39**, p.342.
† D. L. Tilleard: *J. Oil and Colour Chem. Assoc.*, 1937, **20**, pp. 124 149.
‡ D. L. Tilleard: *J. Oil and Colour Chem. Assoc.*, 1942, **25**, p. 226.

pigments are diluted by admixture with a white pigment. Certain pigments possess a very high rate of absorption in a particular region of the spectrum, to such an extent that the rays are completely absorbed within a distance of about a wavelength from the upper surface of the pigment. By a process similar to resonance, the absorbed rays are re-radiated at the same frequency or wavelength, and as the selectively reflected light is mainly directed at the specular reflection angle, the surface acquires a metallic appearance or bronzing.

When a pigment exhibiting the bronzing effect is mixed with a white pigment, the change in colour quality of the surface is a function partly of the ordinary desaturation of the body colour of the pigment by the white particles and partly of the disappearance of the metallic reflection as the proportion of white pigment to coloured pigment is increased. As a result of the interaction of these two effects, the chromaticity locus of different mixtures of the white and coloured pigments departs very widely from the straight line which might be expected. Examples of these dilution curves are illustrated in Fig. 10.8, as plotted in the C.I.E. chart from data kindly supplied in a communication from the Paint Research Station. Four different pigments are represented: Lake green B, Prussian blue, red iron-oxide and a bronze red of the phosphotungsto-molybdic type. The white pigment with which they were mixed was blanc fixe and the mixture was mounted in a silicon-ester medium. The colour of the surface was measured with a Guild colorimeter for 45° illumination and normal viewing, and the curves show the change in chromaticity with decreasing percentage of coloured pigment.

With two pigments, the Lake green and the red iron-oxide, the addition of the white pigment at first produces a change in exactly the reverse direction from what might be expected, namely away from the white point to give a more saturated appearance. This is because the main effect of the white pigment when first added is to reduce the metallic reflection, without producing much change in the body colour; subsequently, the white pigment causes the normal desaturating action. Apart from the inherent interest of the phenomenon, the result is of some commercial importance as showing the most efficient concentration at which the pigments should be used as colouring agents. With the bronze red, the curve indi-

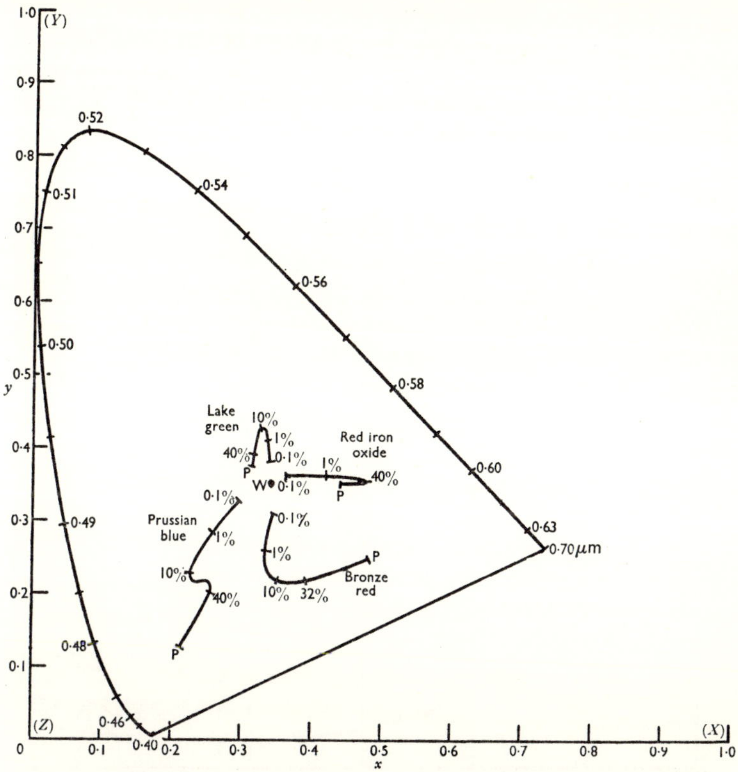

FIG. 10.8 The change in chromaticity of bronzing pigments when mixed with a white pigment and mounted in silicon-ester medium. The figures against the curves give the percentage of bronzing pigment in the mixture, starting from the pure colour *P*. The white illuminant is indicated by *W* (Paint Research Station)

cates that the body colour of the pigment is a bluish purple which becomes apparent at low concentrations but which in the pure form is reddened by the red surface reflection. With the Prussian blue, the pure colour has a brilliant red bronze, but shows the normal desaturation with the initial admixture of the white pigment, as indicated by the curve going straight towards the white point *W*. At a concentration of 60 per cent blanc fixe and 40 per cent Prussian blue, however, the curve shows a sharp inflection due to the sudden reduction in the red reflection. When only 10 per cent of Prussian blue is present in the mixture, the surface becomes matt blue with no trace of bronzing, and subsequent addition of white pigment leads only to progressive desaturation.

The major colorimetric problem in the paint industry is without doubt, however, the establishment of methods by which the colour of a pigment mixture can be related to the spectral reflection

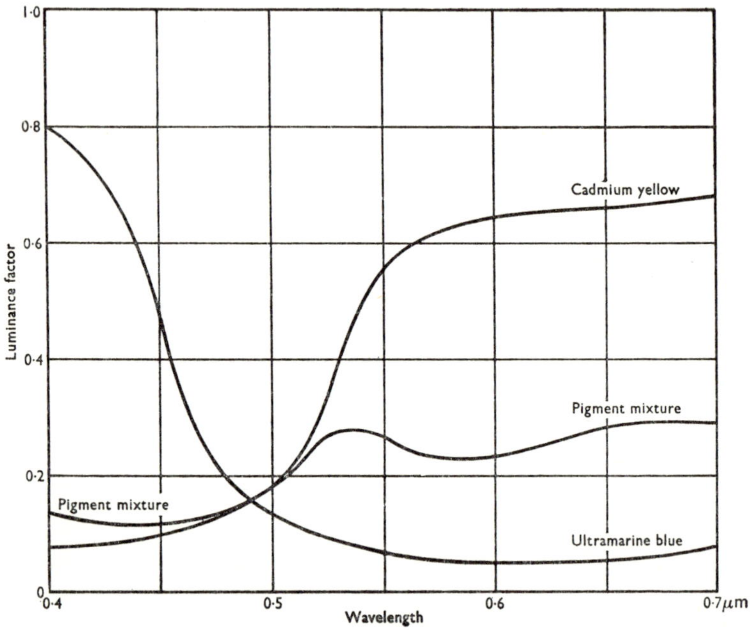

FIG. 10.9 Spectral reflection curves of cadmium-yellow and ultramarine pigments, and of their mixture in the proportion of two parts yellow to one part blue (After Duncan)

characteristics of the component pigments. Thus in Fig. 10.9 the reflection curves of cadmium yellow and ultramarine blue are given, together with the curve found when two parts of the former are mixed with one part of the latter, as taken from an important paper by Duncan on the colour of pigment mixtures.* Now it is immediately apparent that the curve for the mixture corresponds to the laws of neither additive nor subtractive mixture, and a theoretical basis is required from which the mixture curve can be derived.

* D. R. DUNCAN: *J. Oil and Colour Chem. Assoc.*, 1949, **32**, p. 296 and 1962, **45**, p. 300.

As with dyes (p. 263) no rigid solution has yet been formulated, but approximate solutions have been derived in terms of the relative amounts of scattering and absorption occurring in the pigments. These will depend on the dimensions and nature of the pigment particles, their refractive index, the refractive index of the medium in which they are bound, and their state of dispersion in the medium, so that each pigment-medium combination has to be considered as an individual problem.

Duncan shows how the coefficient of absorption α and the coefficient of scatter δ for a number of wavelengths through the spectrum can be deduced, for a given pigment, from transmission and reflection measurements of thin films of the pigment. Their relative magnitudes, which are all that are needed for the prediction of the colour of mixtures, can be calculated more easily from spectral reflection measurements of a thick film of the pigment alone, and of a mixture of the pigment with a standard white pigment in known proportions.

When two or more pigments, $A, B, C \ldots$, are to be mixed in the proportions of $a, b, c \ldots$, the reflection factor ρ can be calculated from

$$\Phi = \frac{2\rho}{(1-\rho)^2}$$

where

$$\Phi = \frac{(a\,.\,\delta_A + b\,.\,\delta_B + c\,.\,\delta_C + \ldots)}{(a\,.\,\alpha_A + b\,.\,\alpha_B + c\,.\,\alpha_C + \ldots)}$$

Calculations of ρ can, of course, be made for each wavelength for which values of α and δ are known, and when ρ is plotted against λ, we have the predicted spectral reflection curve of the mixture. In favourable cases, Duncan has shown that the agreement with the observed values can be as close as 0·005 or 0·010 in ρ, and 0·001 or 0·002 in the derived chromaticity co-ordinates x, y. Yet with darker colours the agreement is less accurate and may break down altogether if the dispersion of the pigments is not uniform.

When we turn to the inverse and very difficult problem of finding the proportions in which pigments should be mixed to match a given colour, Duncan first illustrates how the range of possible colours from a given set of pigments can be derived and then discusses two or three direct methods for calculating the pigment mixtures. Both on account of their complexity and their inaccuracy, he concludes that quantitative applications of the theory may be less important than its qualitative implications. If these provide a better understanding of the relationship which exists between the variables such as tinting strength, mass-tone and undertone with which the craftsman is familiar, and the absorption and scattering coefficients investigated by the scientist, then the theory will have made a very real contribution to the business of manufacturing paint.

However, the outlook has changed since Duncan wrote his paper, owing to the advent of the electronic computer. Some account of the computing procedures for colorant formulation has already been given in this chapter in the section dealing with the dyeing industry, since the Kubelk-Munk and the Duncan type of formulae were applied in the first instance to the prediction of dye mixtures rather than pigment mixtures.* These computation techniques have now been applied to pigments, especially with the Davidson and Hemmendinger COMIC analogue computer,† and there is no doubt this will become a major activity in the pigment industry before long.‡

SIGNAL GLASSES AND ROAD SIGNS

The specification of signal glasses is one of the most notable examples of colour standardization on the C.I.E. system. Coloured signals are of such general use on the roads and railways and for navigation by sea and air, that standardization is essential and agreement must be on a national and international basis. This need was undoubtedly one of the important factors which led to the establishment of the C.I.E. system, and it is natural to find that the value of the system was more speedily appreciated in this type

* T. L. Dawson, A. N. Derbyshire and D. R. Lemin: *J. Soc. Dyers and Colourists*, 1965, **81**, p. 306.
E. Allen: *J. Paint Technology*, 1967, **39**, p. 368.

† H. R. Davidson and H. Hemmendinger: *Official Digest*, 1965, **37**, p. 895; *J. Opt. Soc. Am.*, 1966, **56**, p. 1102.
R. Best: *Proc. Symposium on Colour Measurement in Industry* (Colour Group (G.B.): London), 1967, p. 47.

‡ As an indication of the commercial interest in these developments, we may note that the subject was a major feature on the technical page of the *Financial Times* for May 17, 1968.

of problem than in industries less closely related to photometry.

The selection of the colour required for any given signal depends on several factors, of which the most fundamental is visual appearance. The signal has to be easily identifiable as, say, red or yellow or green, and the particular red, yellow or green should be chosen to be as distinct in appearance as possible, bearing in mind the reduced sensitivity of the eye to colour differences when distant signals are seen as point sources.

In addition to securing a large colour difference between the signals, it is also important that the signal glasses should possess a reasonably high transmission factor if the signals are to have the required intensity. This consideration may operate in the opposite direction from the requirement of maximum colour difference, since a glass giving a saturated colour with a dominant hue either in the extreme red or in the blue-green or blue can transmit only a small fraction of the incident light. Some compromise is therefore called for in which both the hue and saturation of the signal have to be decided in relation to its intensity.

Superimposed on the visual criteria are manufacturing considerations which make the production of a glass to an exact specification a matter of some difficulty. As a consequence, the colour of a signal glass is defined by an area in the chromaticity chart within which the colour of an acceptable glass must fall. The size and shape of this area must be such that when a given type of glass has been manufactured, the percentage of rejections is not too high, provided always that the extreme glasses included within the tolerance area do not differ too greatly from the visual characteristics they are supposed to possess. To restrict the area to smaller dimensions than is necessary from the point of view of the recognition of the signal would merely increase the manufacturing difficulties unnecessarily and raise the cost of the glass. On the other hand, when the glasses can be manufactured to within fairly small limits without much trouble, there is little point in defining an unduly large area, since this merely suggests looseness in drawing up the specification.

It follows that the specification of signal glasses must be to some extent a compromise between visual and manufacturing considerations. The actual specifications which have been drawn up have generally been based on experience with signals which have proved satisfactory over a period of years. For example, the specification of railway signal glasses in British Standard 623: 1940 in effect crystallized the established practice on the railways by specifying colour limits within which the acceptable coloured glasses have from time to time been found to fall.*

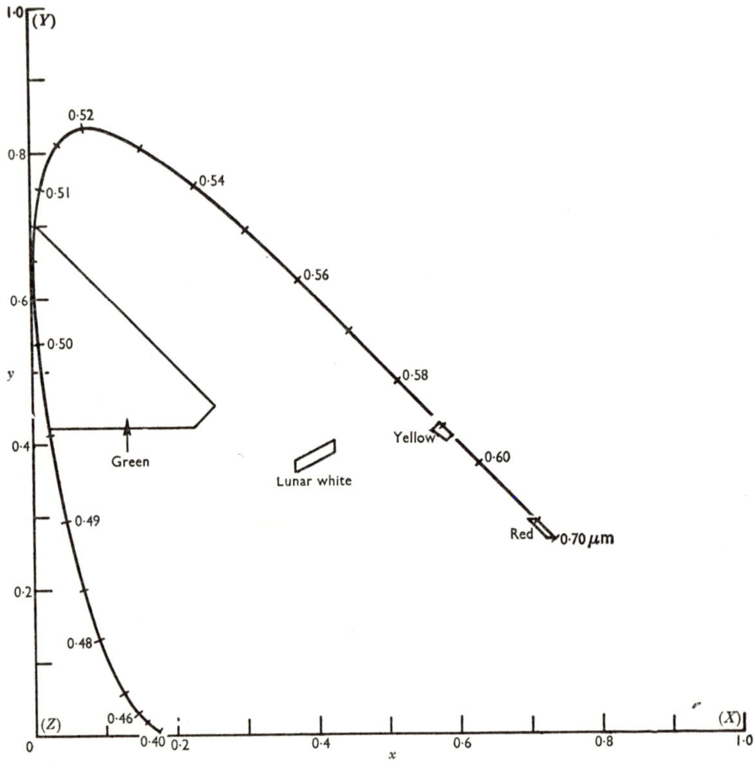

FIG. 10.10 Tolerance areas in the chromaticity chart for red, yellow, green, and lunar white signal glasses for stepped lenses as defined in B.S. 623: 1940

Fig. 10.10 illustrates the tolerance areas on the C.I.E. chart for red, yellow, green and lunar-white signal glasses for stepped lenses as defined in the British Standard. There are two features of these areas which should be noted. The first is that they are bounded by straight lines, which suggests that simplicity in defining the areas rather than any close adherence to manufacturing or visual limits

* See also *C.I.E. Compte Rendu*, 1948, 1951 and 1955.

determined their final shape. The second, and more significant feature, is the very large area specified for the green glass compared especially with the red and yellow areas. This is no doubt in part because red and yellow glasses are more easily reproduced to a given specification by reason of their chemical constitutions and absorption curves. But the great discrepancy in the size of the areas is in part a result of the non-uniformity of the C.I.E. chart, as illustrated, for example, in Fig. 6.3, and in part because the green tolerance area has been extended to the spectrum locus as the most convenient way to define one of the boundaries. In practice none of the green glasses could lie very close to the spectrum locus, because such saturated colours would have too low a transmission.

An investigation in which colorimetry played an essential part was the determination of the colour quality which signals should possess for maximum ease of recognition, assuming that previous working experience was not an adequate guide on the question. Thus, Holmes* carried out an extensive series of observations, using a colour mixer in which eleven different stimuli could be mixed in varying proportions so as to provide a colour signal of adjustable colour quality but with an energy distribution generally similar to that used in full-scale signalling practice. Observers were shown a series of 256 colours obtained with the colour mixer and were asked to name each colour in turn, using only a restricted list of names for the purpose. Some 40 000 observations were analysed to provide contours on the C.I.E. chromaticity chart showing the frequency with which each colour in the chart was recognized as one of the simple colours. These observations were recorded under various conditions of illumination, presence of nearby light sources, etc., and the data were discussed with a view to stating the limits within which signal colours should lie. Holmes concluded that red is the ideal signalling colour, because it can be recognized down to low illuminations and is very easily produced. Blue is not satisfactory unless the colour is highly saturated and the illumination is high, two conditions not easily satisfied simultaneously; green is safe unless blue occurs in the same signalling system, and similarly, both yellow and white are satisfactory provided that they are used separately and not in the same system. These conclusions are in

* J. G. Holmes: *Trans. Ill. Eng. Soc.*, 1941, **6**, p. 71.

accord with the colour confusions associated with small-field tritanopia described on p. 213. This investigation and a parallel one by Hill* provides a further example of the use of the C.I.E. system for the description and analysis of colour tests which have led to conclusions of great scientific interest and practical importance.

More recently, the C.I.E. has been concerned with the design of road signs, many of which are distinguished by their colour as well as their shape. Here again the chromaticity chart is being used in the first place as a means of recording current practice in various countries on the colours in use on existing road signs. Subsequently, international agreement will be sought on the surface colours that are recommended for use on road signs and their definition within specified areas of the chromaticity chart.†

THE ANALYSIS OF OPTICAL AND VISUAL PHENOMENA

There are certain experiments in physical optics in which one of the important results is the production of a coloured pattern or series of coloured fringes. This applies principally to interference, diffraction and polarization phenomena, in which the colour is produced by the interaction between two or more parts of the wavefront in a beam of light, and in such a manner as to reduce the amplitudes of some wavelengths to a greater extent than others. If the beam of light incident on the optical system is initially white, it may emerge with some parts of the spectrum completely removed and other parts reduced in intensity. The emergent beam will then appear coloured, exactly as if the removal of those regions of the spectrum had been produced by absorption in a colour filter. In contrast, however, to the exact description of all the other quantities entering into the experiment, e.g. wavelength, path difference, phase retardation, refractive index, etc., the final result of the experiment, namely the colour, is generally described in a purely qualitative manner. For some purposes this may be adequate; it may, for example, be sufficient to refer to first-, second- and third-order spectra, thereby indicating colours of decreasing saturation,

* N. E. G. Hill: *Proc. Phys. Soc.*, 1947, **59**, p. 560.
† *C.I.E. Compte Rendu*, 1967.

or to describe a magenta as the sensitive tint, and so on. But as a scientific analysis of the phenomena it is clearly inadequate. Moreover, the colour is of some technical importance in certain types of work, such as crystallography and photo-elasticity, and a more exact discussion is likely to be of interest in these problems.

As an example, we may consider the colours produced in photo-elastic experiments, which were analysed on the trichromatic system many years ago.* In photo-elasticity, celluloid or Bakelite models of a mechanical structure are subjected to external forces and the strains introduced are investigated by viewing the model

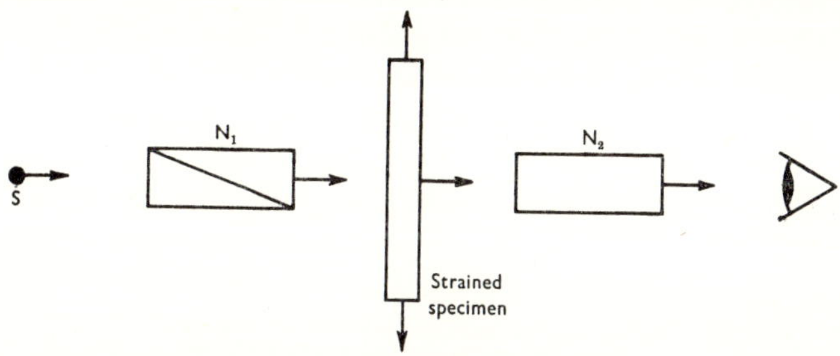

FIG 10.11 Elementary optical system used in photo-elasticity

between crossed polarizing prisms. An elementary arrangement is illustrated in Fig. 10.11. White light from a source S passes through a Nicol prism N_1 and emerges as a plane-polarized beam. The model is mounted at right angles to the direction of the light and is subjected to external forces, e.g. extension, compression, bending, etc., which cause the specimen to become anisotropic. If the light vibration incident at any point in the model is resolved into two perpendicular vibrations in the directions of the principal stresses at that point, one of the components will travel through the model with a higher velocity than the other and give rise to a path difference D between them. D is given by the equation

$$D = k \,.\, (p - q) \,.\, d$$

where k is a constant depending on the material of which the model is composed, p and q are the principal stresses in a lamina

* R. V. BAUD and W. D. WRIGHT: *J. Opt. Soc. Am.*, 1930, **20**, p. 381.

of unit thickness at the point under consideration, and d is the thickness of the model. The path difference D gives rise to a phase difference ϕ between the two vibrations, where

$$\phi = \frac{2\pi \,.\, D}{\lambda}$$

In general, therefore, the light emerges elliptically polarized, and the intensity of the light transmitted through the second Nicol N_2 will be proportional to $\sin^2\phi/2$, assuming that N_2 is crossed with respect to N_1. Since ϕ depends on λ, the fraction of the light transmitted through N_2 will vary through the spectrum, so that the emergent beam will no longer be white but coloured. The energy distribution of the transmitted beam will be a measure of D and therefore of $(p - q)$, and the locus of a particular colour in the model will be a locus of constant stress. The stress distribution in the model can therefore be plotted, and comparison of the colours at different points in the model can be used to relate the magnitude of the stresses at those points.

In practice the system is more elaborate than that shown in Fig. 10.11: lenses can be introduced to illuminate a large area of the model and a projection lens can be used to throw an image of the model on a screen; in particular, it is an advantage to mount crossed quarter-wave plates on either side of the model and hence to examine the specimen in circularly polarized light. The colour analysis, however, is not thereby affected, and the variation of chromaticity with increasing retardation can be calculated by evaluating the energy distribution for each retardation and then multiplying the energy curve by the C.I.E. distribution curves in the usual way. The variation of chromaticity with retardation, as illustrated in Fig. 10.12, shows the results of Baud and the author transformed into C.I.E. units, for values of D from 0 to 2·0 μm and for an illuminant with a colour temperature of 3000° K.

Two or three points of interest in this diagram may be noted. By following the curve round for increasing retardations, the normal colour changes can be traced out: starting from a neutral grey, the colour turns yellow, then orange to purple and blue, then to green, and back to yellow again to start the second-order spectrum. With each successive cycle the loops become smaller as the colours get more desaturated, until they eventually approach asymptotically

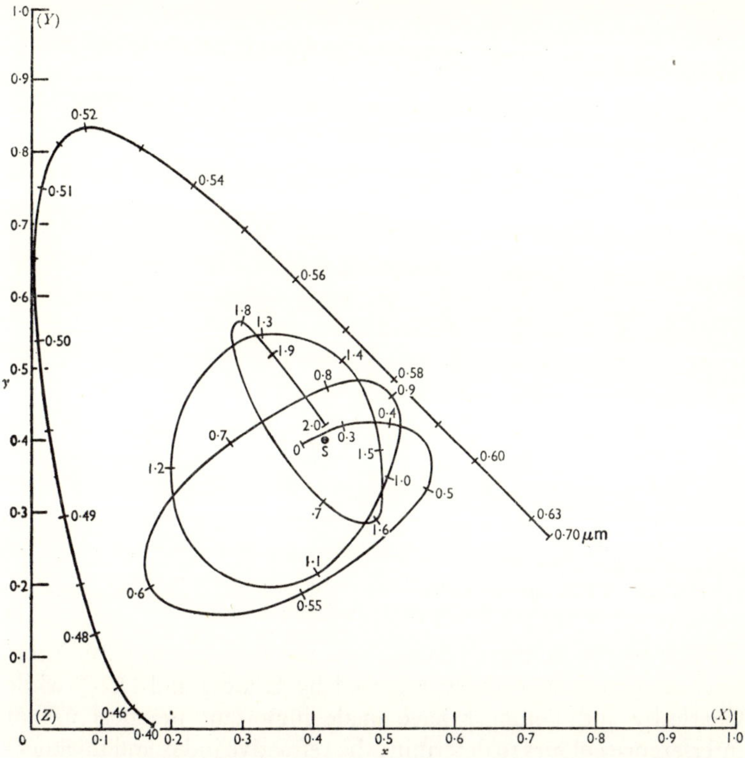

FIG. 10.12 The locus in the chromaticity chart of the colours observed in photo-elastic experiments with increasing optical retardation. The figures against the locus indicate the retardation in μm. *S* represents the colour of the original source (Baud and Wright)

to the point S representing the colour of the incident light. It is interesting to note that S is not the starting-point of the curve; the first light which gets through the system when the specimen is only slightly strained is tinged with blue relative to the colour of the source. Again, the existence of the sensitive tint when the retardation is equal to 550 nm is clearly brought out. A change in retardation of only 50 nm changes the purple colour either to an orange or to a blue, according as the retardation is decreased or increased.

The complete description of the fringes must also include data on their variation in luminance as well as in chromaticity. These can be calculated from the spectral energy curves at each retardation in the usual manner, i.e. by multiplying the energy curve by the $\bar{y}$ distribution curve. Fig. 10.13 shows the periodic type of variation which is found; comparison with Fig. 10.12 shows that the maximum luminance corresponds to the fringes which are yellow in hue and the minimum to the bluish-purple fringes.

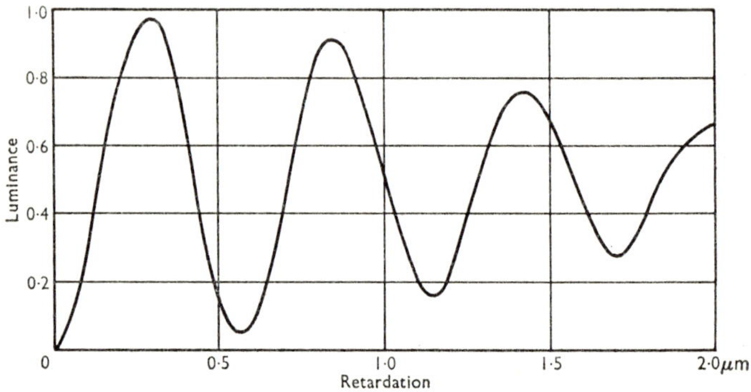

FIG. 10.13 The variation in luminance of the colours observed in photoelastic experiments with increasing retardation

More recent studies on these lines, and their extension to interference colours, have been reported by Kubota and Ose,* while Charlesby and Polling † have made important practical use of interference colours to determine the refractive index and thickness of thin oxide films on tantalum. Brilliantly coloured oxide films can be produced on tantalum by electrolysis, and information was required about the optical characteristics of the films, because of their influence on the surface properties of the metals. By using the spectrophotometer described on p. 247, the wavelengths for which the reflection was a minimum were determined for various angles of incidence for a number of oxide films, and from the variation of wavelength with incidence angle, the refractive index was calculated for each specimen. The absolute thickness could then be derived from the optical retardation, after a correction had been applied for the phase change at the oxide-metal interface.

There is also an immensely interesting field of study in the structural colours which occur in biological material through

* H. KUBOTA and T. OSE: *J. Opt. Soc. Am.*, 1955, **45**, p. 89.

† A. CHARLESBY and J. J. POLLING: *Proc. Roy. Soc.*, *A*, 1955, **227**, p. 434.

optical effects such as interference and scattering.* Thus the blue and green colours of certain feathers, for example the Macaw, are due to light scatter in the minute air cavities in the barbules of the feathers; the iridescent colours of humming-bird feathers are due to multilayer interference as the light is reflected from platelets stacked 7 to 15 deep in the barbules of the feathers;† the blue-green and violet colours in the 'eye' of the peacock's wing is also due to multiple interference, and similarly the metallic green reflections of a variety of beetles.

Recording of the colours of such material by colorimetry or spectrophotometry can be of value and interest for purposes of identification and comparison, but it can also be a means of analysing the structures producing the colour. One of the more remarkable specimens for study is the South American *Morpho* butterfly whose wings are a brilliant iridescent blue. Under the microscope the wing is seen to be covered with parallel rows of scales with an array of very fine ribs 1 or 2 μm apart running across them. Anderson and Richards‡ have deduced a rib structure as shown in Fig. 10.14 from examination with the electron micro-

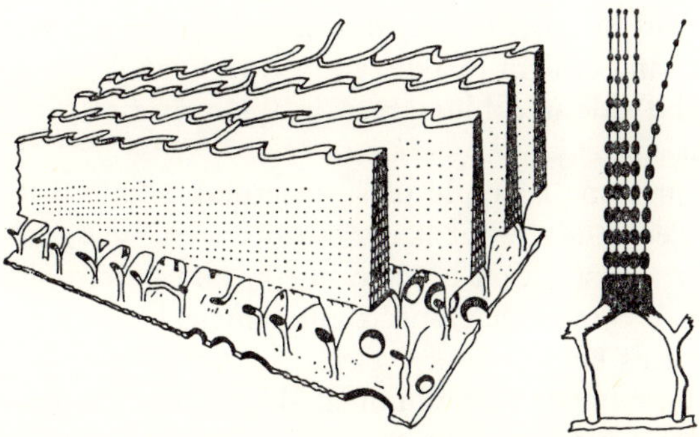

FIG. 10.14 Structure of ribs on scale of *Morpho* butterfly wing as deduced by Anderson and Richards (1942) from electron microscope studies

scope. The coloration may then arise from multiple reflections in the twelve or so longitudinal mullions in the ribs, although it seems

* D. L. FOX: *Animal Biochromes and Structural Colours.* (Cambridge University Press), 1953.

† C. H. GREENEWALT, W. BRANDT and D. D. FRIEL: *J. Opt. Soc. Am.*, 1960, **50**, p. 1005.

‡ T. F. Anderson and A. G. RICHARDS: *J. Applied Physics*, 1942, **13**, p. 748.

likely that each rib must act as a three-dimensional diffraction grating. The author has made some spectral reflection measurements on the wing for different angles of incidence and reflection and has also developed a simple hemispherical viewer* shown in Fig. 10.15. The sample is located at the centre of the hemisphere, which is whitened on the inside, and is illuminated by a pencil of

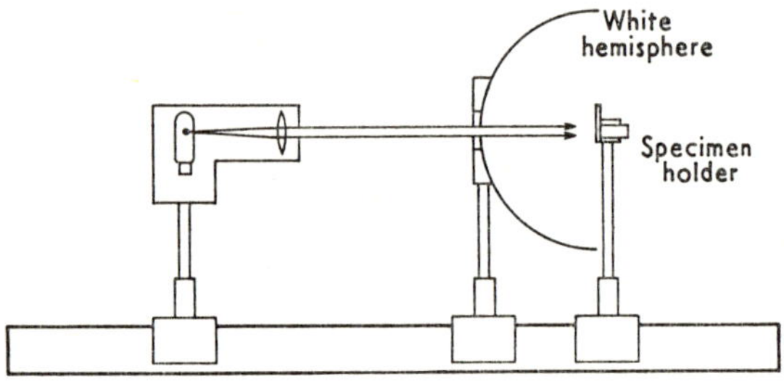

FIG. 10.15. Hemispherical viewer for studying the polar distribution of the light eflected from the *Morpho* butterfly wing and other specimens (Wright, 1963)

light passing through a small aperture in the mid-point of the hemisphere. The polar distribution of the light reflected by the specimen can then be seen from the light pattern on the inside of the sphere; in the case of the Morpho wing, this pattern consists of a bright band of light in a direction at right angles to the ribs on the wing and coloured blue-green in the centre of the band and violet at the ends.

This is just one example of a structural colour that deserves more detailed study,† but it is typical of the biological material suitable for investigation by colorimetry and spectrophotometry. The normal commercial spectrophotometer may not be well adapted to this kind of measurement partly because of the small area that may have to be illuminated and partly because of the need in some cases to make the measurements for specified and varied angles of incidence and reflection. There is scope here for the design of special illuminating and viewing units to be attached to standard spectrophotometers and colorimeters.

Research on vision itself is another area in which colour measurement can make an important contribution to the understanding of the visual process. Some examples of this have already been dis-

* W. D. WRIGHT: *Nature*, 1963, **198**, p. 1239. *Colour and Life* (Inst. Biology: London), 1964.

† P. K. C. PILLAI: *J. Opt. Soc. Am.*, 1968, **58**, p. 1019.

cussed in earlier chapters. Thus, the distribution of the white points in the chromaticity chart of Fig. 4.3 is related to the degree of yellow pigmentation of the 36 observers who took part in the observations; the convergence of the tritanopic confusion loci towards the blue corner of the chromaticity chart, as illustrated in Fig. 8.12, implies that this colour vision defect is due to the absence of the blue colour receptors in the retina; the striking colour changes that are observed with certain colour patterns such as that shown in Plate 6 have been matched and measured with a colorimeter as an aid to their interpretation in terms of contrast, light scatter and chromatic aberration of the eye.

A recent study of the colour of after-images byPadgham* may be quoted as a further illustration of the use to which the C.I.E. system can be put in visual research. After-images occur when an observer fixates on a bright source for a short period and then looks away. A slightly blurred pattern of the same shape as the source is then seen projected against the background at which the observer subsequently looks. If this is a light area, the after-image will be darker than the background, although it is likely also to be coloured. If the background is dark, then the after-image may be quite bright and again it is likely to be coloured. The after-image may persist for several seconds or even minutes, and during this time its appearance may run through a cycle of colour changes which may be repeated as the after-image gradually fades away. The colour at any given moment is a function of the colour and intensity of the original stimulus, the time the observer has fixated on it, and the luminance and colour of the background against which the after-image is viewed.

The physiological processes responsible for these after-image effects are likely to be partly photo-chemical and partly neural. They represent one of the more elusive and fugitive visual phenomena, but they nevertheless need to be studied. They have, in fact, been investigated on many occasions, but much of the information that has been reported has been merely qualitative and descriptive. Padgham, however, has been a pioneer in his attempts to make measurements of the intensity and colour of the after-images. Thus Fig. 10.16, which is taken from his most recent paper, shows the C.I.E. chromaticity chart with the colour

* C. A. Padgham: *Vision Research*, 1968, **8**, p. 939.

sequence of an after-image extending over a period of about two minutes, as measured by the binocular matching technique using a specially designed colorimeter. In this technique the after-image was induced in the left eye and its colour was measured by matching its appearance against the colorimeter field seen in the right eye. The analysis of the curve in Fig. 10.16 and of similar curves recorded for other observing conditions has led to interesting

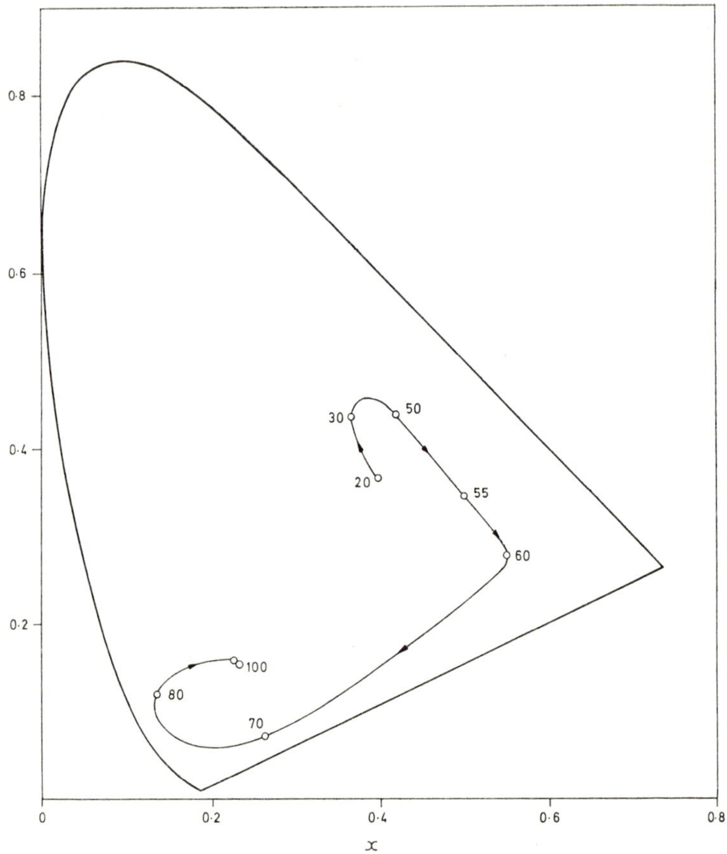

FIG. 10.16. Locus in C.I.E. chromaticity chart of the colour sequence of an after-image produced by exposure to a 5° diameter circular white stimulus viewed foveally for 1 sec as measured using a binocular matching technique. Retinal illumination of stimulus 2×10^6 trolands, colour temperature 3000° K. Numbers on curve indicate time in seconds from beginning of stimulus exposure (Padgham)

deductions about the growth and decay of the after-activities in the three receptor processes. Colorimetric measurements clearly provide a very necessary tool in the study of visual effects of this kind.

METEOROLOGY AND ASTRONOMY

The colour of an object seen from a distance depends on the light incident on its surface, on its spectral reflection characteristics, on the atmospheric absorption between the object and the observer, and on the air-light caused by the scattering in the atmos-

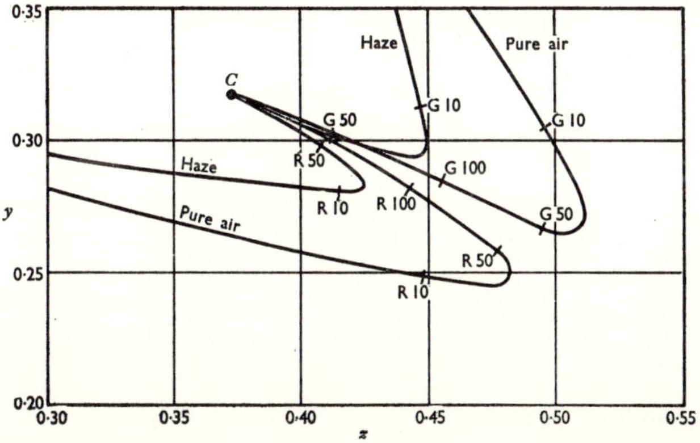

FIG. 10.17 A portion of the C.I.E. chromaticity chart showing the variation in chromaticity of red (R) and blue-reen (G) objects seen at various distances through a slight haze and through dry air. The figures against the curves give the distance from object to observer in kilometres (Middleton)

phere. Since the visibility and recognition of an object are affected by its apparent colour, the subject is of some importance in meteorology and has been analysed by Middleton * in terms of the C.I.E. system.

As an illustration of this type of application, we may refer to Fig. 10.17, due to Middleton, in which the change in the colour of red and blue-green objects when seen at various distances through pure air and through a slight haze are recorded. The diagram represents a portion only of the C.I.E. chromaticity chart, plotted relative to the y and z co-ordinates, and the figures marked against

* W. E. K. MIDDLETON: *Visibility in Meteorology* (University Press: Toronto), 1941.

the loci indicate the distance (in kilometres) of the object from the observer. The initial colours of the two objects, (*R*) and (*G*) are given by the equations

$$(R) \equiv 0{\cdot}489\,(X) + 0{\cdot}317\,(Y) + 0{\cdot}194\,(Z)$$
$$(G) \equiv 0{\cdot}150\,(X) + 0{\cdot}450\,(Y) + 0{\cdot}400\,(Z)$$

and the changes illustrated in Fig. 10.17 show that with increasing distance the colours at first move towards the blue corner of the chart and then reverse their direction, becoming whiter and approaching the point *C*. This point represents the quality of the daylight illumination, which was assumed by Middleton to correspond to the C.I.E. S_C illuminant. For the pure dry air, scattering according to the Rayleigh formula was postulated, while the haze was defined by assuming an extinction coefficient typical for slightly hazy air. The calculations themselves involved, first, the derivation of the tristimulus equations for the coloured objects when allowance had been made for the atmospheric absorption at each distance, and secondly, the tristimulus equation for the air-light alone for each distance. The sum of the two equations gives the tristimulus equation for the object as seen by the observer, from which the luminance and the chromaticity co-ordinates can be derived.

As an indication of the practical utility of the investigation, comparison with the visual range (i.e. the maximum distance at which a given object can be seen in any particular circumstances) of black, grey and white objects, shows that before the limiting range of an uncoloured object is reached, e.g. 375 km for a grey object in pure air or 144 km in the haze (Middleton), the chromaticity of the coloured object has approached so nearly to that of *C* that the difference is less than the saturation threshold. Consequently the coloured object appears grey, and in the determination of the visual range no complication in the way of a second threshold for colour is involved.

Another application in this field is to the colours observed during displays of the aurora. The author's interest in this problem was aroused by Professor S. Chapman in connection with the International Geophysical Year 1957–8, and it was for the purpose of attempting colour measurements on the aurora that the colorimeter described on p. 228 was constructed with the aid of a grant from the Royal Society. Only a few observations were, in fact, possible, but Dr M. Gadsden, who was associated with the project, used one model at the Auroral Station at Omakau, New Zealand, where he worked for some years.

The aurora provides a peculiarly apt application of trichromatic colorimetry, since the main radiations come from three distinct parts of the spectrum, namely, a group of lines in the red in the wavelength region 630–640 nm, a green line at a wavelength of 557·7 nm and a blue group in the region 430–470 nm. If, therefore, the chromaticity of an aurora, or part of an aurora, could be located in the C.I.E. chart, its spectral composition could be uniquely determined on the assumption that only these three parts of the spectrum were involved in its appearance. Actually, of course, other, fainter, lines are present in its spectrum and some background illumination must also be expected. Nevertheless, for want

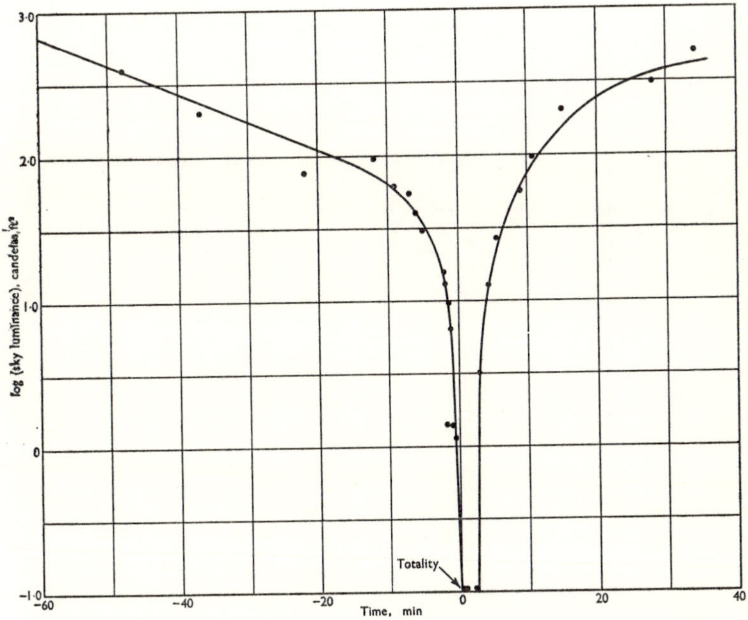

FIG. 10.18 Variation in sky luminance during the total eclipse of the sun on 30 June 1954. as observed from Grebbestad, Sweden. The sky was cloudy and was observed at an altitude of 40° above the horizon in an azimuth plane 90° west from the sun

of measurement, a great deal of potentially useful information is being lost during every colourful display of the aurora, and while the initial function of the portable colorimeter which has been described will no doubt be to make a general study of the range of chromaticities and luminances which occur, it is to be hoped that by one means or another more extensive colour measurements will become possible.

As a field test of the colorimeter, observations were made on the colour of the sky during the total solar eclipse of the 30th June, 1954. One colorimeter was operated by Mr B. McInnes of Edinburgh University in an aeroplane flying along the track of the eclipse near Iceland at a height of about 9000 ft, but for a number of technical reasons, the observations were largely unsuccessful. The author used a second model at Grebbestad in Sweden, which lay near the centre line of the totality zone, and recorded the sky colour in an azimuth plane 90° round from the sun and at a height

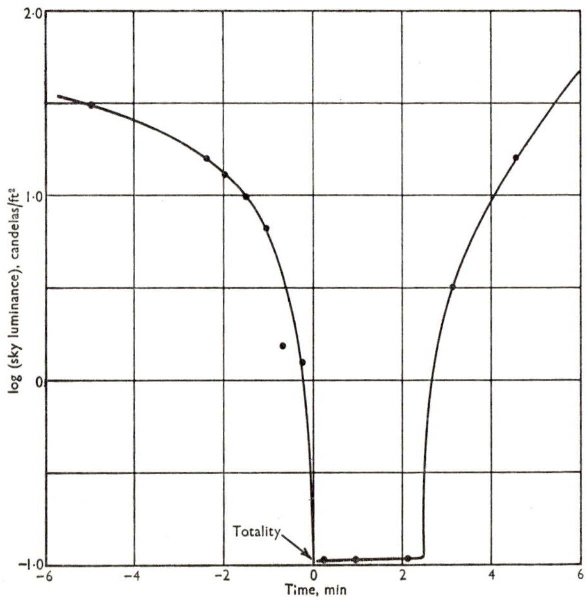

FIG. 10.19. Part of Fig. 10.18 on an enlarged scale. Variation in sky luminance shortly before, during, and immediately after the total solar eclipse of 30 June 1954

of 40° above the horizon. As the sky was cloudy, the observations were of little astronomical value, but Figs. 10.18 and 10.19 show the variation in luminance of the sky before, during and after the eclipse (a colorimeter may also prove a very effective photometer in experiments of this kind), and Fig. 10.20 shows the chromaticity changes that occurred, together with an indication of the chromaticity recorded by McInnes from the air.

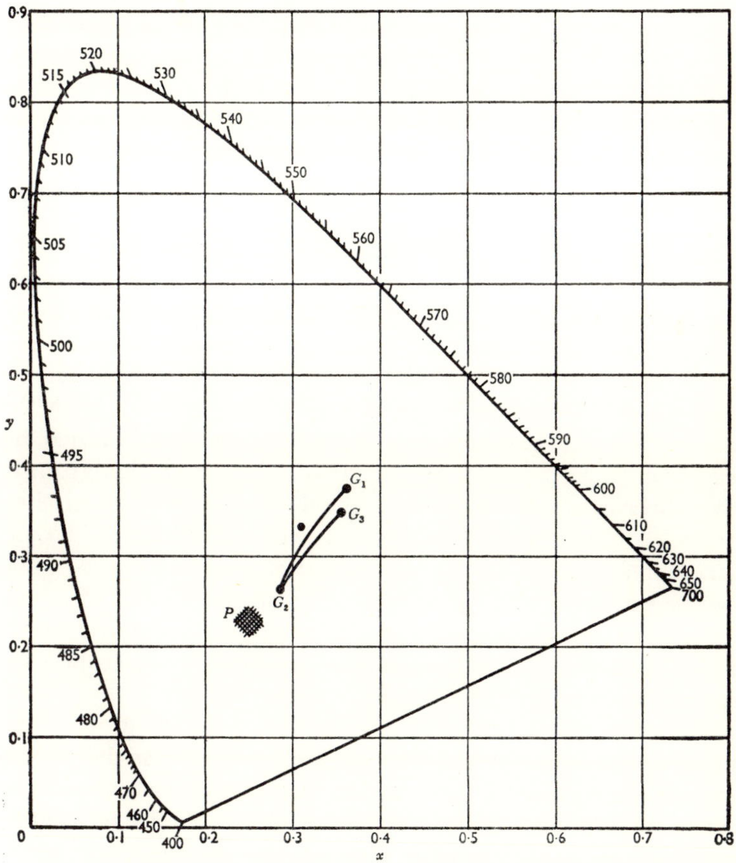

FIG. 10.20 Chromaticity changes in sky colour during the total solar eclipse of 30 June 1954. G_1 and G_3 are the chromaticities observed at Grebbestad at the beginning and the end of the partial phase, G_2 is the chromaticity during totality, and P is the chromaticity observed from an aeroplane at a height of 9000 feet along the totality track near Iceland

With a clear sky the results could be applied to the study of the light scatter in the sky, although with a single instrument the results would be limited to one area of the sky at a time. In a phenomenon such as an eclipse, where changes of luminance and colour occur very rapidly and extend over large areas, a comprehensive investigation might be planned in which visual observations taken with a number of colorimeters pointing to various parts of the sky were combined with continuous photographic records covering large areas of sky. The colorimetric observations could then be used to calibrate the colours of the corresponding points on the photographs, and thus permit interpolation in both time and space to give a much more complete and accurate record than would otherwise be possible by using colorimetry or photography alone.

DICHROISM

The term *dichroism* is often used in the general sense of describing phenomena which show a marked change in hue with change in the experimental situation, e.g. quality of illumination, dilution of absorbing medium, polarization, etc. We shall consider here an example of the change which is sometimes observed when the thickness or concentration of a transparent medium is varied, an effect which can produce an alteration in hue from blue-green in the thin layer to reddish-purple in the thick layer, as with malachite green, or from green to red, as with chromic chloride. The observation appears at first sight to be one of those odd effects which can be explained only by reference to some peculiarity of vision, but in fact the phenomenon of dichroism has a physical origin which is revealed when calculations are made to determine the spectral composition of the light reaching the eye.

We saw in Fig. 1.6 on p. 16 how the relative contributions of the light from different parts of the spectrum will alter as the colouring medium is diluted. Reduction in thickness or concentration of the medium leads to an increased transmission at all wavelengths, but the increase is relatively very much greater at those wavelengths which are most absorbed. When the transmission curve contains one main transmission band at, say, a wavelength of λ_1 and a subsidiary band at λ_2, then the relative proportions of the λ_1 and λ_2 radiations will vary with the thickness of the medium and will almost inevitably give rise to a change in hue. Or again, when the medium transmits freely at λ_1 and absorbs heavily at λ_2, then, in thick layers, only the λ_1 radiations will be transmitted in any quantity, and hence the hue will be determined by these radiations alone. But in thin layers, all the radiations will be freely transmitted except those at λ_2, which will still be slightly absorbed; as a result the medium will be tinged with the hue complementary to that of λ_2.

The colour of blood provides an instructive and important illustration of the phenomenon,* since the transmission curve of haemoglobin shows both a region of high transmission and a strong absorption band. The complete transmission curves of oxyhaemoglobin (1 per cent concentration) for thicknesses ranging from 1 cm to 0·01 mm have been calculated from data given by Heilmeyer† and are illustrated in Fig. 10.21. It is seen from this diagram that for the thickest layers, the only radiations transmitted in any strength are those at the red end of the spectrum, and it is this

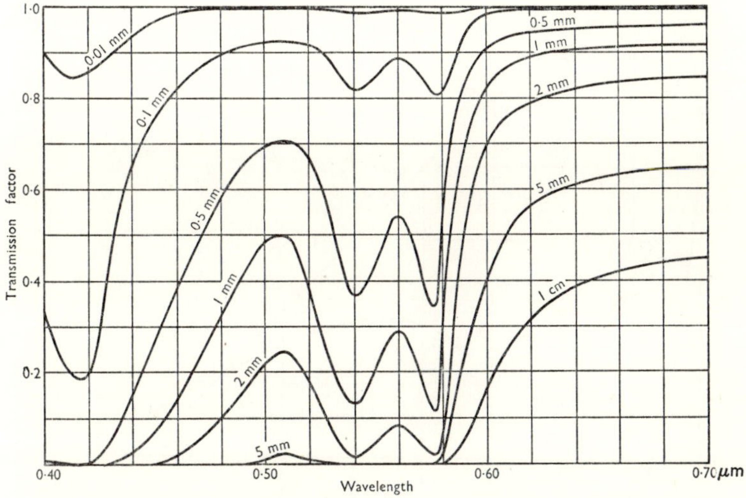

FIG. 10.21 Spectral transmission curves for different thicknesses of oxyhaemoglobin (1 per cent concentration) when reflection losses at the surfaces of the medium are neglected. The thickness of each layer is shown against the curves

* W. D. WRIGHT: *Nature*, 1944, **153**, p. 9.

† *Spectrophotometry in Medicine* by L. HEILMEYER, translated by A. JORDAN and T. L. TIPPELL (Hilger), 1943.

transmission which is responsible for the deep red colour normally observed. For layers of intermediate thickness the transmission around 0·51 μm in the green and blue-green becomes comparable to that in the red and will cause some change in the hue towards the orange or yellow, assisted also by the transmission at 0·56 μm. For very thin layers, on the other hand, there is practically complete transmission of all wavelengths except those in the violet where some significant absorption still occurs. This will have the effect of tingeing the layer with the hue complementary to violet, namely yellow, and will therefore introduce a change of hue in the same direction as that caused by the transmission in the green. The similarity in the trend of the two effects is, however, a coincidence.

The change in hue which accompanies the desaturation can be

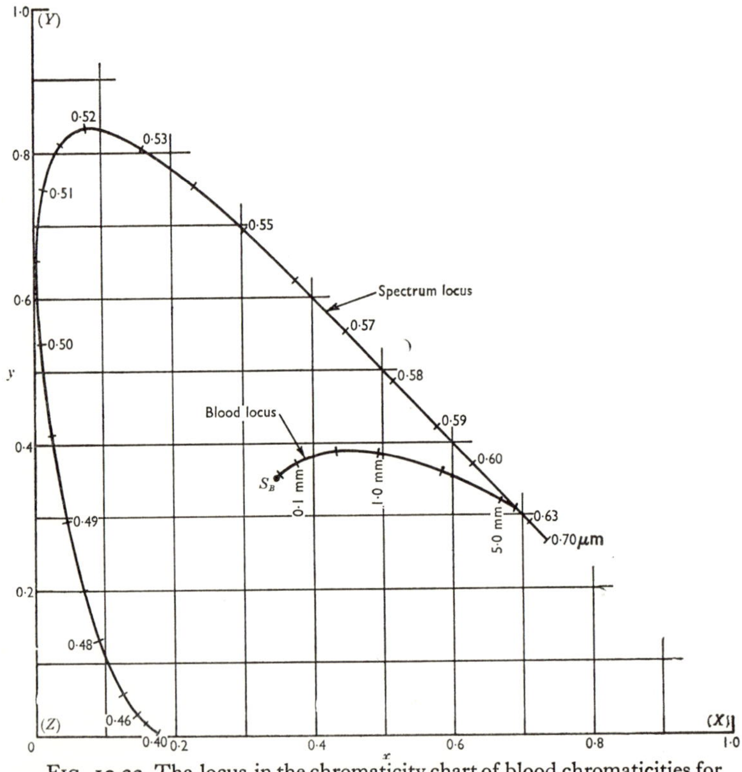

FIG. 10.22 The locus in the chromaticity chart of blood chromaticities for different thicknesses of blood using the illuminant S_B. The thicknesses of the haemoglobin layers are shown against the blood locus

effectively illustrated by calculating the chromaticity co-ordinates for each layer. Assuming the source S_B has been used as the illuminant, the blood locus in the chromaticity chart as shown in Fig. 12.22 is obtained. It is seen that, for thick layers, the locus starts from the spectrum near the wavelength 0·62 μm and then with diminishing thickness follows a curved path which terminates, for extremely thin layers, at S_B. If we draw straight lines from S_B through the points representing each layer and extend them to in-

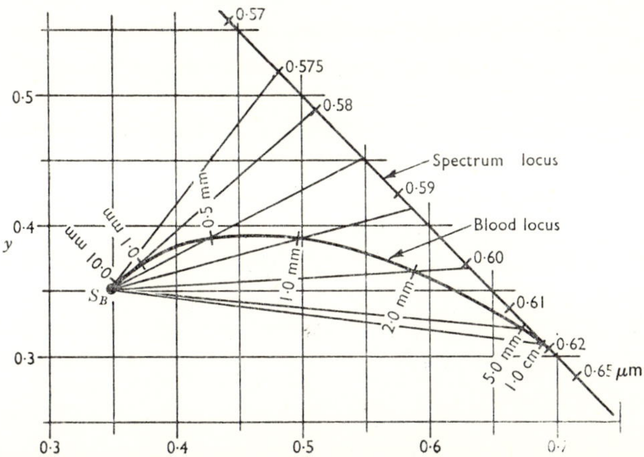

FIG. 10.23. Part of Fig. 10.22 on an enlarged scale to show the variation of dominant wavelength with thickness

tersect the spectrum locus, as in the enlarged diagram (Fig. 10.23) the points of intersection will give the dominant wavelength of each layer. These vary from 0·62 μm, an orange-red, for the thickest layer, to 0·575 μm, a yellow, for very thin layers, and we thus arrive at the numerical evidence for the change of hue which we could deduce from general considerations from Fig. 10.21. If there had been no change of dominant wavelength with thickness, the blood locus would, of course, have been a straight line from the region of 0·62 μm to S_B. It is true, as has been mentioned in Chapter 7 in connection with the Munsell system, that constancy of dominant wavelength does not necessarily imply exact constancy in the apparent hue of a medium, and this subjective dichroism, as we may call it, must be taken into account in making a final estimation of the change in appearance of the medium. From the loci of constant hue given in Fig. 7.5 the subjective dichroism tends to compensate the physical dichroism to a small extent in the case of haemoglobin, but for other media it might be equally prone to accentuate the physical effect.

The subject of dichroism is important apart from its intrinsic interest, more especially in connection with colour tests applied to microscopical objects,* since the identification of a thin section of a specimen may be more difficult if its hue differs appreciably from that of a thick layer of the medium. Thus blood corpuscles are a light straw-yellow when seen under the microscope and their relation to a drop of blood is by no means obvious. The type of colour change to be expected can, however, be estimated if a dilute solution of the medium can be prepared, since change in thickness and change in concentration are equivalent to each other. There is no doubt that the phenomenon is more widespread than is generally imagined and, strictly speaking, every colorant exhibits some degree of dichroism. In dyeing, as noted by Paterson,† dyestuffs which are strongly dichroic are not very satisfactory owing to the change of hue of the lighter tints, and because there will generally be a change of hue towards the red when the dyed material is transferred from daylight to tungsten light (as discussed on p. 261). Dichroism is also evidenced in certain pigments when they are greatly extended with white, as the undertone may reveal a change of hue from that of the mass of pigment.

* J. R. Baker: *Nature,* 1943, **152**, p. 331.
D. A. Webb: *Nature,* 1943, **152**, p. 476.

† D. Paterson: *Textile Colour Mixing* (Benn), 1927.

APPENDIX I

ILLUMINANTS

PART I

In 1931 the Commission Internationale de l'Éclairage specified three illuminants for use in colorimetry, known as sources *A*, *B* and *C*, or S_A, S_B and S_C and based on a report published by Davis and Gibson in 1931.* S_A was to represent an average artificial illuminant, while S_B and S_C were to correspond to two phases of daylight. When a coloured sample is to be measured, the illuminant most appropriate to the nature and use of the sample should be chosen. The selected source must be stated when quoting the specification. The sources have the following colour temperatures: S_A—2854° K, S_B—4900° K, S_C—6700° K, but the energy distributions of S_B and S_C depart significantly from that of a Planckian radiator.

1931 Sources S_A, S_B *and* S_C

S_A. A gas-filled tungsten lamp operated at a colour temperature of 2854° K.

S_B. The source S_A used in conjunction with a colour filter consisting of two solutions B_1 and B_2, each one centimetre in thickness and contained in a double cell made of colourless optical glass. The solutions are described below.

S_C. The source S_A used in conjunction with a colour filter consisting of two solutions C_1 and C_2, each one centimetre in thickness and contained in a double cell made of colourless optical glass. The solutions are described below.

Specification of Solutions

	B_1	C_1
Copper sulphate ($CuSO_4 \cdot 5H_2O$)	2·452 g	3·412 g
Mannite [$C_6H_8(OH)_6$]	2·452 g	3·412 g
Pyridine (C_5H_5N)	30·0 c.c.	30·0 c.c.
Water (distilled) to make	1000 c.c.	1000 c.c.

* R. Davis and K. S. Gibson: *Bur. Stand. Misc. Pub.*, No. 114, 1931.

	B_2	C_2
Cobalt ammonium sulphate $[CoSO_4 . (NH_4)_2SO_4 . 6H_2O]$	21·71 g	30·580 g
Copper sulphate ($CuSO_4 . 5H_2O$)	16·11 g	22·520 g
Sulphuric acid (sp. gr. 1·835)	10·0 c.c.	10·0 c.c.
Water (distilled) to make	1000 c.c.	1000 c.c.

The spectral energy distributions, P_A, P_B and P_C, for the three sources are as shown in Table 1 (p. 311) from the data given by Davis and Gibson. The data give only the relative distributions of energy, the actual values having been adjusted to fit as closely as possible the energy distribution of the ideal Planckian sources which they were designed to represent, the latter having an energy of 100·00 at a wavelength of 560 nm.

In 1951 the C.I.E. extended the table of energy values for S_B and S_C down to a wavelength of 320 nm and these additional values have been included in Table 1.

1964 Daylight Distributions D_{5500}, D_{6500} and D_{7500}

On the basis of several new studies on the spectral composition of daylight extending over a wide range of wavelengths from the ultra-violet to the far red end of the spectrum, the C.I.E. Colorimetry Committee recommended in 1963 the adoption of new standard spectral distributions to represent daylight at various correlated colour temperatures.* This recommendation was officially endorsed by the C.I.E. in 1964. For general use, a correlated colour temperature of 6500° K has been selected as typical of average daylight. Standardized formulae and tables were also recommended for the calculation of daylight distributions at other correlated colour temperatures. The energy distribution for the daylight illuminant D_{6500} is given in Table 2, together with distributions for D_{5500} and D_{7500} as representative of slightly yellower and bluer phases of daylight. It is expected that in due course D_{6500} will replace S_C as the illuminant for which C.I.E. colour specifications are given under daylight conditions.

* *C.I.E. Compte Rendu*, 1963.

TABLE 1. ENERGY DISTRIBUTION OF THE ILLUMINANTS S_A, S_B AND S_C

Wave-length (nm.)	P_A	P_B	P_C	Wave-length (nm.)	P_A	P_B	P_C
320		0·02	0·01	570	107·18	102·60	102·30
330		0·5	0·4	575	110·80	101·90	100·15
340		2·4	2·7	580	114·44	101·00	97·80
350		5·6	7·0	585	118·08	100·07	95·43
360		9·6	12·9				
370		15·2	21·4				
380	9·79	22·40	33·00	590	121·73	99·20	93·20
385	10·90	26·85	39·92	595	125·39	98·44	91·22
390	12·09	31·30	47·40	600	129·04	98·00	89·70
395	13·36	36·18	55·17	605	132·70	98·08	88·83
400	14·71	41·30	63·30	610	136·34	98·50	88·40
405	16·15	46·62	71·81	615	139·99	99·06	88·19
410	17·68	52·10	80·60	620	143·62	99·70	88·10
415	19·29	57·70	89·53	625	147·23	100·36	88·06
420	21·00	63·20	98·10	630	150·83	101·00	88·00
425	22·79	68·37	105·80	635	154·42	101·56	87·86
430	24·67	73·10	112·40	640	157·98	102·20	87·80
435	26·64	77·31	117·75	645	161·51	103·05	87·99
440	28·70	80·80	121·50	650	165·03	103·90	88·20
445	30·85	83·44	123·45	655	168·51	104·59	88·20
450	33·09	85·40	124·00	660	171·96	105·00	87·90
455	35·41	86·88	123·60	665	175·38	105·08	87·22
460	37·82	88·30	123·10	670	178·77	104·90	86·30
465	40·30	90·08	123·30	675	182·12	104·55	85·30
470	42·87	92·00	123·80	680	185·43	103·90	84·00
475	45·52	93·75	124·09	685	188·70	102·84	82·21
480	48·25	95·20	123·90	690	191·93	101·60	80·20
485	51·04	96·23	122·92	695	195·12	100·38	78·24
490	53·91	96·50	120·70	700	198·26	99·10	76·30
495	56·85	95·71	116·90	705	201·36	97·70	74·36
500	59·86	94·20	112·10	710	204·41	96·20	72·40
505	62·93	92·37	106·98	715	207·41	94·60	70·40
510	66·06	90·70	102·30	720	210·36	92·90	68·30
515	69·25	89·65	98·81	725	213·26	91·10	66·30
520	72·50	89·50	96·90	730	216·12	89·40	64·40
525	75·79	90·43	96·78	735	218·92	88·00	62·80
530	79·13	92·20	98·00	740	221·66	86·90	61·50
535	82·52	94·46	99·94	745	224·36	85·90	60·20
540	85·95	96·90	102·10	750	227·00	85·20	59·20
545	89·41	99·16	103·95	755	229·58	84·80	58·50
550	92·91	101·00	105·20	760	232·11	84·70	58·10
555	96·44	102·20	105·67	765	234·59	84·90	58·00
560	100·00	102·80	105·30	770	237·01	85·40	58·20
565	103·58	102·92	104·11	775	239·37	86·10	58·50
570	107·18	102·60	102·30	780	241·67	87·00	59·10

TABLE. 2 ENERGY DISTRIBUTION OF THE ILLUMINANTS D_{5500}, D_{6500} AND D_{7500}

Wave-length (nm)	P_{5500}	P_{6500}	P_{7500}	Wave-length (nm)	P_{5500}	P_{6500}	P_{7500}
300	0·02	0·03	0·04	570	97!2	96·3	95·6
310	2·1	3·3	5·1	580	97·7	95·8	94·2
320	11·2	20·2	29 8	590	91·4	88·7	87·0
330	20·6	37·1	54·9				
340	23·9	39·9	57·3	600	94·4	90·0	87·2
				610	95·1	89·6	86·1
350	27·8	44·9	62·7	620	94·2	87·7	83·6
360	30·6	46·6	63·0	630	90·4	83·3	78·7
370	34·3	52·1	70·3	640	92·3	83·7	78·4
380	32·6	50·0	66·7				
390	38·1	54·6	70·0	650	88·9	80·0	74·8
				660	90·3	80·2	74·3
400	60·9	82·8	101·9	670	93·9	82·3	75·4
410	68·6	91·5	111·9	680	90·0	78·3	71·6
420	71·6	93·4	112·8	690	79·7	69·7	63·9
430	67·9	86·7	103·1				
440	85·6	104·9	121·2	700	82·8	71·6	65·1
				710	84·8	74·3	68·1
450	98·0	117·0	133·0	720	70·2	61·6	56·4
460	100·5	117·8	132·4	730	79·3	69·9	64·2
470	99·9	114·9	127·3	740	85·0	75·1	69·2
480	102·7	115·9	126·8				
490	98·1	108·8	117·8	750	71·9	63·6	58·6
				760	52·8	46·4	42·6
500	100·7	109·4	116·6	770	75·9	66·8	61·4
510	100·7	107·8	113·7	780	71·8	63·4	58·3
520	100·0	104·8	108·7	790	72·9	64·3	59·1
530	104·2	107·7	110·4				
540	102·1	104·4	106·3	800	67·3	59·5	54·7
				810	58·7	52·0	47·9
550	103·0	104·0	104·9	820	65·0	57·4	52·9
560	100·0	100·0	100·0	830	68·3	60·3	55·5

PART 2

In 1931 the C.I.E. recommended that surface colours should be measured with the surface illuminated by light incident at an angle of 45° and viewed in a direction normal to the surface. In practice various other geometrics of illuminating and viewing have been used in different designs of colorimeter and spectrophotometer. In particular many instruments use an integrating sphere for diffuse illumination or light collection. Discrepancies in measurements undoubtedly occur because of these varieties in instrument design, especially the effectiveness with which the specular component of the reflected light is included or excluded in the measurement.

The C.I.E. has recognized the need for various types of geometry but in an attempt to reduce the inconsistencies in experimental data, the Colorimetry Committee recommended in 1967 that these should be limited to four alternative arrangements and has laid down tolerances to the angular conditions of illumination and viewing. These are given below and it is hoped that instrument manufacturers will in future design their instruments to meet these conditions.

Illuminating and Viewing Conditions

It is recommended that the colorimetric specification of opaque specimens be given so as to correspond to one of the following illuminating and viewing conditions:

(*a*) 45°/*normal*: The specimen is illuminated by one or more beams whose axes are at an angle of $45° \pm 5°$ from the normal to the specimen surface. The angle between the direction of viewing and the normal to the specimen should not exceed 10°. The angle between the axis and any ray of an illuminating beam should not exceed 5°. The same restriction should be observed in the viewing beam.

(*b*) *Normal*/45°: The specimen is illuminated by a beam whose axis is at an angle not exceeding 10 degrees from the normal to the specimen. The specimen is viewed at an angle of $45° \pm 5°$ from the normal. The angle between the axis and any ray of the illuminating beam should not exceed 5°. The same restriction should be observed in the viewing beam.

(*c*) *Diffuse/normal*: The specimen is illuminated diffusely using an integrating sphere. The angle between the normal to the specimen and the axis of the viewing beam should not exceed 10°. The integrating sphere may be of any diameter provided the total area of the ports does not exceed 10·0 per cent of the internal reflecting sphere area. The angle between the axis and any ray of the viewing beam should not exceed 5°.

(*d*) *Normal/diffuse*: The specimen is illuminated by a beam whose axis is at an angle not exceeding 10° from the normal to the specimen. The reflected flux is collected by means of an integrating sphere. The angle between the axis and any ray of the illuminating beam should not exceed 5°. The integrating sphere may be of any diameter provided the total area of the ports does not exceed 10·0 per cent of the internal reflecting sphere area.

Note 1: For the conditions 'normal/diffuse' and 'diffuse/normal' the influence of the specular component of specimens with mixed reflection may be reduced by the use of a gloss trap. If a gloss trap is used, details of its size and position should be given.

Note 2: In the 'normal/45°' and the 'normal/diffuse' conditions, specimens with mixed reflection should not be measured with strictly normal illumination.

Note 3: The '45°/normal' condition gives the radiance factor $\beta_{45/0}$. The 'normal/45°' condition gives the radiance factor $\beta_{0/45}$. The 'diffuse/normal' condition gives the radiance factor $\beta_{d/0}$. The 'normal/diffuse' condition gives the reflectance $\beta_{0/d} = \rho$.

Note 4: If it is required to deviate from the recommended standard illuminating and viewing conditions, it is important that the particular conditions used be specified.

APPENDIX II

THE 1931 C.I.E. SYSTEM OF COLOUR SPECIFICATION

This system is based on colour matching data obtained for foveal vision using a field subtending 2°. Colour specifications on this system therefore apply to colours observed foveally and can be used for field sizes up to 4°. The reference stimuli are (X), (Y) and (Z), chosen to satisfy the conditions described in Chapter 4 and defined by the following unit trichromatic equations:

$$C_{\lambda_1} \equiv 0{\cdot}73467(X) + 0{\cdot}26533(Y) + 0{\cdot}00000(Z)$$
$$C_{\lambda_2} \equiv 0{\cdot}27376(X) + 0{\cdot}71741(Y) + 0{\cdot}00883(Z)$$
$$C_{\lambda_3} \equiv 0{\cdot}16658(X) + 0{\cdot}00886(Y) + 0{\cdot}82456(Z)$$
$$S_E \equiv 0{\cdot}33333(X) + 0{\cdot}33333(Y) + 0{\cdot}33333(Z)$$

where $\lambda_1 = 700$ nm, $\lambda_2 = 546{\cdot}1$ nm and $\lambda_3 = 435{\cdot}8$ nm, while S_E is the equal-energy stimulus. The unit equations for the three illuminants S_A, S_B and S_C are as follows:

$$S_A \equiv 0{\cdot}44757(X) + 0{\cdot}40745(Y) + 0{\cdot}14498(Z)$$
$$S_B \equiv 0{\cdot}34842(X) + 0{\cdot}35161(Y) + 0{\cdot}29997(Z)$$
$$S_C \equiv 0{\cdot}31006(X) + 0{\cdot}31616(Y) + 0{\cdot}37378(Z)$$

Table 3 gives the chromaticity co-ordinates x, y and z of each of the tabulated monochromatic radiations through the visible spectrum, and Table 4 gives the distribution coefficients $\bar{x}_\lambda$, $\bar{y}_\lambda$ and $\bar{z}_\lambda$ for the equal-energy stimulus through the spectrum. These values apply to the C.I.E. standard observer as defined in the 1931 resolutions. Tables 5, 6 and 7 give the data prepared by Smith and Guild* in which the distribution coefficients have been weighted by the energy at each wavelength for the illuminants S_A, S_B and S_C respectively, but adjusted to make $\Sigma(P_\lambda \,.\, \bar{y}_\lambda) = 100{\cdot}00$ in each case.

The relative luminous efficiencies $V_{(X)}$, $V_{(Y)}$ and $V_{(Z)}$ of 1 T-unit of (X), (Y) and (Z) are zero, unity and zero respectively.

* T. Smith and J. Guild: *Trans. Opt. Soc.*, 1931, **33**, p. 73.

TABLE 3

Wave-length (nm)	Chromaticity co-ordinates			Wave-length (nm)	Chromaticity co-ordinates		
	x	y	z		x	y	z
380	0·1741	0·0050	0·8209	580	0·5125	0·4866	0·0009
385	0·1740	0·0050	0·8210	585	0·5448	0·4544	0·0008
390	0·1738	0·0049	0·8213	590	0·5752	0·4242	0·0006
395	0·1736	0·0049	0·8215	595	0·6029	0·3965	0·0006
400	0·1733	0·0048	0·8219	600	0·6270	0·3725	0·0005
405	0·1730	0·0048	0·8222	605	0·6482	0·3514	0·0004
410	0·1726	0·0048	0·8226	610	0·6658	0·3340	0·0002
415	0·1721	0·0048	0·8231	615	0·6801	0·3197	0·0002
420	0·1714	0·0051	0·8235	620	0·6915	0·3083	0·0002
425	0·1703	0·0058	0·8239	625	0·7006	0·2993	0·0001
430	0·1689	0·0069	0·8242	630	0·7079	0·2920	0·0001
435	0·1669	0·0086	0·8245	635	0·7140	0·2859	0·0001
440	0·1644	0·0109	0·8247	640	0·7190	0·2809	0·0001
445	0·1611	0·0138	0·8251	645	0·7230	0·2770	0·0000
450	0·1566	0·0177	0·8257	650	0·7260	0·2740	0·0000
455	0·1510	0·0227	0·8263	655	0·7283	0·2717	0·0000
460	0·1440	0·0297	0·8263	660	0·7300	0·2700	0·0000
465	0·1355	0·0399	0·8246	665	0·7311	0·2689	0·0000
470	0·1241	0·0578	0·8181	670	0·7320	0·2680	0·0000
475	0·1096	0·0868	0·8036	675	0·7327	0·2673	0·0000
480	0·0913	0·1327	0·7760	680	0·7334	0·2666	0·0000
485	0·0687	0·2007	0·7306	685	0·7340	0·2660	0·0000
490	0·0454	0·2950	0·6596	690	0·7344	0·2656	0·0000
495	0·0235	0·4127	0·5638	695	0·7346	0·2654	0·0000
500	0·0082	0·5384	0·4534	700	0·7347	0·2653	0·0000
505	0·0039	0·6548	0·3413	705	0·7347	0·2653	0·0000
510	0·0139	0·7502	0·2359	710	0·7347	0·2653	0·0000
515	0·0389	0·8120	0·1491	715	0·7347	0·2653	0·0000
520	0·0743	0·8338	0·0919	720	0·7347	0·2653	0·0000
525	0·1142	0·8262	0·0596	725	0·7347	0·2653	0·0000
530	0·1547	0·8059	0·0394	730	0·7347	0·2653	0·0000
535	0·1929	0·7816	0·0255	735	0·7347	0·2653	0·0000
540	0·2296	0·7543	0·0161	740	0·7347	0·2653	0·0000
545	0·2658	0·7243	0·0099	745	0·7347	0·2653	0·0000
550	0·3016	0·6923	0·0061	750	0·7347	0·2653	0·0000
555	0·3373	0·6589	0·0038	755	0·7347	0·2653	0·0000
560	0·3731	0·6245	0·0024	760	0·7347	0·2653	0·0000
565	0·4087	0·5896	0·0017	765	0·7347	0·2653	0·0000
570	0·4441	0·5547	0·0012	770	0·7347	0·2653	0·0000
575	0·4788	0·5202	0·0010	775	0·7347	0·2653	0·0000
580	0·5125	0·4866	0·0009	780	0·7347	0·2653	0·0000

TABLE 4

Wave-length (nm)	Distribution coefficients for equal-energy stimulus			Wave-length (nm)	Distribution coefficients for equal-energy stimulus		
	$\bar{x}_\lambda$	$\bar{y}_\lambda$	$\bar{z}_\lambda$		$\bar{x}_\lambda$	$\bar{y}_\lambda$	$\bar{z}_\lambda$
380	0·0014	0·0000	0·0065	580	0·9163	0·8700	0·0017
385	0·0022	0·0001	0·0105	585	0·9786	0·8163	0·0014
390	0·0042	0·0001	0·0201	590	1·0263	0·7570	0·0011
395	0·0076	0·0002	0·0362	595	1·0567	0·6949	0·0010
400	0·0143	0·0004	0·0679	600	1·0622	0·6310	0·0008
405	0·0232	0·0006	0·1102	605	1·0456	0·5668	0·0006
410	0·0435	0·0012	0·2074	610	1·0026	0·5030	0·0003
415	0·0776	0·0022	0·3713	615	0·9384	0·4412	0·0002
420	0·1344	0·0040	0·6456	620	0·8544	0·3810	0·0002
425	0·2148	0·0073	1·0391	625	0·7514	0·3210	0·0001
430	0·2839	0·0116	1·3856	630	0·6424	0·2650	0·0000
435	0·3285	0·0168	1·6230	635	0·5419	0·2170	0·0000
440	0·3483	0·0230	1·7471	640	0·4479	0·1750	0·0000
445	0·3481	0·0298	1·7826	645	0·3608	0·1382	0·0000
450	0·3362	0·0380	1·7721	650	0·2835	0·1070	0·0000
455	0·3187	0·0480	1·7441	655	0·2187	0·0816	0·0000
460	0·2908	0·0600	1·6692	660	0·1649	0·0610	0·0000
465	0·2511	0·0739	1·5281	665	0·1212	0·0446	0·0000
470	0·1954	0·0910	1·2876	670	0·0874	0·0320	0·0000
475	0·1421	0·1126	1·0419	675	0·0636	0·0232	0·0000
480	0·0956	0·1390	0·8130	680	0·0468	0·0170	0·0000
485	0·0580	0·1693	0·6162	685	0·0329	0·0119	0·0000
490	0·0320	0·2080	0·4652	690	0·0227	0·0082	0·0000
495	0·0147	0·2586	0·3533	695	0·0158	0·0057	0·0000
500	0·0049	0·3230	0·2720	700	0·0114	0·0041	0·0000
505	0·0024	0·4073	0·2123	705	0·0081	0·0029	0·0000
510	0·0093	0·5030	0·1582	710	0·0058	0·0021	0·0000
515	0·0291	0·6082	0·1117	715	0·0041	0·0015	0·0000
520	0·0633	0·7100	0·0782	720	0·0029	0·0010	0·0000
525	0·1096	0·7932	0·0573	725	0·0020	0·0007	0·0000
530	0·1655	0·8620	0·0422	730	0·0014	0·0005	0·0000
535	0·2257	0·9149	0·0298	735	0·0010	0·0004	0·0000
540	0·2904	0·9540	0·0203	740	0·0007	0·0003	0·0000
545	0·3597	0·9803	0·0134	745	0·0005	0·0002	0·0000
550	0·4334	0·9950	0·0087	750	0·0003	0·0001	0·0000
555	0·5121	1·0002	0·0057	755	0·0002	0·0001	0·0000
560	0·5945	0·9950	0·0039	760	0·0002	0·0001	0·0000
565	0·6784	0·9786	0·0027	765	0·0001	0·0000	0·0000
570	0·7621	0·9520	0·0021	770	0·0001	0·0000	0·0000
575	0·8425	0·9154	0·0018	775	0·0000	0·0000	0·0000
580	0·9163	0·8700	0·0017	780	0·0000	0·0000	0·0000
				Totals	21·3713	21·3714	21·3715

TABLE 5

Wave-length (nm)	Distribution coefficients weighted by energy values of illuminant S_A			Wave-length (nm)	Distribution coefficients weighted by energy values of illuminant S_A		
	$P_A \cdot \bar{x}_\lambda$	$P_A \cdot \bar{y}_\lambda$	$P_A \cdot \bar{z}_\lambda$		$P_A \cdot \bar{x}_\lambda$	$P_A \cdot \bar{y}_\lambda$	$P_A \cdot \bar{z}_\lambda$
380	0·0006	0·0000	0·0029	580	4·8594	4·6139	0·0090
385	0·0011	0·0000	0·0053	585	5·3549	4·4668	0·0077
390	0·0024	0·0000	0·0113	590	5·7896	4·2704	0·0062
395	0·0047	0·0001	0·0224	595	6·1403	4·0379	0·0058
400	0·0097	0·0003	0·0463	600	6·3518	3·7733	0·0048
405	0·0174	0·0004	0·0825	605	6·4299	3·4855	0·0037
410	0·0356	0·0010	0·1699	610	6·3346	3·1780	0·0019
415	0·0694	0·0020	0·3319	615	6·0877	2·8622	0·0013
420	0·1308	0·0039	0·6283	620	5·6865	2·5358	0·0013
425	0·2269	0·0077	1·0974	625	5·1267	2·1901	0·0007
430	0·3246	0·0133	1·5840	630	4·4902	1·8523	0·0000
435	0·4055	0·0207	2·0036	635	3·8779	1·5529	0·0000
440	0·4632	0·0306	2·3236	640	3·2791	1·2812	0·0000
445	0·4976	0·0426	2·5484	645	2·7004	1·0344	0·0000
450	0·5155	0·0583	2·7173	650	2·1681	0·8183	0·0000
455	0·5230	0·0788	2·8621	655	1·7078	0·6372	0·0000
460	0·5097	0·1052	2·9254	660	1·3141	0·4861	0·0000
465	0·4690	0·1380	2·8539	665	0·9850	0·3625	0·0000
470	0·3882	0·1808	2·5581	670	0·7241	0·2651	0·0000
475	0·2998	0·2375	2·1979	675	0·5368	0·1958	0·0000
480	0·2138	0·3108	1·8179	680	0·4022	0·1461	0·0000
485	0·1372	0·4004	1·4575	685	0·2877	0·1041	0·0000
490	0·0799	0·5196	1·1622	690	0·2019	0·0729	0·0000
495	0·0387	0·6813	0·9308	695	0·1429	0·0515	0·0000
500	0·0136	0·8960	0·7545	700	0·1047	0·0377	0·0000
505	0·0070	1·1878	0·6191	705	0·0756	0·0271	0·0000
510	0·0285	1·5398	0·4843	710	0·0549	0·0199	0·0000
515	0·0934	1·9518	0·3585	715	0·0394	0·0144	0·0000
520	0·2127	2·3855	0·2627	720	0·0283	0·0097	0·0000
525	0·3849	2·7859	0·2012	725	0·0198	0·0069	0·0000
530	0·6069	3·1609	0·1547	730	0·0140	0·0050	0·0000
535	0·8631	3·4987	0·1140	735	0·0101	0·0041	0·0000
540	1·1567	3·7999	0·0809	740	0·0072	0·0031	0·0000
545	1·4904	4·0618	0·0555	745	0·0052	0·0021	0·0000
550	1·8660	4·2841	0·0375	750	0·0032	0·0010	0·0000
555	2·2887	4·4701	0·0255	755	0·0021	0·0010	0·0000
560	2·7550	4·6110	0·0181	760	0·0021	0·0010	0·0000
565	3·2564	4·6974	0·0130	765	0·0011	0·0000	0·0000
570	3·7853	4·7285	0·0104	770	0·0011	0·0000	0·0000
575	4·3259	4·7002	0·0092	775	0·0000	0·0000	0·0000
580	4·8594	4·6139	0·0090	780	0·0000	0·0000	0·0000
				Totals	109·8472	100·0000	35·5824

TABLE 6

Wave-length (nm)	Distribution coefficients weighted by energy values of illuminant S_B			Wave-length (nm)	Distribution coefficients weighted by energy values of illuminant S_B		
	$P_B . \bar{x}_\lambda$	$P_B . \bar{y}_\lambda$	$P_B . \bar{z}_\lambda$		$P_B . \bar{x}_\lambda$	$P_B . \bar{y}_\lambda$	$P_B . \bar{z}_\lambda$
380	0·0015	0·0000	0·0070	580	4·4218	4·1984	0·0082
385	0·0028	0·0001	0·0135	585	4·6790	3·9030	0·0067
390	0·0063	0·0001	0·0301	590	4·8644	3·5880	0·0052
395	0·0131	0·0003	0·0626	595	4·9701	3·2684	0·0047
400	0·0282	0·0008	0·1340	600	4·9736	2·9546	0·0037
405	0·0517	0·0013	0·2455	605	4·8999	2·6561	0·0028
410	0·1083	0·0030	0·5163	610	4·7185	2·3672	0·0014
415	0·2139	0·0061	1·0236	615	4·4415	2·0882	0·0009
420	0·4058	0·0121	1·9495	620	4·0700	1·8149	0·0009
425	0·7017	0·0238	3·3944	625	3·6031	1·5392	0·0005
430	0·9916	0·0405	4·8394	630	3·1000	1·2788	0·0000
435	1·2134	0·0621	5·9951	635	2·6296	1·0530	0·0000
440	1·3446	0·0888	6·7448	640	2·1871	0·8545	0·0000
445	1·3878	0·1188	7·1067	645	1·7765	0·6804	0·0000
450	1·3718	0·1551	7·2308	650	1·4074	0·5312	0·0000
455	1·3229	0·1993	7·2399	655	1·0929	0·4078	0·0000
460	1·2269	0·2531	7·0422	660	0·8273	0·3060	0·0000
465	1·0807	0·3181	6·5769	665	0·6085	0·2239	0·0000
470	0·8589	0·4000	5·6599	670	0·4381	0·1604	0·0000
475	0·6365	0·5044	4·6670	675	0·3177	0·1159	0·0000
480	0·4348	0·6323	3·6980	680	0·2323	0·0844	0·0000
485	0·2667	0·7784	2·8332	685	0·1617	0·0585	0·0000
490	0·1475	0·9590	2·1449	690	0·1102	0·0398	0·0000
495	0·0672	1·1826	1·6156	695	0·0758	0·0273	0·0000
500	0·0221	1·4538	1·2242	700	0·0540	0·0194	0·0000
505	0·0106	1·7976	0·9370	705	0·0378	0·0135	0·0000
510	0·0403	2·1798	0·6856	710	0·0267	0·0097	0·0000
515	0·1246	2·6052	0·4785	715	0·0185	0·0068	0·0000
520	0·2707	3·0361	0·3344	720	0·0129	0·0044	0·0000
525	0·4735	3·4272	0·2476	725	0·0087	0·0030	0·0000
530	0·7291	3·7973	0·1859	730	0·0060	0·0021	0·0000
535	1·0186	4·1292	0·1345	735	0·0042	0·0017	0·0000
540	1·3445	4·4168	0·0940	740	0·0029	0·0012	0·0000
545	1·7042	4·6445	0·0635	745	0·0020	0·0008	0·0000
550	2·0915	4·8016	0·0420	750	0·0012	0·0004	0·0000
555	2·5006	4·8840	0·0278	755	0·0008	0·0004	0·0000
560	2·9200	4·8872	0·0192	760	0·0008	0·0004	0·0000
565	3·3360	4·8122	0·0133	765	0·0004	0·0000	0·0000
570	3·7359	4·6669	0·0103	770	0·0004	0·0000	0·0000
575	4·1019	4·4568	0·0088	775	0·0000	0·0000	0·0000
580	4·4218	4·1984	0·0082	780	0·0000	0·0000	0·0000
				Totals	99·0930	100·0000	85·3125

TABLE 7

Wave-length (nm)	Distribution coefficients weighted by energy values of illuminant S_C			Wave-length (nm)	Distribution coefficients weighted by energy values of illuminant S_C		
	$P_C . \bar{x}_\lambda$	$P_C . \bar{y}_\lambda$	$P_C . \bar{z}_\lambda$		$P_C . \bar{x}_\lambda$	$P_C . \bar{y}_\lambda$	$P_C . \bar{z}_\lambda$
380	0·0022	0·0000	0·0101	580	4·2084	3·9958	0·0078
385	0·0041	0·0002	0·0197	585	4·3859	3·6585	0·0063
390	0·0093	0·0002	0·0447	590	4·4920	3·3133	0·0048
395	0·0197	0·0005	0·0938	595	4·5265	2·9767	0·0043
400	0·0425	0·0012	0·2018	600	4·4745	2·6581	0·0034
405	0·0782	0·0020	0·3716	605	4·3617	2·3644	0·0025
410	0·1647	0·0045	0·7850	610	4·1622	2·0882	0·0013
415	0·3263	0·0092	1·5611	615	3·8863	1·8272	0·0008
420	0·6192	0·0184	2·9743	620	3·5349	1·5763	0·0008
425	1·0672	0·0363	5·1628	625	3·1074	1·3275	0·0004
430	1·4986	0·0612	7·3139	630	2·6548	1·0952	0·0000
435	1·8165	0·0929	8·9747	635	2·2358	0·8953	0·0000
440	1·9874	0·1312	9·9687	640	1·8468	0·7216	0·0000
445	2·0182	0·1728	10·3351	645	1·4909	0·5711	0·0000
450	1·9578	0·2213	10·3194	650	1·1743	0·4432	0·0000
455	1·8499	0·2786	10·1235	655	0·9058	0·3380	0·0000
460	1·6811	0·3469	9·6497	660	0·6807	0·2518	0·0000
465	1·4539	0·4279	8·8481	665	0·4965	0·1827	0·0000
470	1·1360	0·5291	7·4860	670	0·3542	0·1297	0·0000
475	0·8281	0·6562	6·0719	675	0·2548	0·0929	0·0000
480	0·5563	0·8088	4·7305	680	0·1846	0·0671	0·0000
485	0·3348	0·9773	3·5571	685	0·1270	0·0459	0·0000
490	0·1814	1·1790	2·6369	690	0·0855	0·0309	0·0000
495	0·0807	1·4197	1·9396	695	0·0581	0·0209	0·0000
500	0·0258	1·7004	1·4319	700	0·0408	0·0147	0·0000
505	0·0121	2·0462	1·0665	705	0·0283	0·0101	0·0000
510	0·0447	2·4165	0·7600	710	0·0197	0·0071	0·0000
515	0·1350	2·8223	0·5183	715	0·0136	0·0049	0·0000
520	0·2881	3·2309	0·3559	720	0·0093	0·0032	0·0000
525	0·4982	3·6052	0·2604	725	0·0062	0·0022	0·0000
530	0·7617	3·9671	0·1942	730	0·0042	0·0015	0·0000
535	1·0593	4·2941	0·1399	735	0·0029	0·0012	0·0000
540	1·3924	4·5742	0·0973	740	0·0020	0·0009	0·0000
545	1·7559	4·7853	0·0654	745	0·0014	0·0006	0·0000
550	2·1412	4·9157	0·0430	750	0·0008	0·0003	0·0000
555	2·5414	4·9636	0·0283	755	0·0005	0·0003	0·0000
560	2·9399	4·9204	0·0193	760	0·0005	0·0003	0·0000
565	3·3167	4·7844	0·0132	765	0·0003	0·0000	0·0000
570	3·6613	4·5736	0·0101	770	0·0003	0·0000	0·0000
575	3·9623	4·3051	0·0085	775	0·0000	0·0000	0·0000
580	4·2084	3·9958	0·0078	780	0·0000	0·0000	0·0000
				Totals	98·0705	100·0000	118·2246

APPENDIX III

CONDENSED TABLES FOR THE 1931 C.I.E. SYSTEM

In many instances, sufficient accuracy in the calculation of the C.I.E. tristimulus values from spectrophotometric data can be obtained by using energy values at wavelength intervals of 10 nm instead of the interval of 5 nm employed in the standard tables. Computation can then be carried out by using alternate entries in the standard tables or, better still, by using the condensed tables specially prepared by Smith* for illuminants *A*, *B* and *C*. Smith's tables have the additional advantage that $\Sigma(P_\lambda \,.\, y_\lambda)$ has for each illuminant been adjusted to have a value of 100·00, so that the *Y* tristimulus value of a surface obtained in the usual way as $\Sigma(p_\lambda \,.\, P_\lambda \,.\, y_\lambda)$ will give the percentage reflection of the surface. They are reproduced in Table 8.

Similar tables prepared by Wyszecki and Stiles† for illuminants D_{5500}, D_{6500} and D_{7500} have also been included. The unit equations for these three illuminants on the 1931 C.I.E. system are as follows:

$$D_{5500} \equiv 0{\cdot}3324(X) + 0{\cdot}3475(Y) + 0{\cdot}3201(Z)$$
$$D_{6500} \equiv 0{\cdot}3127(X) + 0{\cdot}3290(Y) + 0{\cdot}3583(Z)$$
$$D_{7500} \equiv 0{\cdot}2990(X) + 0{\cdot}3150(Y) + 0{\cdot}3860(Z)$$

* T. Smith: *Proc. Phys. Soc.*, 1934, **46**, p. 372.

† G. Wyszecki and W. S. Stiles: *Color Science* (Wiley: New York and London), 1967.

TABLE 8. CONDENSED TABLES FOR THE 1931 DISTRIBUTION COEFFICIENTS WEIGHTED FOR ILLUMINANTS S_A, S_B, S_C, D_{5500}, D_{6500} AND D_{7500}

Wave-length (nm)	Distribution coefficients weighted by energy values of illuminant S_A			Distribution coefficients weighted by energy values of illuminant S_B			Distribution coefficients weighted by energy values of illuminant S_C		
	$P_A \cdot \bar{x}_\lambda$	$P_A \cdot \bar{y}_\lambda$	$P_A \cdot \bar{z}_\lambda$	$P_B \cdot \bar{x}_\lambda$	$P_B \cdot \bar{y}_\lambda$	$P_B \cdot \bar{z}_\lambda$	$P_C \cdot \bar{x}_\lambda$	$P_C \cdot \bar{y}_\lambda$	$P_C \cdot \bar{z}_\lambda$
380	0·0010	0·0000	0·0048	0·0025	0·0000	0·0113	0·0036	0·0000	0·0164
390	0·0046	0·0001	0·0219	0·0123	0·0003	0·0585	0·0183	0·0004	0·0870
400	0·0193	0·0005	0·0916	0·0558	0·0014	0·2650	0·0841	0·0021	0·3992
410	0·0688	0·0019	0·3281	0·2091	0·0057	0·9970	0·3180	0·0087	1·5159
420	0·2666	0·0080	1·2811	0·8274	0·0248	3·9750	1·2623	0·0378	6·0646
430	0·6479	0·0265	3·1626	1·9793	0·0810	9·6617	2·9913	0·1225	14·6019
440	0·9263	0·0609	4·6469	2·6889	0·1768	13·4883	3·9741	0·2613	19·9357
450	1·0320	0·1167	5·4391	2·7460	0·3105	14·4729	3·9191	0·4432	20·6551
460	1·0207	0·2098	5·8584	2·4571	0·5050	14·1020	3·3668	0·6920	19·3235
470	0·7817	0·3624	5·1445	1·7297	0·8018	11·3825	2·2878	1·0605	15·0550
480	0·4242	0·6198	3·6207	0·8629	1·2609	7·3655	1·1038	1·6129	9·4220
490	0·1604	1·0398	2·3266	0·2960	1·9190	4·2939	0·3639	2·3591	5·2789
500	0·0269	1·7956	1·5132	0·0437	2·9133	2·4552	0·0511	3·4077	2·8717
510	0·0572	3·0849	0·9674	0·0810	4·3669	1·3694	0·0898	4·8412	1·5181
520	0·4247	4·7614	0·5271	0·5405	6·0602	0·6709	0·5752	6·4491	0·7140
530	1·2116	6·3230	0·3084	1·4555	7·5959	0·3705	1·5206	7·9357	0·3871
540	2·3142	7·5985	0·1625	2·6899	8·8322	0·1889	2·7858	9·1470	0·1956
550	3·7329	8·5707	0·0749	4·1838	9·6060	0·0840	4·2833	9·8343	0·0860
560	5·5086	9·2201	0·0357	5·8385	9·7722	0·0378	5·8782	9·8387	0·0381
570	7·5710	9·4574	0·0209	7·4723	9·3341	0·0206	7·3230	9·1476	0·0202
580	9·7157	9·2257	0·0170	8·8406	8·3947	0·0154	8·4141	7·9897	0·0147
590	11·5841	8·5430	0·0130	9·7329	7·1777	0·0109	8·9878	6·6283	0·0101
600	12·7103	7·5460	0·0096	9·9523	5·9086	0·0075	8·9536	5·3157	0·0067
610	12·6768	6·3599	0·0044	9·4425	4·7373	0·0033	8·3294	4·1788	0·0029
620	11·3577	5·0649	0·0020	8·1290	3·6251	0·0014	7·0604	3·1485	0·0012
630	8·9999	3·7122	0·0000	6·2135	2·5629	0·0000	5·3212	2·1948	0·0000
640	6·5487	2·5587	0·0000	4·3678	1·7066	0·0000	3·6882	1·4411	0·0000
650	4·3447	1·6389	0·0000	2·8202	1·0638	0·0000	2·3531	0·8876	0·0000
660	2·6234	0·9706	0·0000	1·6515	0·6110	0·0000	1·3589	0·5028	0·0000
670	1·4539	0·5327	0·0000	0·8796	0·3223	0·0000	0·7113	0·2606	0·0000
680	0·7966	0·2896	0·0000	0·4602	0·1673	0·0000	0·3657	0·1329	0·0000
690	0·4065	0·1467	0·0000	0·2218	0·0801	0·0000	0·1721	0·0621	0·0000
700	0·2067	0·0744	0·0000	0·1065	0·0384	0·0000	0·0806	0·0290	0·0000
710	0·1108	0·0398	0·0000	0·0538	0·0193	0·0000	0·0398	0·0143	0·0000
720	0·0556	0·0195	0·0000	0·0253	0·0089	0·0000	0·0183	0·0064	0·0000
730	0·0280	0·0100	0·0000	0·0120	0·0043	0·0000	0·0085	0·0030	0·0000
740	0·0144	0·0062	0·0000	0·0058	0·0025	0·0000	0·0040	0·0017	0·0000
750	0·0063	0·0021	0·0000	0·0024	0·0008	0·0000	0·0017	0·0006	0·0000
760	0·0032	0·0011	0·0000	0·0012	0·0004	0·0000	0·0008	0·0003	0·0000
770	0·0011	0·0000	0·0000	0·0004	0·0000	0·0000	0·0003	0·0000	0·0000
Totals	109·8450	100·0000	35·5824	99·0915	100·0000	85·3094	98·0699	100·0000	118·2216

Wave-length (nm)	Distribution coefficients weighted by energy values of illuminant D_{5500}			Distribution coefficients weighted by energy values of illuminant D_{6500}			Distribution coefficents weighted by energy values of illuminant D_{7500}		
	$P_{5500}\bar{x}\lambda$	$P_{5500}\bar{y}\lambda$	$P_{550}\text{-}\bar{z}\lambda$	$P_{6500}\bar{x}\lambda$	$P_{6500}\bar{y}\lambda$	$P_{6500}\bar{z}\lambda$	$P_{7500}\bar{x}\lambda$	$P_{7500}\bar{y}\lambda$	$P_{7500}\bar{z}\lambda$
380	0·004	0·000	0·020	0·007	0·000	0·031	0·009	0·000	0·041
390	0·015	0·000	0·073	0·022	0·001	0·104	0·028	0·001	0·132
400	0·083	0·002	0·394	0·112	0·003	0·532	0·137	0·004	0·650
410	0·284	0·008	1·354	0·377	0·010	1·796	0·457	0·013	2·179
420	0·916	0·027	4·400	1·189	0·035	5·711	1·424	0·042	6·839
430	1·836	0·075	8·959	2·330	0·095	11·370	2·749	0·112	13·414
440	2·838	0·187	14·235	3·458	0·228	17·343	3·964	0·262	19·883
450	3·136	0·354	16·528	3·724	0·421	19·627	4·199	0·475	22·133
460	2·780	0·574	15·959	3·243	0·669	18·614	3·614	0·746	20·746
470	1·858	0·865	12·243	2·124	0·989	13·998	2·336	1·088	15·395
480	0·933	1·357	7·936	1·048	1·524	8·915	1·139	1·655	9·682
490	0·299	1·941	4·341	0·330	2·142	4·791	0·354	2·301	5·146
500	0·047	3·094	2·606	0·051	3·343	2·815	0·054	3·536	2·978
510	0·089	4·819	1·516	0·095	5·132	1·614	0·099	5·371	1·689
520	0·602	6·754	0·744	0·628	7·041	0·775	0·646	7·245	0·798
530	1·641	8·546	0·418	1·687	8·785	0·430	1·717	8·941	0·438
540	2·821	9·267	0·197	2·869	9·425	0·201	2·899	9·523	0·203
550	4·245	9·747	0·085	4·267	9·796	0·086	4·270	9·803	0·086
560	5·656	9·466	0·037	5·625	9·415	0·037	5·584	9·345	0·037
570	7·048	8·804	0·019	6·947	8·678	0·019	6·844	8·550	0·019
580	8·520	8·089	0·016	8·304	7·885	0·015	8·108	7·699	0·015
590	8·927	6·584	0·010	8·612	6·352	0·009	8·387	6·186	0·009
600	9·541	5·668	0·007	9·046	5·374	0·007	8·704	5·171	0·007
610	9·074	4·553	0·003	8·499	4·264	0·003	8·114	4·071	0·002
620	7·658	3·415	0·002	7·089	3·161	0·002	6·710	2·992	0·002
630	5·528	2·280	0·000	5·062	2·088	0·000	4·754	1·961	0·000
640	3·934	1·537	0·000	3·547	1·386	0·000	3·301	1·290	0·000
650	2·397	0·905	0·000	2·147	0·810	0·000	1·993	0·752	0·000
660	1·417	0·524	0·000	1·252	0·463	0·000	1·152	0·426	0·000
670	0·781	0·286	0·000	0·680	0·249	0·000	0·620	0·227	0·000
680	0·401	0·145	0·000	0·347	0·126	0·000	0·315	0·114	0·000
690	0·172	0·062	0·000	0·150	0·054	0·000	0·136	0·049	0·000
700	0·090	0·032	0·000	0·077	0·028	0·000	0·070	0·025	0·000
710	0·047	0·017	0·000	0·041	0·015	0·000	0·037	0·013	0·000
720	0·019	0·007	0·000	0·017	0·006	0·000	0·015	0·005	0·000
730	0·011	0·004	0·000	0·009	0·003	0·000	0·008	0·003	0·000
740	0·006	0·002	0·000	0·005	0·002	0·000	0·005	0·002	0·000
750	0·002	0·001	0·000	0·002	0·001	0·000	0·002	0·001	0·000
760	0·001	0·001	0·000	0·001	0·000	0·000	0·001	0·000	0·000
770	0·001	0·000	0·000	0·001	0·000	0·000	0·001	0·000	0·000
Totals	95·655	100·000	92·102	95·018	100·000	108·845	94·954	100·000	122·520

APPENDIX IV

TABLE 9. WAVELENGTHS (IN μm) FOR THE CALCULATION OF THE TRISTIMULUS VALUES X, Y, Z OF A SURFACE OR FILTER BY THE HARDY SELECTED ORDINATE METHOD*

Number of ordinate	Illuminant S_A			Illuminant S_B			Illuminant S_C		
	X	Y	Z	X	Y	Z	X	Y	Z
1	0·4440	0·4878	0·4164	0·4281	0·4723	0·4148	0·4244	0·4659	0·4141
2	0·5169	0·5077	0·4249	0·4421	0·4945	0·4229	0·4355	0·4894	0·4222
3	0·5440	0·5173	0·4294	0·4541	0·5057	0·4271	0·4439	0·5004	0·4263
4	0·5542	0·5241	0·4329	0·4681	0·5135	0·4303	0·4521	0·5087	0·4294
5	0·5614	0·5298	0·4360	0·5278	0·5196	0·4330	0·4612	0·5151	0·4320
6	0·5671	0·5348	0·4387	0·5433	0·5248	0·4354	0·4740	0·5206	0·4343
7	0·5720	0·5394	0·4413	0·5519	0·5294	0·4377	0·5312	0·5254	0·4365
8	0·5763	0·5437	0·4437	0·5585	0·5337	0·4399	0·5443	0·5298	0·4386
9	0·5802	0·5478	0·4460	0·5640	0·5377	0·4420	0·5524	0·5339	0·4406
10	0·5839	0·5517	0·4483	0·5688	0·5415	0·4440	0·5587	0·5377	0·4425
11	0·5872	0·5554	0·4505	0·5731	0·5451	0·4460	0·5641	0·5414	0·4444
12	0·5905	0·5591	0·4526	0·5771	0·5487	0·4480	0·5689	0·5449	0·4463
13	0·5935	0·5627	0·4547	0·5809	0·5521	0·4500	0·5732	0·5484	0·4482
14	0·5965	0·5663	0·4568	0·5845	0·5555	0·4519	0·5773	0·5518	0·4501
15	0·5994	0·5698	0·4588	0·5880	0·5590	0·4539	0·5813	0·5551	0·4521
16	0·6023	0·5733	0·4608	0·5914	0·5624	0·4558	0·5850	0·5585	0·4540
17	0·6052	0·5769	0·4629	0·5947	0·5658	0·4578	0·5887	0·5619	0·4559
18	0·6080	0·5805	0·4649	0·5981	0·5693	0·4598	0·5924	0·5653	0·4579
19	0·6109	0·5841	0·4670	0·6014	0·5729	0·4618	0·5960	0·5689	0·4599
20	0·6138	0·5879	0·4692	0·6047	0·5767	0·4639	0·5996	0·5725	0·4620
21	0·6169	0·5918	0·4716	0·6081	0·5806	0·4661	0·6033	0·5764	0·4641
22	0·6200	0·5959	0·4741	0·6116	0·5847	0·4684	0·6070	0·5805	0·4663
23	0·6233	0·6001	0·4768	0·6153	0·5891	0·4708	0·6109	0·5848	0·4687
24	0·6269	0·6047	0·4799	0·6191	0·5939	0·4736	0·6150	0·5896	0·4714
25	0·6308	0·6097	0·4834	0·6233	0·5991	0·4766	0·6194	0·5948	0·4743
26	0·6353	0·6152	0·4875	0·6280	0·6050	0·4802	0·6242	0·6008	0·4777
27	0·6405	0·6215	0·4927	0·6334	0·6118	0·4845	0·6298	0·6077	0·4818
28	0·6469	0·6292	0·4993	0·6401	0·6199	0·4902	0·6366	0·6161	0·4872
29	0·6559	0·6397	0·5084	0·6492	0·6309	0·4986	0·6459	0·6273	0·4952
30	0·6735	0·6590	0·5267	0·6663	0·6507	0·5152	0·6630	0·6474	0·5112
Multiplying factors	0·03661	0·03333	0·01185	0·03303	0·03333	0·02842	0·03268	0·03333	0·03938

* See page 133.

APPENDIX V

THE 1964 C.I.E. SYSTEM OF COLOUR SPECIFICATION

This system is based on colour matching data recorded for an annular zone of the retina around the fovea, the external diameter of the field subtending an angle of 10°. In 1964 the C.I.E.* approved the adoption of colour matching functions derived from the results of Stiles and Speranskaya, as described in Chapter 4, to be used as a supplement to the 1931 C.I.E. system whenever more accurate correlation with visual colour-matching of fields of large angular subtense was desired.

The co-ordinate system is basically similar to the 1931 system and the reference stimuli have the symbols (X_{10}), (Y_{10}) and (Z_{10}). As with the 1931 system, the relative luminous efficiencies $V_{(X_{10})}$, $V_{(Y_{10})}$, $V_{(Z_{10})}$ of 1 T. unit of (X_{10}), (Y_{10}) and (Z_{10}) are zero, unity and zero respectively.

The unit equations for the equal-energy stimulus S_E and for S_A, S_B and S_C are as follows:

$$S_E \equiv 0{\cdot}3333(X_{10}) + 0{\cdot}3333(Y_{10}) + 0{\cdot}3333(Z_{10})$$
$$S_A \equiv 0{\cdot}4512(X_{10}) + 0{\cdot}4059(Y_{10}) + 0{\cdot}1429(Z_{10})$$
$$S_B \equiv 0{\cdot}3499(X_{10}) + 0{\cdot}3526(Y_{10}) + 0{\cdot}2975(Z_{10})$$
$$S_C \equiv 0{\cdot}3104(X_{10}) + 0{\cdot}3191(Y_{10}) + 0{\cdot}3705(Z_{10})$$

The unit equations for illuminants D_{5500}, D_{6500} and D_{7500} on the 1964 C.I.E. system are as follows:

$$D_{5500}\cdot \quad 0{\cdot}3341(X_{10}) + 0{\cdot}3487(Y_{10}) + 0{\cdot}3172(Z_{10})$$
$$D_{6500}\cdot \quad 0{\cdot}3138(X_{10}) + 0{\cdot}3310(Y_{10}) + 0{\cdot}3552(Z_{10})$$
$$D_{7500}\cdot \quad 0{\cdot}2996(X_{10}) + 0{\cdot}3173(Y_{10}) + 0{\cdot}3831(Z_{10})$$

Table 10 gives the 10° chromaticity co-ordinates x_{10}, y_{10} and z_{10} for wavelengths from 380 nm to 780 nm at intervals of 5 nm, and

* *C.I.E. Compte Rendu*, 1963.

Table 11 gives the 10° equal-energy distribution coefficients $\bar{x}_{10}$, $\bar{y}_{10}$ and $\bar{z}_{10}$ for the same range of wavelengths.

In Table 12 the distribution coefficients at wavelength intervals of 10 nm have been weighted by the corresponding energy values of the illuminants S_A, S_B, S_C, D_{5500}, D_{6500} and D_{7500} respectively and adjusted for each illuminant so that $\Sigma(P_\lambda \, . \, y_{10}) = 100{\cdot}00$. The totals at the foot of the tables therefore give the 10° tristimulus values of the perfect white diffuser under these illuminants. The values in these tables have been derived by Judd and Wysczeki* and by Wyszecki and Stiles† from the 10° C.I.E. colour mixture data. These tables can be used directly for the calculation of the 10° tristimulus specifications of surface colours under the illuminants S_A, S_B, S_C, D_{5500}, D_{6500} and D_{7500} from the spectral reflection curves of the surfaces.

* D. B. Judd and G. Wyszecki, *Color in Business, Science and Industry*, (Wiley: New York and London,) 2nd. ed., 1963.

† G. Wyszecki and W. S. Stiles: *Color Science* (Wiley: New York and London), 1967.

TABLE 10. 10° CHROMATICITY CO-ORDINATES

Wave-length (nm)	Chromaticity co-ordinates			Wave-length (nm)	Chromaticity co-ordinates		
	x_{10}	y_{10}	z_{10}		x_{10}	y_{10}	z_{10}
380	0·1813	0·0197	0·7990	580	0·5386	0·4614	0·0000
385	0·1809	0·0195	0·7996	585	0·5654	0·4346	0·0000
390	0·1803	0·0194	0·8003	590	0·5900	0·4100	0·0000
395	0·1795	0·0190	0·8015	595	0·6116	0·3884	0·0000
400	0·1784	0·0187	0·8029	600	0·6306	0·3694	0·0000
405	0·1771	0·0184	0·8045	605	0·6471	0·3529	0·0000
410	0·1755	0·0181	0·8064	610	0·6612	0·3388	0·0000
415	0·1732	0·0178	0·8090	615	0·6731	0·3269	0·0000
420	0·1706	0·0179	0·8115	620	0·6827	0·3173	0·0000
425	0·1679	0·0187	0·8134	625	0·6898	0·3102	0·0000
430	0·1650	0·0203	0·8147	630	0·6955	0·3045	0·0000
435	0·1622	0·0225	0·8153	635	0·7010	0·2990	0·0000
440	0·1590	0·0257	0·8153	640	0·7059	0·2941	0·0000
445	0·1554	0·0300	0·8146	645	0·7103	0·2898	0·0000
450	0·1510	0·0364	0·8126	650	0·7137	0·2863	0·0000
455	0·1459	0·0452	0·8088	655	0·7156	0·2844	0·0000
460	0·1389	0·0589	0·8022	660	0·7168	0·2832	0·0000
465	0·1295	0·0779	0·7926	665	0·7179	0·2821	0·0000
470	0·1152	0·1090	0·7758	670	0·7187	0·2813	0·0000
475	0·0957	0·1591	0·7452	675	0·7193	0·2807	0·0000
480	0·0728	0·2292	0·6980	680	0·7198	0·2802	0·0000
485	0·0452	0·3275	0·6273	685	0·7200	0·2800	0·0000
490	0·0210	0·4401	0·5389	690	0·7202	0·2798	0·0000
495	0·0073	0·5625	0·4302	695	0·7203	0·2797	0·0000
500	0·0056	0·6745	0·3199	700	0·7204	0·2796	0·0000
505	0·0219	0·7526	0·2256	705	0·7203	0·2797	0·0000
510	0·0495	0·8023	0·1482	710	0·7202	0·2798	0·0000
515	0·0850	0·8170	0·0980	715	0·7201	0·2799	0·0000
520	0·1252	0·8102	0·0646	720	0·7199	0·2801	0·0000
525	0·1664	0·7922	0·0414	725	0·7197	0·2803	0·0000
530	0·2071	0·7663	0·0267	730	0·7195	0·2806	0·0000
535	0·2436	0·7399	0·0165	735	0·7192	0·2808	0·0000
540	0·2786	0·7113	0·0101	740	0·7189	0·2811	0·0000
545	0·3132	0·6813	0·0055	745	0·7186	0·2814	0·0000
550	0·3473	0·6501	0·0026	750	0·7183	0·2817	0·0000
555	0·3812	0·6182	0·0007	755	0·7180	0·2820	0·0000
560	0·4142	0·5858	0·0000	760	0·7176	0·2824	0·0000
565	0·4469	0·5531	0·0000	765	0·7172	0·2828	0·0000
570	0·4790	0·5210	0·0000	770	0·7169	0·2831	0·0000
575	0·5096	0·4904	0·0000	775	0·7165	0·2835	0·0000
580	0·5386	0·4614	0·0000	780	0·7161	0·2839	0·0000

TABLE 11. 10° DISTRIBUTION COEFFICIENTS

Wave-length (nm)	Distribution coefficients for equal-energy stimulus			Wave-length (nm)	Distribution coefficients for equal-energy stimulus		
	$\bar{x}_{10}$	$\bar{y}_{10}$	$\bar{z}_{10}$		$\bar{x}_{10}$	$\bar{y}_{10}$	$\bar{z}_{10}$
380	0·0002	0·0000	0·0007	580	1·0142	0·8689	0·0000
385	0·0007	0·0001	0·0029	585	1·0743	0·8256	0·0000
390	0·0024	0·0003	0·0105	590	1·1185	0·7774	0·0000
395	0·0072	0·0008	0·0323	595	1·1343	0·7204	0·0000
400	0·0191	0·0020	0·0860	600	1·1240	0·6583	0·0000
405	0·0434	0·0045	0·1971	605	1·0891	0·5939	0·0000
410	0·0847	0·0088	0·3894	610	1·0305	0·5280	0·0000
415	0·1406	0·0145	0·6568	615	0·9507	0·4618	0·0000
420	0·2045	0·0214	0·9725	620	0·8563	0·3981	0·0000
425	0·2647	0·0295	1·2825	625	0·7549	0·3396	0·0000
430	0·3147	0·0387	1·5535	630	0·6475	0·2835	0·0000
435	0·3577	0·0496	1·7985	635	0·5351	0·2283	0·0000
440	0·3837	0·0621	1·9673	640	0·4316	0·1798	0·0000
445	0·3867	0·0747	2·0273	645	0·3437	0·1402	0·0000
450	0·3707	0·0895	1·9948	650	0·2683	0·1076	0·0000
455	0·3430	0·1063	1·9007	655	0·2043	0·0812	0·0000
460	0·3023	0·1282	1·7454	660	0·1526	0·0603	0·0000
465	0·2541	0·1528	1·5549	665	0·1122	0·0441	0·0000
470	0·1956	0·1852	1·3176	670	0·0813	0·0318	0·0000
475	0·1323	0·2199	1·0302	675	0·0579	0·0226	0·0000
480	0·0805	0·2536	0·7721	680	0·0409	0·0159	0·0000
485	0·0411	0·2977	0·5701	685	0·0286	0·0111	0·0000
490	0·0162	0·3391	0·4153	690	0·0199	0·0077	0·0000
495	0·0051	0·3954	0·3024	695	0·0138	0·0054	0·0000
500	0·0038	0·4608	0·2185	700	0·0096	0·0037	0·0000
505	0·0154	0·5314	0·1592	705	0·0066	0·0026	0·0000
510	0·0375	0·6067	0·1120	710	0·0046	0·0018	0·0000
515	0·0714	0·6857	0·0822	715	0·0031	0·0012	0·0000
520	0·1177	0·7618	0·0607	720	0·0022	0·0008	0·0000
525	0·1730	0·8233	0·0431	725	0·0015	0·0006	0·0000
530	0·2365	0·8752	0·0305	730	0·0010	0·0004	0·0000
535	0·3042	0·9238	0·0206	735	0·0007	0·0003	0·0000
540	0·3768	0·9620	0·0137	740	0·0005	0·0002	0·0000
545	0·4516	0·9822	0·0079	745	0·0004	0·0001	0·0000
550	0·5298	0·9918	0·0040	750	0·0003	0·0001	0·0000
555	0·6161	0·9991	0·0011	755	0·0002	0·0001	0·0000
560	0·7052	0·9973	0·0000	760	0·0001	0·0000	0·0000
565	0·7938	0·9824	0·0000	765	0·0001	0·0000	0·0000
570	0·8787	0·9555	0·0000	770	0·0001	0·0000	0·0000
575	0·9512	0·9152	0·0000	775	0·0000	0·0000	0·0000
580	1·0142	0·8689	0·0000	780	0·0000	0·0000	0·0000
				Totals:	23·3294	23·3323	23·3343

TABLE 12. 1964 (10°) DISTRIBUTION COEFFICIENTS WEIGHTED FOR ILLUMINANTS S_A, S_B, S_C, D_{5500}, D_{6500} AND D_{7500}

Wave-length in μ	Distribution coefficients weighted by energy values of illuminant S_A			Distribution coefficients weighted by energy values of illuminant S_B			Distribution coefficients weighted by energy values of illuminant S_C		
	$P_A . \bar{x}_{10}$	$P_A . \bar{y}_{10}$	$P_A . \bar{z}_{10}$	$P_B . \bar{x}_{10}$	$P_B . \bar{y}_{10}$	$P_B . \bar{z}_{10}$	$P_C . \bar{x}_{10}$	$P_C . \bar{y}_{10}$	$P_C . \bar{z}_{10}$
380	0·000	0·000	0·001	0·000	0·000	0·002	0·001	0·000	0·002
390	0·003	0·000	0·011	0·007	0·001	0·029	0·009	0·001	0·043
400	0·025	0·003	0·111	0·070	0·007	0·313	0·103	0·011	0·463
410	0·132	0·014	0·605	0·388	0·040	1·786	0·581	0·060	2·672
420	0·377	0·040	1·795	1·137	0·119	5·411	1·708	0·179	8·122
430	0·682	0·083	3·368	2·025	0·249	9·997	3·011	0·370	14·865
440	0·968	0·156	4·962	2·729	0·442	13·994	3·969	0·643	20·349
450	1·078	0·260	5·802	2·787	0·673	14·997	3·914	0·945	21·058
460	1·005	0·426	5·802	2·350	0·997	13·568	3·168	1·343	18·292
470	0·737	0·698	4·965	1·585	1·500	10·671	2·062	1·952	13·887
480	0·341	1·076	3·274	0·674	2·125	6·470	0·840	2·675	8·144
490	0·076	1·607	1·968	0·137	2·880	3·528	0·167	3·484	4·268
500	0·020	2·424	1·150	0·032	3·822	1·812	0·037	4·398	2·085
510	0·218	3·523	0·650	0·299	4·845	0·894	0·327	5·284	0·976
520	0·750	4·854	0·387	0·927	6·002	0·478	0·971	6·285	0·501
530	1·644	6·086	0·212	1·920	7·103	0·247	1·973	7·302	0·255
540	2·847	7·267	0·104	3·214	8·207	0·117	3·275	8·362	0·119
550	4·326	8·099	0·033	4·711	8·818	0·035	4·744	8·882	0·036
560	6·198	8·766	0·000	6·382	9·025	0·000	6·322	8·941	0·000
570	8·277	9·002	0·000	7·936	8·630	0·000	7·653	8·322	0·000
580	10·201	8·740	0·000	9·017	7·726	0·000	8·444	7·235	0·000
590	11·967	8·317	0·000	9·768	6·789	0·000	8·874	6·168	0·000
600	12·748	7·466	0·000	9·697	5·679	0·000	8·583	5·027	0·000
610	12·349	6·327	0·000	8·935	4·579	0·000	7·756	3·974	0·000
620	10·809	5·026	0·000	7·515	3·494	0·000	6·422	2·986	0·000
630	8·583	3·758	0·000	5·757	2·520	0·000	4·851	2·124	0·000
640	5·992	2·496	0·000	3·883	1·618	0·000	3·226	1·344	0·000
650	3·892	1·561	0·000	2·454	0·984	0·000	2·014	0·808	0·000
660	2·306	0·911	0·000	1·410	0·557	0·000	1·142	0·451	0·000
670	1·277	0·499	0·000	0·751	0·294	0·000	0·598	0·233	0·000
680	0·666	0·259	0·000	0·374	0·145	0·000	0·293	0·114	0·000
690	0·336	0·130	0·000	0·178	0·069	0·000	0·136	0·053	0·000
700	0·167	0·064	0·000	0·084	0·033	0·000	0·062	0·024	0·000
710	0·083	0·033	0·000	0·039	0·015	0·000	0·028	0·011	0·000
720	0·040	0·015	0·000	0·018	0·006	0·000	0·013	0·004	0·000
730	0·019	0·008	0·000	0·008	0·004	0·000	0·005	0·003	0·000
740	0·010	0·004	0·000	0·004	0·002	0·000	0·003	0·001	0·000
750	0·006	0·002	0·000	0·003	0·001	0·000	0·002	0·001	0·000
760	0·002	0·000	0·000	0·001	0·000	0·000	0·001	0·000	0·000
770	0·002	0·000	0·000	0·001	0·000	0·000	0·001	0·000	0·000
Totals	111·159	100·000	35·200	99·207	100·000	84·349	97·298	100·000	116·137

Wave-length (nm)	Distribution coefficients weighted by energy values of illuminant D_{5500}			Distribution coefficients weighted by energy values of illuminant D_{6500}			Distribution coefficients weighted by energy values of illuminant D_{7500}		
	$P_{5500}\bar{x}_{10}$	$P_{5500}\bar{y}_{10}$	$P_{5500}\bar{z}_{20}$	$P_{6500}\bar{x}_{10}$	$P_{6500}\bar{y}_{10}$	$P_{6500}\bar{z}_{10}$	$P_{7500}\bar{x}_{10}$	$P_{7500}\bar{y}_{10}$	$P_{7500}\bar{z}_{10}$
380	0·001	0·000	0·002	0·001	0·000	0·003	0·001	0·000	0·004
390	0·008	0·001	0·035	0·011	0·001	0·049	0·014	0·002	0·062
400	0·102	0·011	0·458	0·136	0·014	0·613	0·165	0·017	0·744
410	0·507	0·053	2·331	0·667	0·069	3·068	0·804	0·084	3·697
420	1·278	0·134	6·078	1·645	0·172	7·824	1·957	0·205	9·308
430	1·866	0·229	9·211	2·349	0·289	11·594	2·753	0·339	13·589
440	2·867	0·464	14·699	3·464	0·561	17·762	3·946	0·639	20·229
450	3·170	0·765	17·061	3·734	0·902	20·095	4·183	1·010	22·511
460	2·650	1·124	15·303	3·066	1·300	17·703	3·395	1·440	19·601
470	1·706	1·615	11·489	1·934	1·831	13·028	2·113	2·001	14·234
480	0·721	2·270	6·911	0·803	2·529	7·701	0·866	2·729	8·308
490	0·139	2·902	3·554	0·152	3·176	3·890	0·162	3·389	4·151
500	0·033	4·048	1·919	0·036	4·337	2·057	0·038	4·558	2·162
510	0·329	5·330	0·984	0·348	5·630	1·039	0·362	5·853	1·081
520	1·027	6·646	0·530	1·062	6·871	0·547	1·085	7·024	0·560
530	2·150	7·957	0·277	2·192	8·112	0·283	2·217	8·203	0·286
540	3·356	8·569	0·122	3·386	8·644	0·123	3·398	8·676	0·124
550	4·759	8·909	0·036	4·744	8·881	0·036	4·716	8·829	0·036
560	6·152	8·700	0·000	6·069	8·583	0·000	5·985	8·463	0·000
570	7·452	8·103	0·000	7·285	7·921	0·000	7·130	7·753	0·000
580	8·648	7·409	0·000	8·360	7·162	0·000	8·109	6·948	0·000
590	8·921	6·201	0·000	8·536	5·933	0·000	8·259	5·740	0·000
600	9·258	5·422	0·000	8·706	5·099	0·000	8·322	4·874	0·000
610	8·553	4·382	0·000	7·945	4·071	0·000	7·536	3·861	0·000
620	7·038	3·272	0·000	6·462	3·004	0·000	6·076	2·825	0·000
630	5·109	2·237	0·000	4·641	2·032	0·000	4·329	1·896	0·000
640	3·476	1·448	0·000	3·109	1·295	0·000	2·874	1·197	0·000
650	2·080	0·834	0·000	1·848	0·741	0·000	1·704	0·684	0·000
660	1·202	0·475	0·000	1·053	0·416	0·000	0·963	0·381	0·000
670	0·666	0·261	0·000	0·576	0·225	0·000	0·521	0·204	0·000
680	0·321	0·125	0·000	0·276	0·107	0·000	0·249	0·097	0·000
690	0·138	0·054	0·000	0·119	0·046	0·000	0·108	0·042	0·000
700	0·069	0·027	0·000	0·059	0·023	0·000	0·053	0·020	0·000
710	0·034	0·013	0·000	0·029	0·012	0·000	0·027	0·010	0·000
720	0·013	0·005	0·000	0·012	0·004	0·000	0·011	0·004	0·000
730	0·007	0·003	0·000	0·006	0·002	0·000	0·005	0·002	0·000
740	0·004	0·001	0·000	0·003	0·001	0·000	0·003	0·001	0·000
750	0·002	0·001	0·000	0·002	0·001	0·000	0·001	0·000	0·000
760	0·000	0·000	0·000	0·000	0·000	0·000	0·000	0·000	0·000
770	0·001	0·000	0·000	0·001	0·000	0·000	0·001	0·000	0·000
Totals	95·815	100·000	91·000	94·827	100·000	107·414	94·444	100·000	120·686

NAME INDEX

SUBJECT INDEX